Innovative Renewable Energy

Series editor

Ali Sayigh
World Renewable Energy Congress, Brighton, UK

The primary objective of this book series is to highlight the best-implemented worldwide policies, projects and research dealing with renewable energy and the environment. The books will be developed in partnership with the World Renewable Energy Network (WREN). WREN is one of the most effective organizations in supporting and enhancing the utilisation and implementation of renewable energy sources that are both environmentally safe and economically sustainable. Contributors to books in this series come from a worldwide network of agencies, laboratories, institutions, companies and individuals, all working together towards an international diffusion of renewable energy technologies and applications. With contributions from most countries in the world, books in this series promote the communication and technical education of scientists, engineers, technicians and managers in this field and address the energy needs of both developing and developed countries.

Each book in the series contains contributions from WREN members and covers the most-up-to-date research developments, government policies, business models, best practices, and innovations from countries all over the globe. Additionally, the series will publish a collection of best papers presented during the annual and bi-annual World Renewable Energy Congress and Forum each year.

More information about this series at http://www.springer.com/series/15925

Ali Sayigh
Editor

Sustainable Building for a Cleaner Environment

Selected Papers from the World Renewable
Energy Network's Med Green Forum 2017

 Springer

Editor
Ali Sayigh
World Renewable Energy Congress
Brighton, UK

ISSN 2522-8927 ISSN 2522-8935 (electronic)
Innovative Renewable Energy
ISBN 978-3-030-06881-3 ISBN 978-3-319-94595-8 (eBook)
https://doi.org/10.1007/978-3-319-94595-8

Printed on acid-free paper

This Springer imprint is published by the registered company Springer International Publishing AG part of Springer Nature.
The registered company address is: Gewerbestrasse 11, 6330 Cham, Switzerland

Introduction

This is the fourth WREN-WREC Med Green Forum and the second one to be held in Florence, teaming with ABITA from the University of Florence and ETA from Renewable Energy Florence.

We received 87 contributions from 50 countries and several students' posters. There were also presentations from well-known Italian industries in the built environment.

The three organizations, WREN, ABITA, and ETA, worked together to highlight the importance of sustainable buildings and renewable energy especially in the regions of abundant sunshine – the Mediterranean Zone.

I thank all the sponsors of this event, especially ISESCO, Springer, and the Department of Architecture, University of Florence, for hosting the Med Green Forum No. 4. We are very grateful to the technical committee and organizing committee for their excellent efforts in making this a successful meeting.

This area of combining buildings with energy conservation, efficiency, and renewable energy is the only way to combat climate change and save energy. The Forum offered a great opportunity where participants could network during the coffee and lunch times, and in the evenings.

This Proceeding consists of 39 papers from 41 countries, covering all the following areas:

1. Sustainable architecture
2. Building construction management and environment
3. Ventilation and air movement in buildings
4. Renewable energy in building and cities
5. Eco materials and technology
6. Policy education and finance
7. Sustainable transport
8. Urban agriculture and soilless urban green space

To encourage readers to read papers from every category, the papers in this Proceeding have not been subdivided into topic sections. We hope that architects, builders, energy specialists, and researchers in the built environment will find all papers interesting and stimulating.

Brighton, UK   Ali Sayigh

Contents

1 Proposing a New Method for Fenestration Shading Design
   in Prefabricated Modular Buildings   1
   Seyedehmamak Salavatian

2 Effectiveness of Occupant Behavioral Ventilation Strategies
   on Indoor Thermal Comfort in Hot Arid Climate   15
   Ali Sedki, Neveen Hamza, and Theo Zaffagnini

3 Effectiveness of Materials, Technologies, and Renewable Energy
   in Educational Buildings Through Cluster Analysis of Energy
   Retrofitting   25
   Francesco Asdrubali, Laura Calcagnini, Luca Evangelisti,
   Claudia Guattari, and Paola Marrone

4 Renewable Energy in Argentina   39
   Carlos Labriola

5 Wind Energy Potential Research in a Low Building
   within an Urban Environment   53
   Jorge Lassig, Claudia Palese, Juan Valle Sosa, Ubaldo Jara,
   and Carlos Labriola

6 Natural Light in Architecture: Use Inspired
   by the Constructive Tradition   63
   Fabio G. S. Giucastro

7 The Cost of Building to the nearly-Zero Energy Building
   Standard: A Financial Case Study   71
   Shane Colclough, John Mernagh, Derek Sinnott, Neil J. Hewitt,
   and Philip Griffiths

8 Policy of Intensification, Diversification, Conservation,
   and Indexation in Pursuing Sustainable Transport   79
   Koesmawan

9   The Problem of Education in Developing Renewable Energy   95
     Ellya Marliana Yudapraja

10   Winter Performance of Certified Passive Houses in a Temperate
     Maritime Climate: nZEB Compliant?   103
     Shane Colclough, Philip Griffiths, and Neil J. Hewitt

11   A Parametric Tool for Assessing Optimal Location of Buildings
     According to Environmental Criteria.   115
     Giacomo Chiesa and Mario Grosso

12   Methodology of Solar Project Managing Through All Stages
     of Development   131
     Andrejs Snegirjovs, Peteris Shipkovs, Kristina Lebedeva,
     Galina Kashkarova, and Dimitry Sergeev

13   Computational BIPV Design: An Energy Optimization
     Tool for Solar Façades   141
     Omid Bakhshaei, Giuseppe Ridolfi, and Arman Saberi

14   Feasibility Study of a Low Carbon House in the United Kingdom . .   153
     Timothy Aird and Hossein Mirzaii

15   Why Do We Need to Reduce the Carbon Footprint in UAE?   165
     Riadh AL-Dabbagh

16   Urban Farming in the Era of Crisis in Greece: The Case Study
     of the Urban Garden of Ag. Anargiri-Kamatero and Fili   179
     Konstadinos Abeliotis and Konstadinos Doudoumopoulos

17   Resilient Urban Design. Belgrade and Florence:
     Reconnect the Waters to the City   187
     Chiara Odolini

18   Strategic Sustainable and Smart Development Based
     on User Behaviour.   199
     Shahryar Habibi and Theo Zaffagnini

19   High|Bombastic: Adaptive Skin Conceptual Prototype
     for Mediterranean Climate   209
     Omid Bakhshaei, Giuseppe Ridolfi, and Arman Saberi

20   Quality of Healthcare: A Review of the Impact of the Hospital
     Physical Environment on Improving Quality of Care.   217
     Jazla Fadda

21   Enhancing Indoor Air Quality for Residential Building
     in Hot-Arid Regions   255
     Ghanim Kadhem Abdul Sada and Tawfeeq Wasmi M. Salih

22 Performance of Solar Window Film with Reference to Energy
Rationalizing in Buildings 265
Kamil M. Yousif and Alan Ibrahim Saeed

23 Visualizing the Infrared Response of an Urban Canyon
Throughout a Sunny Day 277
Benoit Beckers, José Pedro Aguerre, Gonzalo Besuievsky,
Eduardo Fernández, Elena García Nevado, Christian Laborderie,
and Raphaël Nahon

24 Meta-Design Approach to Environmental Building Programming
for Passive Cooling of Buildings 285
Giacomo Chiesa and Mario Grosso

25 Urban and Architectural Sustainability in the Restoration
of Iranian Cities (Strategy and Challenges): Case Study
of Soltaniyeh 297
Nazila Khaghani

26 Influence of the Period of Measurements on Wind Potential
Assessment for a Given Site 311
H. Nfaoui and A. Sayigh

27 Integration Strategies of Luminescent Solar Concentrator
Panels: A Case Study in Florence, Italy 327
Lucia Ceccherini Nelli and Giada Gallo Afflitto

28 Photovoltaic and Thermal Solar Concentrator Integrated
into a Dynamic Shading Device 335
Giulia Chieli and Lucia Ceccherini Nelli

29 A University Master's Course and Training Programme for Energy
Managers and Expert in Environmental Design in Italy 347
Marco Sala, Lucia Ceccherini Nelli, and Alessandra Donato

30 A Project for the NZero-Foundation in the South of Italy 361
Lucia Ceccherini Nelli, Vincenzo Donato, and Danilo Rinaldi

31 Planning Without Waste. 371
Adolfo F. L. Baratta, Laura Calcagnini, Fabrizio Finucci,
and Antonio Magarò

32 Production of ZnO Cauliflowers Using the Spray
Pyrolysis Method. 383
Shadia J. Ikhmayies

33 Evaluating Deep Retrofit Strategies for Buildings
in Urban Waterfronts 391
Nicola Strazza, Piero Sdrigotti, Carlo Antonio Stival, and Raul Berto

34 Enhancing the Thermophysical Properties of Rammed Earth
    by Stabilizing with Corn Husk Ash. 405
    Amina Lawal Batagarawa, Joshua Ayodeji Abodunrin,
    and Musa Lawal Sagada

35 Thermal Monitoring of Low-Income Housings Built
    with Autoclaved Aerated Concrete in a Hot-Dry Climate. 415
    Ramona Romero-Moreno, Gonzalo Bojórquez-Morales,
    Aníbal Luna-León, and César Hernández

36 Renewables Are Commercially Justified to Save Fuel and Not
    for Storage. 427
    Donald T. Swift-Hook

37 Climate Change Adaptation: Assessment and Simulation
    for Hot-Arid Urban Settlements – The Case Study of the Asmarat
    Housing Project in Cairo, Egypt. 437
    Mohsen Aboulnaga, Amr Alwan, and Mohamed R. Elsharouny

38 Ventilation Effectiveness of Residential Ventilation Systems
    and Its Energy-Saving Potential. 451
    Mohammad Reza Adili and Michael Schmidt

39 Assessment of Cardboard as an Environment-Friendly Wall
    Thermal Insulation for Low-Energy Prefabricated Buildings. 463
    Seyedehmamak Salavatian, M. D'Orazio, C. Di Perna,
    and E. Di Giuseppe

Conclusions. 471

Chapter 1
Proposing a New Method for Fenestration Shading Design in Prefabricated Modular Buildings

Seyedehmamak Salavatian

Abstract Use of prefabricated construction in developing countries has been increased lately; (for several purposes i.e. educational, industrial, recreational, commercial, etc. as well as temporary homes in post-disaster situations) while the minimum consideration regarding energy efficiency and bioclimatic design strategies are paid attention in this regard. Typically, these buildings are constructed based on industrial production systems and are installed through a modular design process. Subsequently, their fenestration design mainly follows the modularity of envelope panels -in size and geometry- and impacts of environmental factors as solar gain, natural ventilation, and heat transmission are neglected in design decisions even though they play an important role in building energy consumption scales. This study aims to analyze windows and their shading systems in a mid-range altitude and temperate-humid climate in Iran. According to the comfort zone suggested by ASHRAE, the time intervals of the year which necessitate shadows on the windows surfaces are determined. On the condition that shading is provided for the interior spaces, comfort condition is guaranteed needless of any auxiliary solutions. On the basis of the attained sun/shadow calendar, the matrix of fenestration design alternatives is studied in Ecotect software. Parameters as windows geometrical ratio, shading type (vertical, horizontal, mixed), and proportional shading size are studied in each geographical directions and the optimized solutions are proposed in order to provide shadows in the required time periods. This architect-friendly method tried to equip designers with non-numerical algorithmic programs and assist them in energy efficient modular fenestration design. The obtained prototypes are utilizable in prefabricated modular buildings meanwhile the decision-making procedure could also be applied in other altitudes and climatic regions to gain similar models.

S. Salavatian (✉)
Department of Architecture, Rasht Branch, Islamic Azad University,
Rasht, Iran
e-mail: salavatian@iaurasht.ac.ir

© Springer International Publishing AG, part of Springer Nature 2019   1
A. Sayigh (ed.), *Sustainable Building for a Cleaner Environment*,
Innovative Renewable Energy, https://doi.org/10.1007/978-3-319-94595-8_1

1.1   Introduction

A building designer needs to find the best solution to satisfy various requirements in different design aspects. Regarding air-conditioning solutions, designers prefer to adopt mechanical systems to achieve the required indoor thermal comfort, which results in increased energy consumption in buildings. Thus, buildings are responsible for a substantial part of energy consumption, mostly the result of the heating, cooling, and artificial ventilation systems of a building [1]. Thus, more investigations are needed to determine sustainable alternatives and passive techniques to avoid high rates of energy use.

Solar heat gain is identified as one of the main contributors to overheating in residential buildings. Windows, as the transparent parts of the envelope, have a significant role in the amount of heat gain and the thermal performance of the building. The shape, size, thermal properties, orientation, and shading of windows determine the visual and thermal comfort for the occupants inside buildings [2]. The high cost of advanced glazing types makes them inappropriate strategies for low-cost projects such as prefabricated temporary buildings. Therefore, there is a general perception that sustainable solutions are not cost-effective for temporary types of buildings. This attitude has caused considerable neglect toward energy-efficient temporary buildings, although solar heat gain can be simply controlled by introducing optimized shadings to minimize solar transmission and heat gains through glazed areas [3].

On the other hand, the recent development of building simulation tools has been a revolution in the manual calculation of bio-climatic and solar design; this necessitates an up-to-date systematic approach to organize a step-by-step method assisted by the appropriate simulation software.

This chapter explores the methodology linked with the architectural design process to provide modular buildings with shading devices for hot periods without deprived the occupants of pleasant sunshine in the cold season.

Finally, this work aims to suggest the optimized properties of window shading devices to provide internal spaces with maximum thermal comfort. It also provides designers with the adequate knowledge to design shading devices as an integral part of the fenestration system.

1.2   Research Background

Design of shading devices has been reported in the literature from different aspects, such as illuminance level, visual comfort, building energy consumption, solar gain, and natural ventilation. Indeed, several studies have been carried out to demonstrate the significant effects of appropriate shading on the thermal performance of internal spaces. In the early 2000s, the C.E .Faculty [4], Gugliermetti and Bisegna [5], and Tzempelikos and Athienitis [6] all conducted studies to explore the effects of different shading design strategies on thermal performance improvement in indoor environments and provided the best solutions as design guidelines. A number of researchers attempted to form accurate guidelines for the design of shading devices and to provide interior spaces with the best possible thermal comfort [7]. Other studies combined other fenestration parameters with shading device properties to recognize the most effective items for reduction of building energy consumption [8].

There are also studies that focused on a specific shading type, such as fixed/movable, internal/external, and horizontal/vertical louver shadings by taking advantage of the capabilities of simulation tools; in the primary steps, Datta [9] in 2001 used TRNSYS to study many horizontal shading variables in various locations in Italy. Palmero-Marrero and Oliveira [10] conducted a similar study in many different latitudes and showed the great impact of shading devices on saving energy loads. Hammad and Abu-hijleh [11] investigated the energy consumption of external dynamic louvers, integrated to office building facades, in AbuDhabi. In 2014, the performance of internal shading devices was compared with external installations by Atzeri et al. [12] in terms of heating/cooling loads.

More recent literature has reported optimization algorithmic programs to classify shading devices. Some of these take a multi-objective optimization approach including a shading system whereas others specifically consider shading devices. Manzan [13] used a genetic optimization to identify a possible geometry to achieve the lowest energy impact. Chua and Chuo [14] examined a novel approach based on an established value that measures the envelope thermal performance in high-rise residential buildings to determine the most suitable shading devices for different orientations of the building.

Tahbaz [15] introduced a graphical, geometric, and step-like method, using the "shading mask" and "climatic needs calendar" initially developed by Olgyay [16, 17], and applied this approach to an inadequately shaded outdoor space. By this method, shadow in the necessary periods of the pattern year was provided to modify the inappropriate existing sunshade. In a following study by the same researcher [18], a generalized methodology was suggested for the solar design of buildings in preliminary design stages. In the sequential method suggested by this study, six simple steps are followed to achieve the efficient sunshade for any architectural project. The "climatic needs calendar" in solar design studies was also applied in the work of other researchers [19, 20]. Also, Krüger and Dorigo [21] applied the shading mask procedure to run a daylighting analysis with RADIANCE and ECOTECT in a public school for different time schedules and orientations. The aforementioned efforts that applied the "Olgyay" recommended procedure are mostly accomplished regardless of powerful user-friendly non-numerical simulation software that lets architects analyze solar aspects of the project easily and quickly in a visual interface.

1.3   Methodology

1.3.1   Identifying Shadow Need Periods

Human thermal comfort is related to several factors such as air temperature, air movement, amount of clothing worn, and activity level including the human body itself [22]. Uncomfortable thermal conditions affect a person's productivity, health, and quality of life. According to ASHRAE 55 [23], thermal comfort is defined as "that condition of mind that expresses satisfaction with the thermal environment and is assessed by subjective evaluation." The Givoni bioclimatic chart [24, 25] considers human and climatic measures as well as building envelope effects and, in this chapter, is used as a proper predictor to analyze the climatic needs of interior spaces. Figure 1.1 shows the thermal zones in which providing shadow (blue line) or shadow+ natural ventilation (green line) guarantees indoor thermal comfort. In other words, so long as natural ventilation is considered for internal spaces, through locating appropriate openings in windward and leeward sides, thermal comfort is satisfied for both zones.

On the basis of Givoni's bioclimatic index, equivalent temperature lines within the "climatic need calendar" are drawn. In Fig. 1.2, the "climatic needs calendar" has two perpendicular axes for days and hours, including all periods of a year. It is utilized to distinguish different climatic needs in various time periods: sunshine need, shadow need, shadow + ventilation need, cold conditions, and very hot conditions. Based on the average hourly climatic data (including temperature and relative humidity, which allow us to gain an effective temperature), the equivalent temperature lines are drawn in the software of Surfer 14. Surfer is a powerful mapping program, very practical in various fields of engineering and scientific studies, which creates a grid-based map from an XYZ data file.

This calendar demonstrates time periods at which shadow provision has a significant role in indoor thermal comfort. The "time zone" area enclosed within M curves needs attention regarding shading system design. A shading device must be made such that the glazing surface is protected exclusively in these periods during the year. The periods during sunray penetration must be avoided or be allowed in the interior spaces are determined. Consequently, the sunshade pattern is designed according to periodic shadow needs and, ultimately, shading devices are proposed to balance these two conditions.

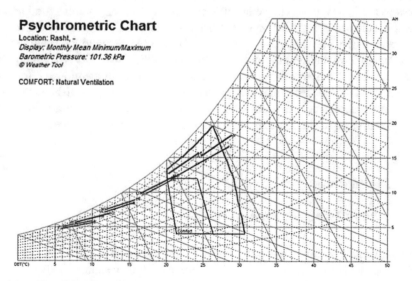

Fig. 1.1 "Givoni" bioclimatic chart of Rasht (Autodesk Ecotect Analysis 2011)

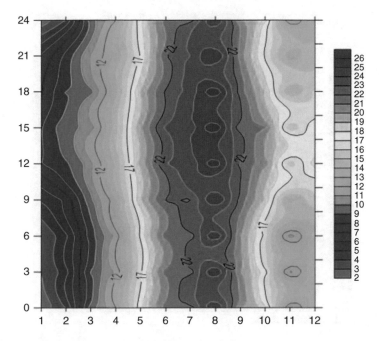

Fig. 1.2 Climatic needs in Calendar of Rasht

1.3.2   Climatic Region

The Guilan Province is located in the northern part of the country. Guilan weather is generally mild, caused by the influences of both Alborz Mountains and the Caspian Sea. This region has a humid temperate and Mediterranean climate with abundant annual rainfall and high relative humidity (between 40% and 100%), and its average temperature is 17.5 °C [26]. The weather data of Rasht, the capital city of the province, were utilized for the simulations as the representative of a moderate humid climate.

The latitude of Rasht is 37°2′ N and 49°6′ E; therefore, the corresponding sun path relevant to the latitude was drawn by solar tool software (Fig. 1.3) that helps designers determine sun location and shadow-casting conditions at any moment of the year. Figure 1.4 demonstrates the transfer of shadow need periods into the sun path diagram. From the shadow angles of a sun protractor (0–90°), the desired horizontal and vertical sunshade angles are estimated (in Fig. 1.5, the black line indicates the shading mask that must be considered in the sunshade design of south-oriented windows). As the proportional sizes of windows are already determined, the sunshade pattern is achievable. This procedure is repeatable for any window orientation.

Climatic data were obtained by the relevant meteorology station, and hourly data were taken from Meteonorm software, converted to wea. format, ready to be applied in the Autodesk weather tool 2011.

S. Salavatian

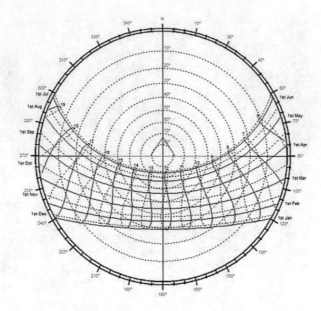

Fig. 1.3 Annual sun path in Rasht

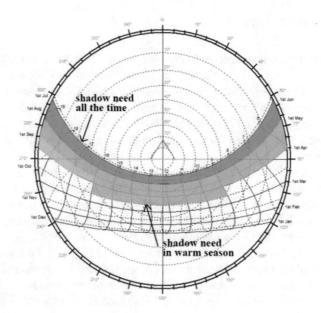

Fig. 1.4 Various shadow needs on the sun path

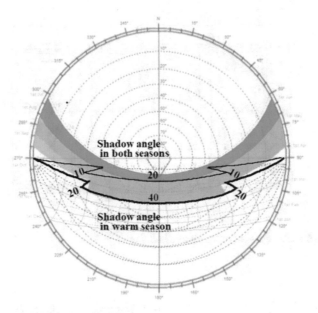

Fig. 1.5 Shadow angle for south-oriented window

1.3.3   Prefabricated Building Prototype

A case building is considered in accordance with architectural and structural considerations: rapidity and simplicity of assembly, expandability and flexibility, and compatibility with the environment. This typically designed residential building model has the
dimensions 9.80 m in length and 5.60 m in width, with a total area of 55 m², equipped
with bathroom, kitchen, technical room, and a multi-purpose space (Fig. 1.6). This
model constitutes three attached modules, two types of 1.80 m × 4.80 m and
2.40 m × 4.80 m, which is the smallest configuration and is planned for occupation by
a couple. Other larger alternatives, including four or five or more modules, are also
feasible to satisfy the needs of larger families (see Fig. 1.7).

   According to the recommended design strategies in this climate, the longitudinal axis of the building is oriented along the east–west direction to maximize heat
gain in winter and heat loss in summer. Because natural ventilation is the other
condition in the second zone to provide indoor thermal comfort, rather than
shadow on the window surface, the Wind Rose pattern of the summer season in
the city of Rasht was considered; the decided orientation is in line with the optimal opening location in the windward and leeward wall sides in the one-layer plan
design of the modular buildings.

   Many different parameters in window properties affect the thermal comfort of
interior spaces. Except for the shading strategies, which were taken as the main

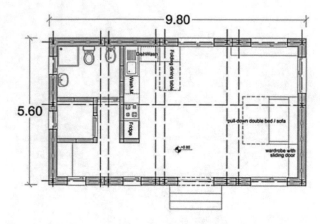

Fig. 1.6 Indoor space layout of prefabricated building

Fig. 1.7 Modularity of building types

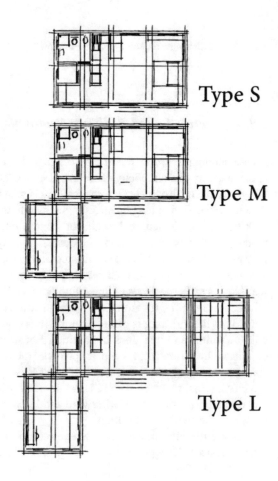

Fig. 1.8 View of
prefabricated building
model in the annual sun
path, Rasht location
(Autodesk Ecotect
Analysis 2011)

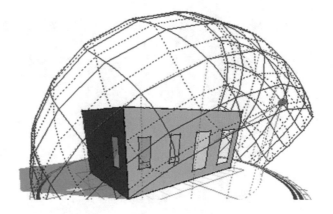

variables, other features are kept identical (e.g., area, geometry, windows design, glazing area, and transmittance ratios).

1.3.4 Simulation Tool

The emergence of a large number of software programs has dramatically changed the manual methods of solar design, leading building engineers and architects to revise the fundamental principles/methods to adjust these with the novel capabilities; among them, Ecotect software has been utilized in various reliable applications in the field of solar studies. For example, Yang et al. [27], Dutta et al. [1], Aldali and Moustafa [28], and Jamaludin et al. [29] used Ecotect for the analysis of solar radiation effects on envelopes.

For the purpose of this chapter, Ecotect was used as the building performance tool, mainly because its pleasant interface makes it easy for application by architects. Simulations were conducted in Ecotect to evaluate the effects of various shading strategies on thermal comfort in a prefabricated building in a temperate climate.

Individual spaces are generated as divided areas in Ecotect, named "zones." Further, annual and daily sunrays over the building are simulated to investigate shadow conditions on the glazing area (Fig. 1.8). Finally, simulations and evaluations can lead to providing shading design recommendations.

1.3.5 Shading Device Variables

Sunshades are the studied variable in this research, and the optimum range of other parameters was assumed according to the studies in the literature. External shadings operate up to 30% more effectively compared to internal shadings.

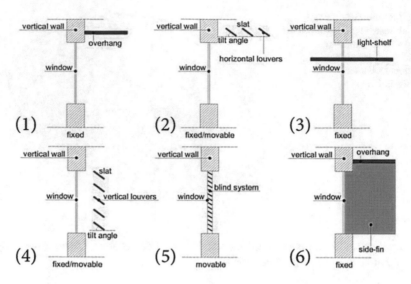

Fig. 1.9 Main shading types [31]

Horizontal external shading is generally recommended for south- and north-oriented windows whereas vertical shading is more appropriate for east- and west-oriented fenestrations [30].

The investigation of this study is limited to shading properties. Shading device types are categorized as (1) overhang, (2) horizontal louvers, (3) light shelf, (4) vertical louvers, (5) blind system, and (6) side fine (see Fig. 1.9) by Bellia et al. [31] in a review research of solar shading systems. In each category, the shape, depth, and length were varied and the shadow obtained on the glazing surface was investigated. The most common practical window types and sizes, which are adjusted to the modular panels, were considered in the analysis. Several configurations were explored to achieve the optimal solution for maximum internal thermal comfort. Because of the wall structure requirements (studs at the usual distance of 60 cm), 70 cm was considered for the panel width, and window dimensions are designated in three different heights and two different widths (Fig. 1.10).

1.4 Results and Discussion

On the basis of the shadow/sun need periods, dimensional and proportional characteristics of shading systems were obtained via Ecotect simulations because the shadow period need is not symmetrical with respect to the solstice. There are periods in warm seasons when shadow is needed (e.g., midday in September) whereas sun is preferred in the corresponding cold season (e.g., in March). This priority is embedded within the subject of climate. In a temperate climate in which humid

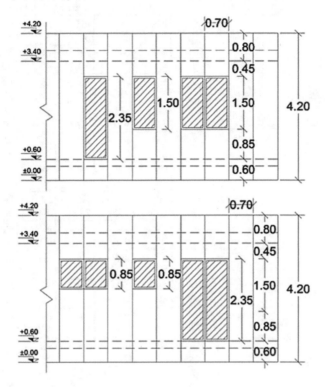

Fig. 1.10 Modular window prototypes

summers get uncomfortable, the priority of thermal comfort goes with the warm seasons. The other consideration in this study is to use the least variety and the most similarity in building elements (in size, geometry, etc.) to be in line with the nature of prefabricated construction. Furthermore, all the sunshades were assumed to be fixed; although movable shading provides more practical solutions, they require higher levels of technology and facilities that are probably unavailable in temporary, low-budget projects.

Window types were investigated in four main orientations. Despite the subtle difference of shadow needs between west and east orientations, the greater shadow need was considered as the dominant criterion to minimize variety and maximize homogeneity in the modular design. According to the shading mask, the required depth was attained for both horizontal and vertical shades; in the case of wider or higher windows, the acquired depth is broken into more than one element. Table 1.1 summarizes the results of simulations for all six window types in main cardinal orientations. For the aim of simplicity, type 1 modeling in the south direction is demonstrated.

On the south side, horizontal and vertical sunshades are needed at the distance of 70 cm. Thus, three, two, and one blades are required for 240 cm, 150 cm, and 85 cm windows, respectively. An angle of 20° allows keeping the shaded area in warm periods and avoiding it in cold seasons. However, no inclination for vertical shades is effective.

Table 1.1 Summary of shading characteristics in four cardinal directions

	Shading simulation for type 1	Type 1	Type 2	Type 3	Type 4	Type 5	Type 6
South		nH 2	nH 2	nH 3	nH 3	nH 1	nH 1
		dH 40	dH 40	dH 40	dH 40	dH 40	dH 40
		lH 170	lH 100	lH 100	lH 170	lH 100	lH 170
		aH −20	aH −20	aH −20	aH −20	aH −20	aH −20
		nV 2	nV 2	nV 2	nV 2	nV 2	nV 2
		dV 10	dV 10	dV 10	dV 10	dV 10	dV 10
		lV 155	lV 155	lV 240	lV 240	lV 90	lV 90
		aV 0	aV 0	aV 0	aV 0	aV 0	aV 0
North		nH 1	nH 1	nH 1	nH 1	nH 1	nH 1
		dH 40	dH 40	dH 40	dH 40	dH 40	dH 40
		lH 140	lH 70	lH 70	lH 140	lH 70	lH 140
		aH 0	aH 0	aH 0	aH 0	aH 0	aH 0
		nV 2	nV 2	nV 2	nV 2	nV 2	nV 2
		dV 40	dV 20	dV 20	dV 40	dV 40	dV 40
		lV 150	lV 150	lV 240	lV 240	lV 70	lV 240
		aV 0	aV 0	aV 0	aV 0	aV 0	aV 0
East & West		nH 3	nH 3	nH 5	nH 5	nH 2	nH 2
		dH 40	dH 40	dH 40	dH 40	dH 40	dH 40
		lH 155	lH 85	lH 85	lH 155	lH 85	lH 155
		aH +45	aH +45	aH +45	aH +45	aH +45	aH +45
		nV 3	nV 2	nV 2	nV 3	nV 2	nV 3
		dV 30	dV 30	dV 30	dV 30	dV 30	dV 30
		lV 155	lV 155	lV 240	lV 240	lV 90	lV 90
		aV +45	aV +45	aV +45	aV +45	aV +45	aV +45

nH number of horizontal sunshades, *dH* depth of horizontal sunshade (cm), *lH* length of horizontal sunshade (cm), *aH* angle of horizontal sunshade (°), *nV* number of vertical sunshades, *dV* depth of vertical sunshade (cm), *lV* length of vertical sunshade (cm), *aV* angle of vertical sunshade (°)

On the west/east sides, the incident ray angle, which is closer to perpendicular, allows the depth of the horizontal blades to be equal to the glazing height; this necessitates a greater number of both shades (Table 1.1). Also, an angle of 45° for both vertical blades (downward) and horizontal blades (clockwise) increases the efficiency of the shading system.

In the north orientation, the major challenge is focused on the warm season, because in the cold period the north side is not the subject of sun radiation in the Northern Hemisphere. In the required periods of the calendar, northern windows are protected by one simple horizontal on the top and two vertical blades on the sides (as the dimensions are determined in Table 1.1).

The overall advantages of this method include (1) providing the ability to control shadow and sunshine in a given period of a year and (2) allowing the design of different alternative shades; it also (3) enables the designer of high mass production projects (i.e., prefabricated buildings) to generalize the analysis of a limited number of fenestration systems to a large number of cases.

1.5   Conclusion

On the basis of the results of this study, although application of shading improves thermal comfort conditions, none of the simulated strategies was sufficiently effective to satisfy the absolute thermal comfort criteria. Therefore, solar shading should be used in line with other strategies to attain thermal comfort inside a building. However, considering temporary modular buildings, the geometrical visual guidelines for a pattern to design their fenestration system in a sample latitude are provided. Obviously, the shape and form of sunshades in any individual project are finalized by the specific aesthetic, construction, and economic issues.

Further studies are recommended for modular buildings in other climatic conditions throughout the world at lower or middle latitudes where shading is an essential passive strategy to improve the thermal performance of buildings. Additionally, considering other parameters as variables provides more comprehensive solutions in terms of climate-responsive design.

References

1. Dutta A, Samanta A, Neogi S (2017) Influence of orientation and the impact of external window shading on building thermal performance in tropical climate. Energy Build 139:680–689
2. Khatami M, Khatami MS (2014) Design optimization of glazing façade by using the GPSPSOCCHJ algorithm. pp 1–9
3. Hashemi A, Khatami N (2016) Effects of solar shading on thermal comfort in low-income tropical housing. Energy Procedia 111:235–244
4. Faculty CE (2003) Whole building energy simulation with complex external shading devices. pp 571–576

5. Gugliermetti F, Bisegna F (2006) Daylighting with external shading devices: design and simulation algorithms. Build Environ 41(2):136–149
6. Tzempelikos A, Athienitis AK (2007) The impact of shading design and control on building cooling and lighting demand. Sol Energy 81(3):369–382
7. Freewan AAY (2011) Improving thermal performance of offices in JUST using fixed shading devices. World Renew Energy Congr Sweden, pp 8–13
8. Aldawoud A (2013) Conventional fixed shading devices in comparison to an electrochromic glazing system in hot, dry climate. Energy Build 59:104–110
9. Datta G (2001) Effect of fixed horizontal louver shading devices on thermal perfomance of building by TRNSYS simulation. Renew Energy 23(3-4):497–507
10. Palmero-Marrero AI, Oliveira AC (2010) Effect of louver shading devices on building energy requirements. Appl Energy 87(6):2040–2049
11. Hammad F, Abu-Hijleh B (2010) The energy savings potential of using dynamic external louvers in an office building. Energy Build 42(10):1888–1895
12. Atzeri A, Cappelletti F, Gasparella. A (2014) Internal versus external shading devices performance in office buildings. Energy Procedia 45:463–472
13. Manzan M (2014) Genetic optimization of external fixed shading devices. Energy Build 72:431–440
14. Chua KJ, Chou SK (2010) Evaluating the performance of shading devices and glazing types to promote energy efficiency of residential buildings. Build Simul 3(3):181–194
15. Tahbaz M (2006) Architecture of shadows, living deserts: is a sustainable urban desert still possible. Arid Hot Reg, pp 9–12
16. Olgyay A, Olgyay V (1957) Solar control and shading devices. Princeton University Press, Princeton
17. Olgyay V (1963) Design with climate. Princeton University Press, Princeton
18. Tahbaz M (2012) Primary stage of solar energy use in architecture: shadow control. J Cent South Univ Technol (English Ed) 19(3):755–763
19. Farajolahi M Abbasi M (2011) Optimization of building orientation in City of 'Qir' affected by sun radiation using cosine equation methods (in Persian). Land 35:43–59
20. Hoseynabadi S, Lashkari H, Salmani Moghadam M (2012) Bio-climatic design of residential building in city of 'Sabzevaar,' considering building orientation and shading depth. Geogr Dev 27:103–116
21. Krüger EL, Dorigo AL (2008) Daylighting analysis in a public school in Curitiba, Brazil. Renew Energy 33(7):1695–1702
22. Randall T (2006) Environmental design. Taylor & Francis, New York
23. ASHRAE (2013) Standard55: thermal environmental conditions for human occupancy
24. Givoni B (1981) Man, climate and architecture. Applied Science Publishers, London
25. Givoni B (1998) Climate consideration in building and urban design. Van Nostrand Reinhold, New York
26. Climatological Research Institute of Iran [online]. Available: http://www.cri.ac.ir/show=251. Accessed 07 Jan 2017
27. Yang L, He BJ, Ye M (2014) Application research of ECOTECT in residential estate planning. Energy Build 72:195–202
28. Aldali KM, Moustafa WS (2016) An attempt to achieve efficient energy design for high-income houses in Egypt: case study: Madenaty City. Int J Sustain Built Environ 5(2):334–344
29. Jamaludin N, Mohammed NI, Khamidi MF, Wahab SNA (2015) Thermal comfort of residential building in Malaysia at different micro-climates. Procedia Soc Behav Sci 170:613–623
30. Simmler H, Binder B (2008) Experimental and numerical determination of the total solar energy transmittance of glazing with venetian blind shading. Build Environ 43(2):197–204
31. Bellia L, Marino C, Minichiello F, Pedace A (2014) An overview on solar shading systems for buildings. Energy Procedia 62:309–317

Chapter 2
Effectiveness of Occupant Behavioral Ventilation Strategies on Indoor Thermal Comfort in Hot Arid Climate

Ali Sedki, Neveen Hamza, and Theo Zaffagnini

Abstract This paper discusses the effectiveness of occupant behavioral strategies of open and close windows on indoor thermal comfort in residential buildings in the hot arid climate of Cairo, Egypt. Based behavioral survey scenarios were deduced from a questionnaire analysis in both winter and summer seasons. The behavioral scenarios were compared to the base case and were categorized into two main groups. The first group includes the based behavioral survey scenarios that were deduced from the questionnaire analysis, and the second group includes the scenarios that were suggested to improve thermal comfort if applied in summer season. Each scenario was applied on the case study and then was simulated through IESVE simulation software program. The effect of each scenario was investigated in winter (represented by the months that need zero cooling demand), summer (represented by the months that need zero heating demand), and spring and autumn months (represented by the months where both cooling and heating are needed).

2.1 Introduction

The aim of this paper is to identify the interrelationship between occupant behavioral different scenarios of natural ventilation and indoor thermal comfort in hot arid climates; these different scenarios were applied on the research case study.

A. Sedki (✉)
Beirut Arab University, Faculty of Architecture – Design, and Built Environment, Beirut, Lebanon

N. Hamza
Newcastle University, Department of Architecture, Newcastle, UK

T. Zaffagnini
University of Ferrara, Department of Architecture, Ferrara, Italy

© Springer International Publishing AG, part of Springer Nature 2019   15
A. Sayigh (ed.), *Sustainable Building for a Cleaner Environment*,
Innovative Renewable Energy, https://doi.org/10.1007/978-3-319-94595-8_2

As acknowledged, wind is considered a very important design factor for architects and air velocity is one of the most significant factors that affect thermal comfort. Naturally ventilated buildings use the adaptive approach of thermal comfort by specifying acceptable operative temperature ranges for naturally conditioned spaces [1]. Thus, natural ventilation design frameworks have got to be more practical and were recognized by international standards as an approach to enhance sustainability, energy efficiency, and comfortable environment in buildings. The airflow throughout a building depends on the area and resistances of the openings and the dissimilarity in air pressure between different areas [2]. This pressure variation probably happens because of wind (wind-driven natural ventilation) or variation in temperature between indoor and outdoor that causes differences in air density (stack effect) or the use of fans to create a pressure dissimilarity in a mechanical way [3].

In hot dry climates, natural ventilation is considered one of the most important passive cooling strategies. It is based on the idea that when the air velocity increases around the human body, then the rate of heat dissipation by evaporation and convection will be significantly accelerated [4]. Accordingly, indoor thermal comfort can be improved by natural ventilation through two different ways [5–7]. The first one is to raise the indoor air velocity by opening the windows letting the air cross inside the spaces and then increasing the cooling effect indoors. The second way is called "nocturnal ventilative cooling" that can indirectly happen by ventilating the spaces at nighttime to flush out heat from the building during the night, and this, consequently, can cool the internal mass of the building and consequently can reduce the indoor temperature rise in the successive day [8].

However, the second way of nocturnal ventilative cooling (or as it is called in other literature "night purge ventilation") is more advisable in hot dry climate regions because opening the windows during the day can significantly increase the indoor temperature and consequently increases discomfort, while during the night, it can diminish the heat that was gained during the daytime [5].

Givoni [6] indicated that, in hot dry climates, nocturnal ventilative cooling is beneficial when diurnal temperature ranges between 32 °C and 36 °C and nocturnal temperature drops down until 20 °C; this process becomes more efficient with high thermal mass buildings. However, in case of the extreme hot diurnal temperature that exceeds 36 °C, nocturnal ventilative cooling is not effective even with high thermal mass buildings. Asimakopoulos and Santamouris [4] stated that an appropriate nocturnal ventilative rate for a building with high thermal mass and all windows closed during the daytime can attain 35–45% diminution in indoor temperature.

Few other studies investigated the role of occupant behavior regarding natural ventilation on indoor comfort in residential spaces in hot dry climates. Liu et al. [9] have conducted a field study in Chongqing, China, to test different types of occupants' behavioral adaptation as a reaction to various thermal conditions throughout the year. Occupants had an active role to be adapted with environmental condition in different seasons as well as in different times of the day.

Indraganti [10] indicated in a field survey conducted in summer 2008 on apartment buildings in Hyderabad in India that 60% of the occupants were uncomfortable in summer and, furthermore, neutral temperature of 29.2 °C and comfort range of 26.0 °C and 32.5 °C were specified by regression analysis, while the outdoor maximum and minimum air temperature were 40.4 °C and 27.3 °C, respectively. In addition, occupants adaptively used the physical environmental controls like windows, balcony, doors, external doors, and curtains to achieve better comfort in the indoor environment.

Gado and Osman [11] conducted a study on the state-funded dwellings in New Al-Minya City in Egypt to examine the efficiency of natural ventilation strategies for this kind of buildings. The study was conducted in two phases. The first phase was a pilot study that discussed the impact of the use of transformations, like reshaping the window and installing vertical and horizontal shading devices, on natural ventilation. The second phase examined the natural ventilation effectiveness during the hottest period of the year using Autodesk Ecotect simulation software program, and it investigated the air movement using computational fluid dynamics software FloVENT. The main finding of the study was that the cross ventilation of nocturnal ventilative cooling is not effective for the case study because it achieved only 4.9% reduction in temperature.

El-Hefnawi [12] conducted a computer simulation parametric analysis for a case study of a typical youth housing dwelling in El-Obour City in the eastern desert of Cairo. The aim of his study was to examine different strategies to ameliorate the thermal performance of these buildings that were used in summer season. The tested strategies are different building materials, wall thicknesses, shading devices, and night purge ventilation. The study recommended using high thermal mass with external wall insulation, reducing window area to be at maximum 40% of the space area, using reflective glass for the windows with a thickness of 6 mm, fixing external shading devices for both west and south orientations, and using night purge ventilation.

The examined behavioral ventilation scenarios in this paper were deduced from a field survey questionnaire (that was distributed on 30 occupied residential units for low-income class (Fig. 2.1)) [13] and were compared to the base case and categorized into two main groups.

The first group includes the based behavioral survey scenarios that were deduced from the questionnaire analysis in winter and summer seasons, and the second group includes the scenarios that were suggested to improve thermal comfort in summer season. Each scenario was applied on the reference case and then was simulated using IESVE simulation software. The effect of natural ventilation was investigated for each month in winter (represented by the months that have zero cooling demand), summer (represented by the months that have zero heating demand), and spring and autumn months (represented by the months where both cooling and heating are needed) (Fig. 2.2).

Fig. 2.1 Residential units for low-income class in 6th of October City, Egypt

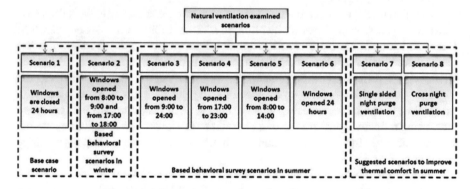

Fig. 2.2 Natural ventilation examined scenarios

2.2 Effect of Natural Ventilation Different Scenarios on Research Case Study

Based on the above literature, natural ventilation as one of important strategies to improve thermal comfort was adopted in this research to be examined on the case study of 6th of October City in Greater Cairo. The research used IESVE software simulation program to examine the effect of natural ventilation different scenarios on indoor thermal comfort. The examined scenarios (Fig. 2.2) were categorized into three main groups. The first group includes only the base case scenario. The second group includes the based behavioral survey scenarios that were deduced from questionnaire analysis in winter and summer seasons. The third group includes the hypothetical suggested scenarios to improve the situation in summer season. Each scenario is explained more in detail below:

- Scenario 1 (base case): windows were closed 24 h.
- Scenario 2: windows were opened from 8:00 to 9:00 at morning and from 17:00 to 18:00 at afternoon – internal doors were opened continuously.
- Scenario 3: windows were opened from 9:00 to 24:00 – internal doors were opened continuously.
- Scenario 4: windows were opened from 17:00 to 23:00 – internal doors were opened continuously.
- Scenario 5: windows were opened from 8:00 to 14:00 – internal doors were opened continuously.
- Scenario 6: windows were opened 24 h– internal doors were opened continuously.
- Scenario 7: single-sided night purge ventilation; all windows were opened during nighttime (from sunrise to sunset) – internal doors were closed continuously.
- Scenario 8: cross night purge ventilation; all the windows were opened during nighttime (from sunrise to sunset) – internal doors were opened.

The effect of each of the abovementioned scenarios is explained in detail in the following sections.

2.3 Effect of Natural Ventilation Different Scenarios on Thermal Comfort for a Living Room in Typical Floor Facing South

The southern apartments of the typical floor were simulated for the whole year according to the abovementioned scenarios for natural ventilation. The following analyses show in percentages the number of hours whereby heating is needed, the number of hours whereby cooling is needed, and the number of hours that are within comfort range wherein neither heating nor cooling is needed. These were specified according to the comfort zone analysis explained in previous studies Sedki et al. [13] that showed the extreme limits of comfort and adaptive comfort for this case study are ranged between 19.6 °C and 29 °C. These were considered as fixed limits of comfort zone for the whole year.

The year was divided into three main groups: winter months (where cooling demand is approximating zero percent from total hours), summer months (where heating demand is approximating zero percent from total hours), and spring-autumn months (where both heating and cooling are needed).

Accordingly, the winter season was represented by the month of January, February, March, November, and December since it was observed that in these months,

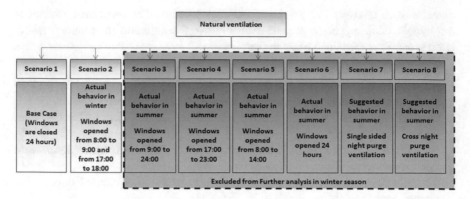

Fig. 2.3 Compared scenarios of natural ventilation in winter season

the cooling demand is approximating 0.0% from the total hours of the whole month. The summer season was represented by the month of June, July, August, and September because in these months the heating demand is approximating 0.0% from the total hours of the whole month. The spring-autumn seasons were represented by the month of April, May, and October. Although these months are considered the most thermally comfortable months in Cairo climate, they have small percentages from total hours that are of both cooling and heating demand according to comfort zone analysis.

In the winter months January, February, March, November, and December (Fig. 2.3), the comparison was made between the base case scenario (scenario 1) and the based behavioral survey scenario that is derived from questionnaire analysis (scenario 2) (see Fig. 2.3 below). Scenario 2 is the most likely to happen in winter as deduced from the questionnaire analysis; this is because the inhabitants prefer opening the external windows only for fresh air intake for little time at morning after they wake up and little time at afternoon after they come back from work (this little time was approximated to be 1 h for simulation purposes).

For the whole period of winter months (the period of zero cooling demand in January, February, March, November, and December) (Fig. 2.4), it was observed that the base case scenario (scenario 1) where windows were closed 24 h achieved higher percentage (35.2%) from total hours within comfort range, while when opening the windows 1 h at morning and 1 h at evening (scenario 2), it reduced the comfort by 1.3% to become 34.1% from total hours within comfort range. As a consequence, even though natural ventilation is undesirable in winter season as it reduces the indoor thermal comfort, it is important for indoor air quality and fresh air intake. So, scenario 2 is preferable to be followed in winter season.

In the summer months June, July, August, and September, the comparison was made among the base case scenario (scenario 1), the based behavioral survey scenarios that are derived from questionnaire analysis (scenario 3, scenario 4, scenario 5, and scenario 6), and the hypothetical suggested scenarios (scenario 7 and scenario 8) (Fig. 2.5).

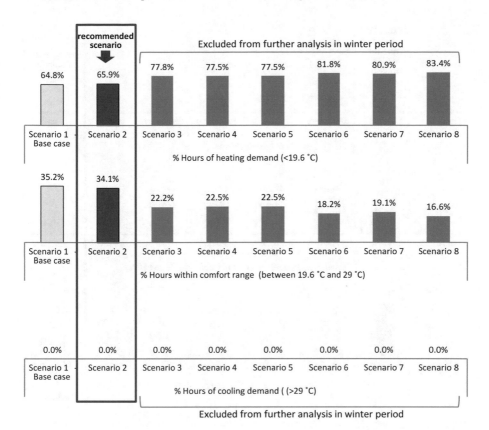

Fig. 2.4 Effect of natural ventilation different scenarios on southern living room in typical floor during the whole winter month's period

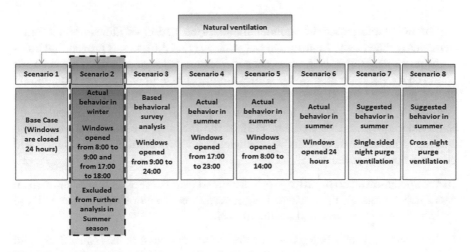

Fig. 2.5 Compared scenarios of natural ventilation in summer season

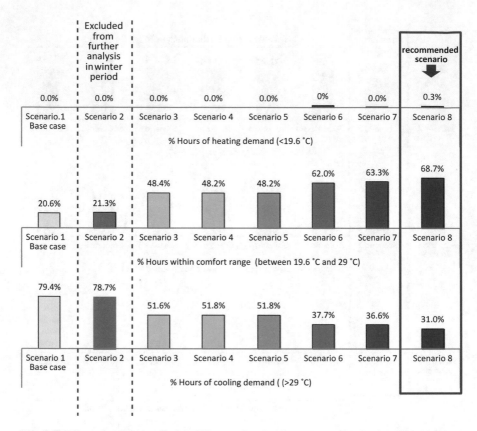

Fig. 2.6 Effect of natural ventilation different scenarios on southern living room in typical floor during the whole summer month's period

For the whole period of summer months (the period of almost zero heating demands in June, July, August, and September) (Fig. 2.6), it was proved that cross night purge ventilation is the best scenario for the summer months. This scenario achieved 68.7% from total hours within comfort range improving the situation by 48.1% from total hours higher than the base case scenario. Although the percentage is different, this result corroborates the findings of Givoni [6], Asimakopoulos and Santamouris [4], and El-Hefnawi [12] in the fact that the nocturnal ventilative cooling is one of the most effective strategies to improve thermal comfort in hot arid climates in summer season.

In the spring-autumn months April, May, and October (Fig. 2.7), the comparison was made among all the presented eight scenarios of natural ventilation as these months have both heating and cooling demand.

In general, for the whole period of the spring-autumn months (April, May, and October) (Fig. 2.6), the occupants should follow scenario 4 (windows opened

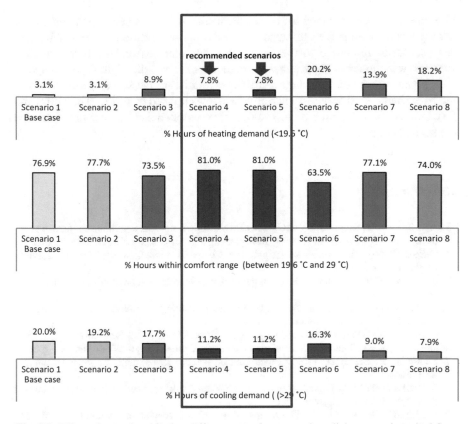

Fig. 2.7 Effect of natural ventilation different scenarios on southern living room in typical floor during the whole period of spring and autumn

from 17:00 to 23:00) or scenario 5 (windows opened from 8:00 to 14:00) because these scenarios have achieved equally the highest percentages from total hours within comfort range (81.0%) improving the situation by 4.1% compared to the base case scenario.

2.4   Conclusion

This paper examined the effectiveness of different scenarios of natural ventilation on indoor thermal comfort for the case study. The scenarios were categorized into three: the base case scenario, the based behavioral survey scenarios that were obtained from questionnaire analysis, and the hypothetical suggested scenarios to improve the situation in summer. The findings divulged that the best scenario in winter season was the base case scenario, but the based behavioral survey scenario in winter was recommended for the fresh air intake. This is because the indoor air quality would be negatively affected if the windows were closed for 24 h in the base case scenario. In regard to the findings of summer season, the recommended scenario was the cross night purge ventilation as it gives the highest percentage of total hours within comfort range. Further, in relation to the spring-autumn period, the based behavioral survey scenarios 4 (windows opened from 17:00 to 23:00) and 5 (windows opened from 8:00 to 14:00) were recommended because they achieved the highest comfort.

References

1. ANSI/ASHRAE Standard 55-2010 (2010) Thermal environmental conditions for human occupancy. American Society of Heating, Refrigerating and Air-Conditioning Engineers, Atlanta
2. CIBSE (2006) Environmental design. The Chartered Institution of Building Services Engineers, London
3. Roaf S et al (2001) Ecohouse: a design guide. Architectural Press, an imprint of Butterworth-Heinemann, Oxford
4. Asimakopoulos D, Santamouris M (1996) Passive cooling of buildings. James & James, Science Publishers, Ltd, London
5. Givoni B (1994) Passive and low energy cooling of buildings. Wiley, New York
6. Givoni B (1998) Climate consideration in building and urban design. Wiley, Vaughan, p 36
7. McMullan R, Seeley IH (2007) Environmental science in building. Palgrave Macmillan, Basingstoke
8. Krishan A et al (2001). Climate responsive architecture: a design handbook for energy efficient buildings. McGraw-Hil, New Delhi
9. Liu J, Yao R, Wang J, Li B (2012) Occupants' behavioural adaptation in workplaces with non-central heating and cooling systems. Appl Therm Eng 35:40–54
10. Indraganti M (2010) Adaptive use of natural ventilation for thermal comfort in Indian apartments. Build Environ 45(6):1490–1507
11. Gado T, Osman M (2009) Investigating natural ventilation inside walk-up housing blocks in the Egyptian desert climatic design region. Int J Vent 8(2):145–160
12. El-Hefnawi AIK (2000) Climatic design for low cost housing in Egypt. Case of the youth housing project in El-Obour City. AEE Architecture Energy & Environment – tools for climatic design – advanced international training program. Lund University, Lund
13. Sedki A, Hamza N, Zaffagnini T (2014) Improving Indoor thermal comfort in residential buildings in hot arid climates through the combination of natural ventilation and insulation retrofitting techniques – case study: existing social housing stocks in greater Cairo. PhD

Chapter 3
Effectiveness of Materials, Technologies, and Renewable Energy in Educational Buildings Through Cluster Analysis of Energy Retrofitting

Francesco Asdrubali, Laura Calcagnini, Luca Evangelisti, Claudia Guattari, and Paola Marrone

Abstract This paper is part of wider research on a stock of 80 schools and their energetic retrofit projects financed by the EU in the Lazio region in 2016 to reduce CO_2 emissions. Because in these procedures the suggested retrofit strategies are often selected on the basis of common best practices (considering average energy savings), but are not supported by proper energy investigations, following the principles of an evidence-based approach, we analysed all the information on the interventions and developed a database to support public administration in future informed financing processes.

All the design choices – including the use of eco-materials, envelope technologies and renewable energy sources – have been studied in relation to context, climate zone, school typology, use, construction quality and costs. Three levels of investigation have been defined to identify significant clusters according to reported data: data analysis, monitoring and field surveys, and observations of technological performance of the interventions. For each cluster, the most representative actual building was identified to allow comparability of different materials and technologies, to control and plan new energy retrofit interventions, thereby reducing inefficiencies. For each representative building, a validating procedure based on dynamic simulation was developed.

In this paper, a picture of the interventions regarding the use of materials and technologies for envelope retrofitting is given.

The first results highlight that the interventions have an effective energy performance overall, but they have many deficiencies in terms of environmental quality of the materials used and nothing has been done to comply with the guidelines of the recent minimum environmental criteria (CAM).

F. Asdrubali · L. Evangelisti · C. Guattari
Roma Tre University, Department of Engineering, Rome, Italy

L. Calcagnini (✉) · P. Marrone
Roma Tre University, Department of Architecture, Rome, Italy
e-mail: laura.calcagnini@uniroma3.it

© Springer International Publishing AG, part of Springer Nature 2019
A. Sayigh (ed.), *Sustainable Building for a Cleaner Environment*,
Innovative Renewable Energy, https://doi.org/10.1007/978-3-319-94595-8_3

Purpose: The comparison between the expected and actual energy performance of the refurbished schools allows indicators to be singled out to support the development of new calls to allocate public funds according to priority intervention areas, especially in view of the obligations recently established by the CAM and confirmed both in the recent National Energy Strategy and in the Sustainable Development Strategy.

An evidence-based approach and cluster analysis are the methodologies used to select the most representative buildings and homogeneous building classes in terms of constructive and energy characteristics of the building stock.

Research is limited to the regional field, but the methodology can be extended on a national and international scale.

The paper examines a significant number of case studies and develops an evaluation methodology for energetic retrofit interventions, in compliance with the latest environmental and energy standards for building envelope materials.

3.1   Reasons Behind the Research

This chapter presents a part of a wider research effort focused on energy renovation projects for educational buildings, funded by EU in the Lazio region in central Italy. The energy renovation interventions, carried out in 2015 and completed in 2016, were studied by analyzing the morphological and technological characteristics of the buildings and the materials and solutions adopted to reduce energy consumption.

Ex-post evaluation of energy retrofitting is aimed at identifying the most effective solutions, taking into account both technical and economic aspects, with the goal to support public administration in subsequent processes of energy renovation financing.

The availability of an important database of 155 energy–environmental renovation projects of public buildings, provided by the Lazio region, allowed a rigorous evaluation of common practices in the field of energy retrofitting in the Lazio territory.

The retrofitting interventions were evaluated according to three different levels: data analysis, monitoring and modeling, and field investigation. The ultimate aim of the research, which is still in progress, is to develop a replicable evaluation methodology for financing energy retrofitting of public assets.

This chapter examines part of the data referring to the actual energy and environmental quality of the interventions for a group of about 80 schools, with particular attention to the materials and the technologies used for envelope retrofitting.

3.2   Trends and Objectives

The research focuses on case studies of school buildings, which represent a relevant number (about 80 buildings) compared to the total number of renovated buildings (155 buildings). Moreover, school retrofitting is an important issue, as the international literature, national legal and operational context, and characterization of the available data demonstrate:

- Several references in the international scientific literature address the issue of school energy consumption and the evaluation of energy retrofitting interventions [1].
- In Italy, the issue of energy consumption by school buildings is relevant in the public assets scene as well as for research on public heritage retrofitting [2–4].
- The national school buildings heritage concerns about 42,000 buildings characterized by a certain obsolescence because of the prevalent age of the buildings, dating from the 1960s to the 1980s [5].
- Statistics for the Lazio region are similar to the national figures: 3% of the buildings were built before 1900, most of them (around 45%) in a period from 1961 to 1980 (Fig. 3.1). Thus, in Italy about 20,000 buildings were built before the first regulations on energy savings were applied and, therefore, they are large energy consumers with inefficient envelopes and building systems [5].
- According to PREPAC (Program for Energy Retrofitting of Existing Buildings and of Public Buildings), buildings for nonresidential use in Italy were responsible for 13% of final energy consumption in 2015, compared to a building sector responsibility of 30% [6].
- The energy efficiency program developed by PREPAC establishes that the energy consumption of at least 3% of public buildings should be reduced on an annual basis [7].
- According to Italian Law 102/2014, € 355 million has been assigned for the period 2014–2020 for the energy renovation of Public Administration buildings [7].
- As 52% of the school buildings have been renovated [5], it can be assumed that retrofitting for at least 10,000 buildings is now needed.

These data show the importance of school assets both for typological and constructive homogeneity characteristics and for interest in completing energy use renovation for public assets.

The research database was built considering about 80 school buildings, so the amount of data was rather daunting because the data were separated into different categories, from architectural data to technological performance (e.g., thermal insu-

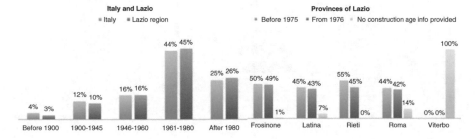

Fig. 3.1 Construction age of school buildings in the Lazio region and [5] the Lazio provinces of Italy [5]

lation of the roofing) and economic concerns. In this chapter we present the outcomes associated with interventions on the building envelope, which involved most of the cases and the greatest amount of invested funds.

3.3 Methodology

Following the outcome of the call for proposals POR FERS 2007–2013, the Lazio region made available to the research partners (Roma Tre University and Enea) a final proposal of the interventions on 77 buildings for school use. Three different levels of investigation were defined, with the dual aim of validating the path made by the region and developing critical instruments for future calls (Fig. 3.2):

1. Analysis of the data on all available final proposals
2. Monitoring (2a) the status and effects of the rehabilitation interventions and dynamically modeling the ante- and post status (2b)
3. Field investigation of technological and energy performance aspects of the interventions

With the aim of evaluating project performances and comparing these to those actually achieved, we managed the amount of data and all the levels of investigation as follows:

- The first level of investigation, the data analysis of the final proposals, was applied to all buildings (77 buildings).

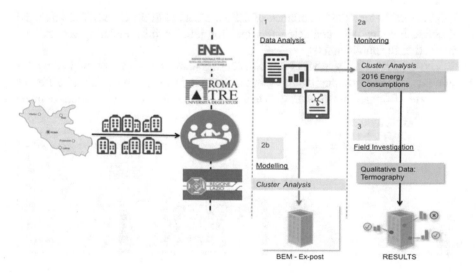

Fig. 3.2 Research methodology. The research effort was organized into three phases: 1, data analysis (77); 2, monitoring (a) and modeling (b) (7); and 3, field investigation (7)

- The second level, the monitoring and modeling of buildings, was accomplished on a limited number of case studies that could exemplify both the constructive and the energy characteristics of the school building heritage of the region (7 buildings).
- The third level, field investigation and data analysis, involved an evaluation of the technological and material qualities, which was performed on all buildings (77 buildings), and an instrumental analysis that was performed on the reference buildings (7 buildings).

The cluster analysis tool was used to select the most representative buildings to carry out the second and third phases of the research. This tool was used to obtain building classes of homogeneous characteristics according to their technological and energetic aspects. In view of our objectives, building classes were determined according to the following parameters: gross heated volume (V), the ratio between the external surface and the gross heated volume, called ratio (S/V), primary energy consumption ex ante (before the interventions), declared energy savings, and energy-class improvement (jumps) expected at the design stage. The cluster analysis also defined a "centroid" for each homogeneous building class, that is, the most representative building of each class.

3.4   Phase 1. Data Analysis

To analyze the data, we prepared a general database of buildings and, for each building, a construction and energy analysis data sheet (Fig. 3.3).

The buildings are quite evenly distributed within the five provinces of the Lazio territory, falling into three different climatic zones (C, D, E) according to the Italian zoning that classifies, depending on degree-days, the territory in six climatic zones (DPR 412/1993), from the hottest climate (zone A) to the coldest one (zone F).

Fig. 3.3 Data sheets. A data sheet was made for each building

Fig. 3.4 Distribution and cost of building energy redevelopment: 50% are interventions on the building envelope, 20% on building systems, and 29% on renewable energy sources (RES)

The interventions are classified into seven types (Fig. 3.4):

1. On the building envelope (three kinds: replacement of windowframes, thermal insulation of opaque vertical closures, thermal insulation of opaque horizontal closures)
2. On the building systems (two kinds: requalification of both the heating plant and the cooling plant)
3. On the renewable energy sources (two kinds: solar thermal installation, solar photovoltaic installation)

The cost of all the interventions on the school buildings, financed by European funds, totals approximately 19.3 million euros: about 13.2 million euros have been spent on energy redevelopment of the building envelope and just over 6 million euros for interventions on building systems and renewable energy sources. More than 50% of the interventions concern the building envelope, which accounts for approximately 70% of the costs.

The first observations of these data drive us to deepen the evaluation of effectiveness of the interventions: in building envelope renovation, the material choices and the comprehension of the work have important impacts on both the environmental and technological quality of buildings.

The cost of renovation per floor area is mainly less than 500 €/m² (Fig. 3.5). This figure is significant for an ordinary cost, which could represent the use of traditional technological standards, linked to common (not best) practices and, supposedly, to traditional materials.

In terms of cost/benefits, the declared average saving is 18%,[1] 5.2 euros per kWh saved (Table 3.1). The lowest cost possible per kWh saved is one of the evaluation criteria of the energy retrofitting funding for public administration buildings.[2] The PANZEB (National Action Plan for Nearly Zero Energy Buildings improvement)

---

[1] On 57 buildings.

[2] The other criteria are the shortest time for the construction time and the percentage of co-funding [8].

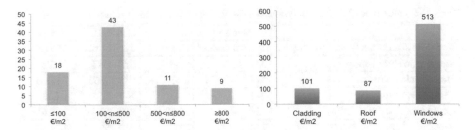

Fig. 3.5 Number of interventions for different costs per floor area and cost per square meter of intervention surface area

Table 3.1 Energy savings and relative costs analysis

Energy saving (without RES) [kWh]	Total energy consumption in 2012 [kWh]	Average energy savings [%]	Total cost for all interventions [€]	Cost per energy saved [€/ kWh]	Energy saved per heated volume [kWh/m³a]
4,357,863.60	24,208,966.42	18	22,785,086.53	5.23	8.95

estimated, for the energy renovations of existing school buildings and up to 2020, a potential energy savings per heated volume per year equal to 9, 19, and 21 kWh/m³ year, respectively, for climate zones C, D, and E; our average data for the three climate zones are 8.95 kWh/m³ with 50% of the buildings in climate zone D³ [8].

3.5 Phase 2. Monitoring and Modeling

This phase was carried out on selected buildings, the most representative of the 77 school buildings, determined by the cluster analysis methodology.[4] By this method, we defined two classes of buildings, with two buildings closer to the centroids that are called from here on the "reference buildings."

The two homogeneous classes, group 39 and 18 buildings, respectively, with the range of characteristics that define the class and the centroid values are given in Table 3.2.

The reference buildings, n. 2 and n. 58, were the subject of both energy consumption monitoring (for the 2016 year, considering that all the work had been completed by December 2016) and the dynamic simulation of the interventions to close the validation path of the design process.

Documentary energy consumption monitoring was carried out on a sample of approximately 10% of the buildings, which corresponds to seven buildings. An analysis of the savings obtained through the dynamic modeling of buildings was

---

[3] The distribution for climate zone of the buildings is 23 in zone C, 37 in zone D, and 17 in zone E.
[4] It was carried out on 57 buildings instead of 77 for more congruent and reliable data.

Table 3.2 Outcomes of cluster analysis and variables that determined the classes

	Cluster 1		Cluster 2	
Variables determinant for clustering	Typical values	Centroid	Typical values	Centroid
Shape ratio S/V [1/m]	0.29–0.74	0.46	0.51–0.91	0.75
Heated volume [m³]	2,682–41,633	8,535.77	685–4,156	1,906.00
Class jumps [number]	0–7	3.43	1–7	4.72
Cost of saved kWh [€/kWh]	1.33–34.97	12.57	5.76–69.89	25.61
Normalized consumption [kWh/m³ N]	7.82–49.79	25.26	6.97–58.56	24.59
Expected savings [kWh]	0.24–15.48	6.12	1.85–27.60	10.09

Table 3.3 Comparison between energy savings in the design proposal (ex ante), in the dynamic simulation, and after the realization (ex post)

		Energy consumption for thermal use [KWh/year]			
		Year 2012 (ex ante)	Year 2016 (ex post)	Dynamic modeling	Energy savings (%) after interventions (comparing 2012 and 2016 energy consumption)
Cluster 1	Building n. 2 (reference building)	75,146.58	59,853.96	56,933.00	20%
	Building n. 72	655,655.58	222,949.08		66%
	Building n. 60	51,840.36	44,855		13%
Cluster 2	Building n. 58 (reference building)	96,604,8	64,389.92	61,884.80	33%
	Building n. 43	46,095	42,815.52		7%
	Building n. 55	24,202,98	15,588.36		35%
	Building n. 61	36,118,44	21,144		41%

carried out on the two most representative buildings. Table 3.3 shows the savings data actually obtained by comparing energy consumption ex ante and ex post; it also contains the results of modeling on the reference buildings.

The conclusion of this phase is that, in terms of energy savings, the interventions made for the buildings considered – the most significant ones and the two reference buildings – imply an energy savings that corresponds to that stated in the project. In terms of energy performance only, the financing, design, and construction processes seem to have been efficient and praiseworthy.

3.6   Phase 3. On Field Investigation

Once having determined the effectiveness of interventions in terms of energy performance, we moved to this further level of investigation focused on the material and technological aspects of the project and of the realization.

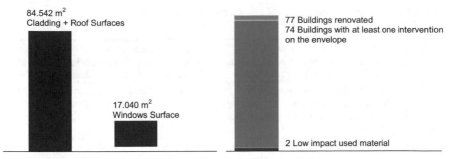

Fig. 3.6 Intervention surfaces and first material evidence

The interventions consist mostly of work done according to common practices, certainly not the best, and with no eco-friendly materials (Fig. 3.6):

- Coat insulation of the vertical cladding with 6–10 cm insulation in polyurethane or polystyrene
- Application of 6–10 cm roof insulation in polyurethane or polystyrene
- Replacement of existing windows with LE and safety double-glazed windows with aluminum frames

The 83% of the upgraded surfaces (about 85,000 m²) corresponds to opaque (horizontal and vertical) closures, but the overall cost for the window substitutions is greater than that of opaque surfaces. Seventy-four of 77 buildings with energy redevelopment are buildings with at least one intervention on the enclosure, but low environmental impact and/or ISO 14000 product labels were chosen in only 2 buildings.

The use of eco-friendly materials is therefore almost negligible; the reduction of the environmental impact in the whole of the interventions would seem to be attributed only to the energy savings and not to material choice.

In addition to these three levels of analysis, we chose to evaluate the interventions, from the material and technological point of view, according to the latest material and energy environment legislation: the Minimum Environmental Criteria (CAM). Even if this legislation comes after the interventions, a comparison (Table 3.4) could be useful to trace future trends and movements toward a more effective respect of environmental criteria, which are not completed or are partially met by the case studies as presented in the comparative picture of the criteria and interventions [9].

Finally, to verify the actual, in situ quality of interventions, thermal imaging was carried out on the selection of the seven representative buildings obtained from the cluster analysis.

Thermal imaging was carried out starting from the knowledge of the execution planning of the interventions: the investigation process followed examination of the specific technical documentation, thermal imaging of the building enclosure (for building elements), and evaluation of defects.

Table 3.4 Comparison between CAM criteria and project solutions

CAM criteria	Our data
New external windows should have a U value <1.8–2.1 W/m² K depending on climate region (1.8 for zone C, 2.0 for zone D, 2.1 for zone C)	Of 45, 24 interventions respect the CAM; the remaining 21 are without value in the final proposal and as-built documentation
New external windows with no renewable materials should have a percentage of recycled material; if in wood they should have at least a FSC or PEFC label	No information provided in all the 77 final proposals or in the as-builts
Old external windows should be recovered and recycled through a clear process and operators	No information provided in all the 77 final proposals or in the as-builts
It is necessary to realize a Green Roof can absorb emissions when an insulating roof intervention is expected	Absent in all the 44 interventions of roof insulation
In insulating roof interventions, materials with Solar Reflectance Index (SRI) > 29 for pitched roof and >76 when slope <15% should be used	No information provided in all the 77 final proposals or in the as-builts

The Solar reflectance is a measure of the constructed surface's ability to reflect solar heat, as shown by a small temperature rise

Fig. 3.7 Thermal imaging on the north-east façade of the building n. 72

Defect assessment was based on the categories of damage that represent the possible visible defects of a building enclosure on an existing building [10], that is, the air infiltration eventually present in the windows and in the joints or in the building elements connections, the uneven insulation, and the humidity and leaks.

Some examples of the investigations and criticalities found at this stage are provided in Figs. 3.7, 3.8, and 3.9.

The thermal imaging of Fig. 3.7 follows the intervention of windows substitution: an unusual thermal density of opaque vertical frames and cold glazing is evident, and furthermore significant thermal gradients are evident at the connection with the horizontal structures not involved in the renewal projects.

Figures 3.8 and 3.9 present two buildings with more than one intervention on the envelope; specifically, Fig. 3.8 is related to a building with interventions of both coat insulating on the vertical cladding and of windows substitution, and thermal imaging shows the evidence of a significant thermal gradient at ground connection.

Fig. 3.8 Thermal imaging on the north façade of the building n. 61

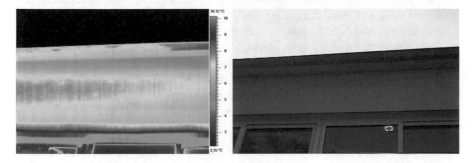

Fig. 3.9 Thermal imaging on the north façade of building n. 43

Figure 3.9 shows the thermal imaging of a building characterized by three interventions on the envelope: roof insulation, coat insulation of the vertical cladding, and window substitution. In Fig. 3.8, significant thermal gradients in the connection nodes between external windows and opaque vertical closures are evident.

Generally, in both the design and the realization of the interventions, the building enclosure is not considered as a set of elements (i.e., roof, cladding, windows) linked together but as isolated elements; the result is low quality in the interventions, which do not take into account the connections between the elements and determine, for the most part, uneven insulation defects.

3.7   Conclusions

The aim of this contribution was to highlight, through the studies of retrofitting of schools, some aspects concerning the existing and typical constructive features of the school buildings heritage in terms of existing materials and technologies used for energy improvement. The aim was to evaluate performances obtained and potentially obtainable from the energy–environmental profile and the technological quality of the realization. On the basis of the accomplished analysis and referring to the 77 case studies we can state that:

- Retrofitting interventions were mainly aimed at the thermal insulation of the opaque closures if compared to the interventions on systems and renewable energy sources, accounting for more than 70% of the costs of construction.
- The interventions are globally effective in terms of energy performance (because of the PA process).
- The interventions made for the reference buildings analyzed within the Phase 2 research imply an energy savings that corresponds to that stated in the project. The design and construction processes were reliable in terms of energy performance prevision.
- The average cost for energy retrofitting for the educational buildings in the Lazio region is 16,625 euro/m² of internal heated building surface.
- The quality of the work is rather low because the interventions are pointless and do not address the existing construction nodes and the connection between the building elements (floor-to-wall, wall-to-windows, etc.).
- Only the energy savings is responsible for the environmental impact reduction. Low-impact building material choices are limited to only 2 interventions with environmentally friendly material solutions of 74 realized.
- There are also many deficiencies in the environmental quality of the materials used in view of complying with the recent minimum environmental criteria (CAM) guidelines.

The performance picture obtained from the database, the comparison between expected/declared performance and the actual performance, and the considerations from the on-field investigation could be a useful framework to draw guidelines and support an effective 'green' policy for future calls for the allocation of public funding for schools retrofitting, especially in view of the obligations recently established by CAMs and reaffirmed both in the recent National Energy Strategy and in the National Strategy for Sustainable Development under preparation.

References

1. Dias Pereira L, Raimondo D, Corgnati SP, Gameiro Da Silva M (2014) Energy consumption in schools – a review paper. Renew Sust Energy Rev 40:911–922
2. Corgnati SP, Corrado V, Filippi M (2008) A method for heating consumption assessment in existing buildings: a field survey concerning 120 Italian schools. Energy Build 40:801–809
3. Dall'O G, Sarto L (2013) Potential and limits to improve energy efficiency in space heating in existing school buildings in northern Italy. Energy Build 67:298–308
4. Ascione F, Bianco N, De Masi RF, Mauro GM, Vanoli GP (2017) Energy retrofit of educational buildings: transient energy simulations, model calibration and multi-objective optimization towards nearly zero-energy performance. Energy Build 144:303–319
5. Miur. Dalla presentazione dell'Anagrafe edilizia scolastica del 7 Agosto 2015. http://www.istruzione.it/ediliziascolastica/anagrafe.shtml. Last visited on 12.04.2017
6. PREPAC Linee guida alla presentazione dei progetti per il Programma per la Riqualificazione Energetica degli edifici della Pubblica Amministrazione Centrale (D.M. 16 Settembre 2016). Maggio 2017

7. MISE. http://www.sviluppoeconomico.gov.it/index.php/it/energia/efficienza-energetica/pubblica-amministrazione. Last visited on 14.08.2017
8. ENEA, RAEE Rapporto Annuale Efficienza Energetica. Analisi e risultati delle policy di efficienza energetica del nostro paese, 2017
9. CAM Criteri ambientali minimi per l'affidamento di servizi di progettazione e lavori per la nuova costruzione, ristrutturazione e manutenzione di edifici per la gestione dei cantieri della pubblica amministrazione. Adottati con Decreto Ministeriale 11 gennaio 2017 (G.U. Serie Generale n. 23 del 28 gennaio 2017) e Criteri ambientali minimi serramenti esterni. Adottati con DM 25 luglio 2011 (G.U. n. 220 del 21 settembre 2011)
10. Lanzoni D Il quadro normativo nel settore della

Chapter 4
Renewable Energy in Argentina

Carlos Labriola

Abstract Since 2010, Argentina has had a revival of interest in the use of renewable energy sources (RES). In particular, with the GEN-REN plan, an impulse was given to the wind and solar energy systems in farm-type installations and biomass (ethanol or biodiesel) used in transport and agroindustries.

The number of bidders exceeded expectations, but the economic conditions of energy price in the wholesale electricity market brought about for a few of them to be installed.

In 2016, after the new administration took office, there was a call for the RENOVAR plan; 1000 MW were tendered but more than 6000 MW were offered. Market conditions were much better, and by September 2017, all selected projects will be implemented.

During 2010 to 2016, legal conditions for low voltage users were approved, which permit them to be energy generators, and these laws allow today the opening of a large market of wind and solar energy systems for domestic installations. In particular, several universities and institutes of renewable energy sources are researching on the use of wind turbines in tall buildings to supply the energy for common uses (elevators, water pumps, hallway lighting). It is also possible to adapt new building projects to the use of mini-hydro turbines by the accumulation of gray water.

In Neuquén Province, northwest Patagonia, a delegation of the National Institute of Industrial Technology is in Cutral Có City. That institution has a test bench for small wind turbines to certify its operation. In the Faculty of Engineering of the National University of Comahue is the Center for Study and Analysis of Applications of Renewable Energy Sources composed of three consolidated research groups. They are developing and testing different wind turbines of their own projects or local entrepreneurs. Also they developed and tested micro- and mini-hydro turbines in the Laboratory of Mini Hydro Power Plants. They made a project of energy efficiency on a building of about 10 floors, where it can obtain 20–25% more wind

C. Labriola (✉)
Faculty of Engineering, Universidad Nacional del Comahue, Buenos Aires, Argentina

© Springer International Publishing AG, part of Springer Nature 2019   39
A. Sayigh (ed.), *Sustainable Building for a Cleaner Environment*,
Innovative Renewable Energy, https://doi.org/10.1007/978-3-319-94595-8_4

speed than an average one and better radiation without obstacles. Also, using wind, solar (on terraces), and mini-hydro (on gray water pipes) energies, it can supply between 30 and 40% of energy for common services of a building especially in the hours of maximum demand, during the day and night. This kind of projects would allow to reduce the demand in the low voltage grid and to distribute generation in urban places. It represents for the users of the building a monthly RE equipment amortization fee during a few years.

4.1 Introduction

4.1.1 Beginning of the Use of the Renewable Energy Sources in Argentina

During 1984 Argentina implemented the first national institutions related to renewable energy sources. These were the Regional Centers for the Development of Renewable Energy Sources (RES) (Table 4.1):

The interest on RES of private companies in Argentina began a few years earlier. In 1977–1978 the San Miguel Institute, a part of the University of Salvador, joined some experts to work on photovoltaic cells and train in solar energy. Then some of them were absorbed by the National Commission of Space Research, dependent of the Argentine Air Force at that time. Then, during 1985, Neuquén began to work on geothermal energy prospecting in Caviahue-Copahue, and during 1988, it developed a rural electrification plan for dispersed inhabitants without access to the services, by means of PV systems. In addition, the province of Chubut, since the beginning of 1990, has been developing its rural electrification plan for dispersed inhabitants by means of installation of wind turbines and photovoltaic panels.

Table 4.1 Regional Centers for the Development of Renewable Energy Sources in Argentina 1984

Centers	Location	Characteristics
Regional Wind Energy Center (RWEC)	Rawson, Chubut http://organismos.chubut.gov.ar/cree	It continues working and developing the wind map of the Argentine Republic. It is also an international consultant for the study of the wind resource and installation of wind farms
Regional Geothermal Energy Center (RGEC)	Neuquén, Provincial Development Council	The first geothermal development in South America is in "Las Mellizas" lagoons (Caviahue-Copahue) where a power plant of 670 kW is installed. It is attended by the provincial energy company
Regional Micro Hydraulic Center (RMHC)	Obera, Misiones	Absorbed by other institutions, the National University of Misiones continues in developing micro-hydro turbines Banki and Pelton
Regional Solar Energy Center (RSEC)	Salta INENCO	Represented by the research Institute of non-Conventional Energy of the National University of Salta that develops solar thermal equipment

Since 1995, a national electrification plan for the dispersed rural population has been projected, evaluating how many inhabitants there were in that condition (4,000,000 inhabitants). Private companies were created, particularly in the Northwest of Argentina, which installed and maintained photovoltaic systems, with a subsidized rate of 50% on electricity tariff, but in 2001 the companies began to be deficient. Rural electrification throughout the country was continued in 2004 by means of the World Bank funds, which financed the Renewable Energy for Rural Markets Project (PERMER) [1].

In the case of wind farms, since the early 1990s, Argentina has been installing wind equipment in Patagonia. The first state wind farms had the problem that they were not properly maintained, resulting in a short service life (Table 4.2).

The national biofuels industry began during the 1980s. The government by means of the National Company of Fuels (YPF) promoted the consumption of "alconafta" in cars from 1984 to 1991 using bioethanol from sugarcane. Then, during the 1990s, the production of "alconafta" was suspended because it was not profitable.

In the mid-1990s, as a result of the opening to the international market, a new impetus was given to wind energy where several cooperatives and municipalities began to install wind turbines of European origin (Neg-Micon, Enercon, Bonus, etc.). These installations are made by cooperatives and municipal services connected at medium voltage to the Argentine Interconnection System (SADI). Most of these facilities are still in service today and have been the private pioneers in wind energy in the country (see Table 4.3).

Table 4.2 First wind farms in Argentina

Year	Wind farm	Turbines, type, total power	N°	Power	Situation
1995	Pico Truncado, Santa Cruz	VENTIS 20-100, 100 kW	10	1 MW	Disassembled
1990	Río Mayo, Chubut	Aeroman, 30 kW	4	120 kW	Disassembled

Table 4.3 Wind farms from 1994 to 2002 of cooperatives and municipalities

Location	Province	Service since	N° turb.	P. turb. (kW)	Total (kW)	Cumul. (kW)
C. Rivadavia	Chubut	July 1, 1994	2	250 – Micon	500	500
Cutral Có	Neuquén	Oct. 1, 1994	1	400 – Micon	400	900
Punta Alta	B. Aires	Feb. 1, 1995	1	400 – Micon	400	1300
Tandil	B. Aires	May 1, 1995	2	400 – Micon	800	2100
Rada Tilly	Chubut	Mar. 1, 1996	1	400 – Micon	400	2500
C. Rivadavia	Chubut	Sep. 1, 1997	8	750 – Vestas	6000	8500
Mr. Buratovich	B. Aires	Oct. 1, 1997	2	600 – Bonus	1200	9700
Darregueira	B. Aires	Oct. 1, 1997	1	750 – Vestas	750	10,450
Punta Alta	B. Aires	Dec. 1, 1998	3	600 – Bonus	1800	12,250
Claromeco	B. Aires	Jan. 1, 1999	1	750 – Vestas	750	13,000
P. Truncado	Santa Cruz	Nov. 1, 2000	2	600 – Enercon	1200	14,200
C. Rivadavia	Chubut	Jul. 1, 2001	16	660 – Gamesa	10,560	24,760
Gral. Acha	La pampa	Mar.1, 2002	2	900 – Vestas	1800	26,560

4.1.2   Training in FER

During the 1990s, the lack of specialized workforce, mainly in maintenance of renewable energy devices, was critical. From 1977 to 1990, courses were given on solar energy from the Solar Research Group of the University of Salvador with the support of the Argentine Air Force. This research group had been formed to develop photovoltaic cells but this project did not prosper. Eventually wind energy courses were given, and the Argentine Navy manufactured a wind turbine of 16 kW whose project was directed by Dr. Bastianon [2].

Since 1998, a postgraduate training in renewable energy has been developed in Argentina. The first institution to do it was INENCO at the University with Workshop methodology. Then this course was given in different places (Buenos Aires, Comodoro Rivadavia), expanding training possibilities.

But the reality on workforce market was that there were no technicians for large-scale RES device installations except for Big Hydro installations.

The author of this work returned to his country in 2000 after his postgraduate in renewable energy at the University of Reading and realized that there is no skilled workforce for installation, operation, and maintenance for RES equipment. That is why in 2003 he created the "Technicature" of Renewable Energy and Environment (REE) in the National Technological University [3], located in Plaza Huincul, which allowed to have the first 55 technicians in REE of the country in 2005. From this event other institutions and provincial education councils have been created, "Technicatures" in renewable and nonrenewable sources. Today practically most of the provinces with renewable resources have courses to be RE technicians.

From 2008, there were more postgraduate trainings offered in national universities such as the National University of Rosario, National Technological University (Buenos Aires), and National University of Comahue. The first two institutions developed a Master's Degree in Renewable Energy and the third a Specialization in Wind Energy. The National University of Rosario has been dedicated especially to the application of solar and wind energy sources in buildings and industrial installations.

In the case of the National University of Southern Patagonia, Santa Cruz, the Academic Unit Caleta Olivia dictates a degree course in Electromechanical Engineering with Renewable Energy Orientation, including the vector hydrogen as a vector of energy accumulation.

Also in Pico Truncado, Santa Cruz, is the first Hydrogen Generation Station by Wind Energy in South America (Fig. 4.1) which offered international courses of H generation [4], security, supply facilities, etc. Argentina has been a pioneer in Hydrogen (Santa Cruz) and Geothermal (Neuquén) technology applications in South America.

4.1.3   Legal Framework for Promotion and Regulation of RES

From the mid-1990s onwards, awareness about caring for the environment began in Argentina. It is initiated in primary and secondary education and then in state universities, also highlighting the advantages of renewable energy sources (RES).

Fig. 4.1 Pico Truncado plant of H [4]

Already in 1984, Argentina stopped using coal and oil to generate electricity in the country. Argentina begun to use only natural gas for electricity generation. Argentina, in the late 1990s, commissioned IPCC experts to study the influence of terrestrial overheating in the country for the next 100 years. It is the only country in America to deal with this type of study. These actions permit Argentina to begin to have a legal framework, which is summarized in the following table (Table 4.4):

4.2   The Great Takeoff of RES in Argentina

Law 26,190/2006 [9] of RES promotion allowed to initiate the national private wind industry that was impulsed by import protections and permitted to design and to produce wind turbines in the country by two companies: Pescarmona Industries (IMPSA) and the consortium NRG (Fig. 4.2). The first one developed a wind turbine of 2.5 MW with multipolar generator with Nd-Fe-Bo permanent magnets, without multiplier and pitch-controlled rotor. The second is a consortium of oil service companies that bought a 1.5 MW turbine design, built a prototype, tested it, and had it certified by international standards. In addition, NRG developed strongly the production of spare parts for wind turbines for international market.

From 1998 to 2014, Argentina suffered a lack of investments in exploration and exploitation of hydrocarbons and dry years for Big Hydro Energy plants. This situation produced an imbalance between energy demand and availability. So, Argentina had to import gas and oil. In 2010 the Argentinean government promoted GEN-REN plan [12] of 1000 MW for Wind, Solar, Biofuels and Mini-/Micro-Hydro implementation. Fifty-one projects with 1436.5 MW offered in this plan. During the middle of 2010, 895 MW were awarded, where 84% were wind energy (17 wind farms shared in Chubut, Santa Cruz, and Buenos Aires), 12% were thermal energy using biofuels (Buenos Aires, Corrientes, and Santa Fe), 1% were Mini-Hydro (Jujuy, Catamarca, Mendoza), and 2% of PV (San Juan).

Table 4.4 Laws related to RES

Law/year	Topic	Characteristics	Restrictions
25,019/1998 [5] (cooperatives impulse)	Wind and solar energy promotion	Tax advantages for investments in solar and wind energy. Creation of promotion funds for RE investments	Only for energy companies of distribution or generation plants
26,093/2006[a] [6] (biofuels impulse)	Biofuels promotion for production and use	Biodiesel promotion by means of soya oil. Twenty percent of national production should be used by distribution oil companies	Instability of local and external markets of biofuels
26,334/2008[a] [7] (ethanol impulse)	Bioethanol promotion	Ethanol legal framework obtained. Eight percent should be used in the local market	Instability of the local market
26,123/2006[b] [8]	H promotion to development, use, and applications as energy	National fund of H promotion and tax advantages for H industries. Five percent should be included in the pipeline	No regulation approved
27,191/2015–16 Review of 26,790	National promotion for RES use and applications	Argentina opens to international RE market. It permits a minimum 30% of national purchasing	National market is stable

[a]Legal framework was established with Laws 26,093/2006 [6] and 26,334/2008 [7]. Biofuels had the major production impulse with regulating its production and use: 5% of bioethanol should be included in gasoline and 7% of biodiesel in gas oil. This regulation guaranteed at the beginning the use of 20% of the local production of biodiesel in oil refinery companies. Biodiesel surplus is exported to Europe and other countries where Argentina is the largest producer of soybean derivatives in the world, including oil

[b]Argentina was a pioneer in Ibero-America in the development of technology, use, and an application of H as energy vector, but it has not been possible to develop its use because the H Law was not regulated

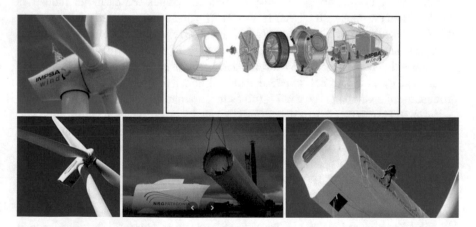

Fig. 4.2 Wind turbines manufactured in Argentina (Pescarmona Industries S.A., Mendoza [10];. NRG Patagonia S.A., Comodoro Rivadavia, Santa Cruz [11])

But in that time, the tariff conditions of Wholesale Electricity Market (MEM) did not give a capital return to amortize installation and operation. That is why few projects were implemented; for example, of about 754 MW wind farms, only 130 MW (20%) was installed. The restrictions on generation required special provisions for cogeneration, for example, specific events of large electric charges (such as spectacles, football matches, etc.) have to use autonomous generators because of the lack of energy and low reliability of the electric distribution in the low voltage system.

In spite of this, there is notable approval of provincial policies, for example, of San Juan province (2008), which promoted the installation of solar farms (2% of GEN-REN's) and a solar panel factory (2012) to produce wafers for PV cells (2014) (Table 4.5).

New tariff conditions were created for new power plant installations by "Energy Plus" plan because of the successive energy crises since 2004. Some thermal power plants (combined cycle) and wind farms (see Table 4.6) were built after GEN-REN 2010.

Figure 4.3 (left) shows wind farms that were in production in 2013. It can be seen that most of them are in Patagonia and south of Buenos Aires province. The Veladero (Barrick Gold mine) (7 in Fig. 4.3) in San Juan has the highest wind installation in the world (3000 m), 2.5/1.6 MW.

Table 4.5 Wind farms offered at GEN-REN 2010 (755 MW were offered, 130 MW were installed) [12, 13]

Wind farm	Company	MW	It began installation on
Malaspina I	IMPSA	50	
Puerto Madryn Oeste	Energías Sustentables S.A.	20	
Malaspina II	IMPSA	30	
Puerto Madryn II	Emgasud Renovables S.A.	50	
Puerto Madryn I	Emgasud Renovables S.A.	50	
Rawson I	Emgasud Renovables S.A.	50	OK – September 2011
Rawson II	Emgasud Renovables S.A.	30	OK – January 2012
Puerto Madryn Sur	Patagonia wind energy S. A.	50	
Puerto Madryn Norte	International new energies S.A.	50	
Koluel Kaike I	IMPSA	50	
Koluel Kaike II	IMPSA	25	
Loma Blanca I	ISOLUX S.A.	50	OK – July 2013
Loma Blanca II	ISOLUX S.A.	50	
Loma Blanca III	ISOLUX S.A.	50	
Loma Blanca IV	ISOLUX S.A.	50	
Loma Blanca I Básica	Sogesic S. A.	49,5	
Loma Blanca II Básica	Sogesic S. A	49,5	

Table 4.6 Wind farms by "Energy Plus" plan

Wind farm	Company	MW	It began operation on
Arauco – La Rioja	IMPSA	50.4	2011–2013
Diadema – Chubut	Compañía Argentina de Petróleo S. A.	6.3	2011

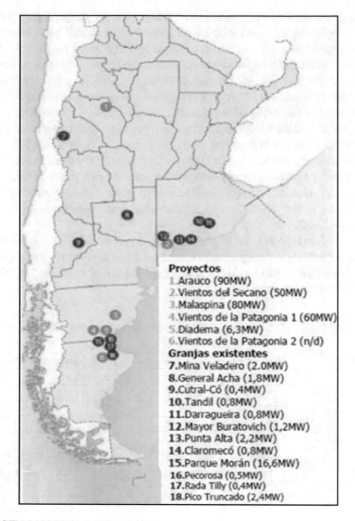

Fig. 4.3 GEN-REN 2010 plan (green) [13]

Facilities 8–18 correspond to cooperatives; these did not exceed 3 MW per wind farm, except Antonio Morán Park in Comodoro Rivadavia, Chubut, of 16.6 MW (15 in Fig. 4.3). Since 2001, several cooperatives entered into financial crisis and did not perform the corresponding maintenance, and some machines are out of service today (15 in Fig. 4.3).

Changes in the electricity tariff of Wholesale Energy Market during 2016 and the application of Law 27,191 (Law 26,190 review) permitted RENOVAR plan [14, 15] was implemented by the present government (Fig. 4.4). This plan will permit to have 8% of RE into the grid in 2018 and 20% in 2025. The first stage of RENOVAR-1 required 1000 MW of RE installations. It had 123 offers with a total of 6343 MW. The government accepted 103 offers (5209 MW) for wind farms (3468 MW), solar

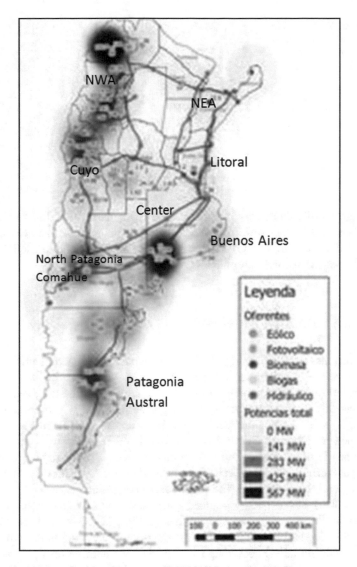

Fig. 4.4 500 kV national grid and location of RENOVAR-1 projects [15]

farms (2811 MW), Biofuels and Biogas plants (11 MW), and Mini-Hydro plants (5 MW). Eighteen offers were not accepted for 598 MW Wind, 506 MW Solar, and 30 MW Biomass. It may be that in the future, the RE proposal will be increased more than 1000 MW (RENOVAR 2, during 2017–2018).

RENOVAR-1 plan gave a real and great impulse to RES in Argentina, and it allowed to continue projects of GEN-REN 2010 which were standstill. In the case of Patagonia because of the transmission limit on a 500 kV network, the power to be installed is limited to 600 MW in Comahue and 400 MW in Southern Patagonia.

Table 4.7 Offered prices in RENOVAR-1 and reference ones for the government (blue) [15]

Tecnología	PRECIO USD/MWh		
	OFERTADO		MÁXIMO DE ADJUDICACIÓN
	Mínimo	Promedio*	
EÓLICA	49,1	69,5	82
SOLAR	59,0	76,2	90
BIOMASA	110,0	114,6	110
BIOGÁS	118,0	177,8	160
PAH	111,1	114,5	105

Average Wholesale Energy Market prices offered in RENOVAR-1 for Wind and Solar farms (Table 4.7) were lower than the government reference ones. During 2017, 59 RE projects of RENOVAR-1 will be finishing its installation. RENOVAR-2 will be during the second half of 2017.

4.2.1 RES Pending Development in Argentina

In Argentina, as described, energy facilities have been developed by means of wind energy (since 1990), solar energy (since 1985), mini- and micro-hydro power (since 1960), and biomass energy (since 1984). In the case of geothermal energy, during 1985, Neuquén province began studies of geothermal energy prospection at "Las Mellizas," Caviahue-Copahue Geothermal Park, and install in 1990 the first geothermal power station of South America (670 kW). It works until 1997 when it began to be out of service because of steam well decay. Today three provinces are making geothermal energy prospection: Neuquén on the volcano Domuyo (North Comahue), and Salta and Jujuy in North West of Argentina.

During 2010 ocean energy sources (tidal and wave) began to be studied in Argentina by several National Universities (UNPA, UTN, etc.). Technology has been developed (marine and fluvial hydrokinetic turbines, oscillating column systems, and nodding tubes) according to available data, but no prototypes have been installed in situ until now. At present, efforts are being made to unify a database on wave and tidal characteristics and ocean currents not only on the coast but also on the high seas. This data will be able to have details of the oceanic resources to have microscale maps and to be able to design appropriate prototypes.

4.3 Domestic Energy Generation Situation

Since the 1980s, several provinces of Patagonia began rural electrification using PV panels and wind turbines in Neuquén (Fig. 4.5) and Chubut. Then, during the 1990s, a rural electrification plan was put in practice, which allowed isolated rural

Fig. 4.5 Rural electrification plan in Neuquén (Mina La Continental, left; Currumil Quillen, School. N°65, right) [16]

Fig. 4.6 INTI testing plant in Cutral Có, Neuquén [17]

electricity users served by a rural electrical equipment company with a subsidized rate in most of the rest of the Argentine provinces.

In 2004, the PERMER National Plan [1] financed all the provincial rural electrification plans by means of PV panels. During 2014 and 2015, laws which promoted domestic electric power generation by RE were developed in provincial governments, where Neuquén, Santa Fe, Córdoba, and San Juan were the pioneers. In addition, since 2008, the University of Córdoba developed a laboratory for matching PV panels and the National Institute of Industrial Technology (INTI) installed a test bench in Cutral Có, Neuquén to verify characteristics of low-power wind turbines. This test bench is relevant to test wind turbines in isolated sites in the provincial rural electrification plans (Chubut, Neuquén, etc.) and in remote communication stations, particularly in hard winds of Patagonia. Figure 4.6 shows a Cutral Có wind turbine testing plant, which is able to certify turbines up to 10 kW.

Since 2010, architecture and civil engineering companies began to apply RE systems in buildings. The opening of the international market during 2016 permitted domestic solar and wind applications to be integrated in houses and buildings up to 20 floors, which are located in cities in the most critical areas of electrical demand. Figure 4.7 shows technology which can be applied for domestic and rural use.

Fig. 4.7 RE devices developed and/or available in Argentina for domestic and rural applications Wind turbines developed in UNCo (right); solar thermal and PV technology (center); micro-hydro (left)

4.4 Conclusions and Recommendations

It can be seen that the development of the applications of RES devices in Argentina has been very dependent on the economic situation and sometimes the country's politics. But in spite of this, the last 40 years has had a great growth, especially in the last decade, generating a legal and tariff framework that allows RE applications in constant evolution with greater installed power reaching 2500 MW in 2018 (10% of total demand, 25,000 MW).

Argentina has been a pioneer in South America in the use of different renewable energy products and devices: domestic photovoltaics (1980), geothermal (1985), private wind farms (1990), biomass (1983 ethanol, 1995 soybean biodiesel), and micro-hydro (1940). In the last three decades and with more emphasis in the last 5 years, wind farms, thermal power plants using biomass, and PV farms have been included in the Argentine System of Electrical Interconnection (SADI), giving the bases in the next future to export Energy to Chile (Hydrocarbons dependent) and Brazil (Big Hydro dependent).

Development Centers and Universities give the possibility to obtain the necessary Human Resources in the next future to designing, construction, installation, operation, and maintenance of RE systems.

RE converters developed and tested in Argentine Institutions (micro-hydro and wind turbines), and solar thermal and PV (Fig. 4.7) applied on projects of energy efficiency on buildings of about 10 floors, permit supply between 30 and 40% of energy for common services of a building, especially in peak demand hours, during the day and night. These developments will allow to reduce the demand in the low voltage grid and to distribute generation in urban distribution networks.

In Argentina, the development of RE technologies and exchange of experiences had been possible thanks to research and development between European and Argentine Universities and Institutes, taking experience by means of European RE equipment installed in Argentina.

References

1. Schmukler M, Garrido S (2016) ELECTRIFICACIÓN RURAL EN ARGENTINA. ADECUACIÓN SOCIO-TÉCNICA DEL PROGRAMA PERMER EN LA PROVINCIA DE JUJUY. Universidad Nacional de Quilmes, ASADES, 2016
2. Bastianon (1987) First Wind Turbine developed in Argentina, Argentine Navy, 1985–1987
3. TSERMA (2003–2005) "Technicature" on renewable energy and environment, first course to be technicians in RE applications, National Technological University, 2003
4. Course H (2006) International course of H production and applications, H Plant of Pico Truncado, Santa Cruz, September 2006
5. Law N° 25.019 (1998) REGIMEN NACIONAL DE ENERGIA EOLICA Y. Objeto Alcance Ámbito de aplicación Autoridad de aplicación Políticas Régimen de inversiones Beneficiarios Beneficios Sanciones Fondo Fiduciario de Energías Renovables Sancionada April 12 1998
6. Law N° 26.093 (2006) Régimen de Regulación y Promoción para la Producción y Uso Sustentables de Biocombustibles. Aprobación 19/04/2006; promulgación 12/05/2006
7. Biofuels (2010) "Impacto de la Ley Nacional N° 26.093de Biocombustibles sobre la producción y el uso de los recursos en el sector agropecuario. Instituto de Economía y Finanzas – UNC, Aug 10 2010
8. Law N° 26.123 (2006) PROMOCION DEL HIDROGENO: "Desarrollo de la tecnología, la producción, el uso y aplicaciones del hidrógeno como combustible y vector de energía". Objetivos. Sujetos. Autoridad de Aplicación. Infracciones y Sanciones. Creación del Fondo Nacional de Fomento del Hidrógeno. Régimen Fiscal Promocional. Disposiciones complementarias. Sancionada: Agosto 2 de 2006. Promulgada de Hecho: Agosto 24 de 2006
9. Law N° 26.190 (2006) "Régimen de Fomento Nacional para el uso de fuentes renovables de energía destinada a la producción de energía eléctrica". Objeto Alcance Ámbito de aplicación Autoridad de aplicación Políticas Régimen de inversiones Beneficiarios Beneficios Sanciones Fondo Fiduciario de Energías Renovables Sancionada: Diciembre 6 de 2006 Promulgada de Hecho: Dec 27 2006
10. IMPSA (2014) Pescarmona Industries S.A. first wind turbine of 2.5 MW manufactured in Argentina
11. NRG (2015) NRG Patagonia S.A., local manufacturer of Turbines Type II of 1,5 MW in Argentina since 2015
12. GEN-REN (2010) Renewable energy plan 2010, National Ministry of Energy, Argentina
13. Gareis M, Ferraro (2011) Revisión del estado de la Energía Eólica en la Argentina, período 1994–2010. Universidad Nacional de Mar del Plata, HYFUSEN, 2011
14. RenovAr (2016) Convocatoria Nacional e Internacional, Ronda 1, Ministerio de Energía y Minería de la Nación Argentina, July, 2016
15. Renovar Plan (2016) El Gobierno obtuvo precios mínimos récord" Jorgelina do Rosario, INFOBAE, Septiember 30, 2016
16. EPEN (2017) Rural electrification plan 1985–1995. Provincial Development Council, Neuquen
17. INTI (2017) National Institute of Industrial technology, wind turbine test bench, Cutral Co test Plant, Neuquén, 2012

Chapter 5
Wind Energy Potential Research in a Low Building within an Urban Environment

Jorge Lassig, Claudia Palese, Juan Valle Sosa, Ubaldo Jara,
and Carlos Labriola

Abstract The use of wind energy for distributed generation in urban environments, where the intensity of turbulence begins to be a decisive factor, is currently one of the main challenges within renewable energies.

Wind tunnel tests have evaluated shaped buildings that can capture wind within the urban atmospheric boundary layer by directing them to wind turbines located in the building so that they produce electricity. In particular, this work evaluates a low building within the city of Neuquén, in Patagonia, Argentina.

The methodology consisted in constructing a 1:200 scale model, performing wind tunnel tests to characterize the flow pattern on the roof of the tunnel. Wind measurements were also performed for two years with an automatic weather station on the roof of the building. The possible power generation was modeled with two prototypes of small wind turbines: one of horizontal axis and one of vertical axis, both developed in the zone. The annual energy savings that could be achieved would be of the order of 20%.

5.1 Introduction

Wind is a resource that exists all over the world and under certain orographic conditions can significantly increase its potential of energy. The so-called concentrator effects can significantly improve the average speed of the wind flow. These effects do not occur only in nature, but also in urban environments, which make it recommendable to study as an energy resource.

The integration of wind turbines into buildings is becoming a new possibility of energy efficiency and has begun to be studied in university research centers on wind energy, such as the Delft University of Technology in the Netherlands.

J. Lassig (✉) · C. Palese · J. V. Sosa · U. Jara · C. Labriola
Faculty of Engineering, Universidad Nacional del Comahue, Buenos Aires,
Neuquén Argentina
e-mail: lassig@uncoma.edu.ar

© Springer International Publishing AG, part of Springer Nature 2019   53
A. Sayigh (ed.), *Sustainable Building for a Cleaner Environment*,
Innovative Renewable Energy, https://doi.org/10.1007/978-3-319-94595-8_5

This university began to study the use of small-scale wind turbines in several Dutch cities such as Amsterdam, The Hague, Tilburg, and Twente, and also in the United Kingdom [1].

Actually, placing a wind turbine above a building is a possibility to obtain electricity, taking advantage of the effect of wind acceleration on it. Wang et al. [2] discuss this possibility. On the other hand Andrew Grant et al. [3] study turbines located at vertical and horizontal axes inside ducts, which are installed in the edges of the terraces of high buildings, taking advantage of the suction that occurs on them. Lin Lu and Ka Yan Ip [4] perform simulation studies in Computational Fluid Dynamics (CFD), with different sets of buildings, and obtain as result the increase of wind speed between those buildings can be between 1.5 and 2 times the average speed of the wind, increasing the available wind power three to eight times. On the other hand Nalanie Mithraratne [5] assesses the possibility of installing small wind turbines on roofs of houses in New Zealand and concludes that it could reduce carbon emissions in that country by 26–81%.

5.2 Methodology

All buildings produce an acceleration of the free wind flow in particular places located near it. Taking distance to the building, the wind speed tends to be the speed of the free wind flow.

In a building with sharp edges at windward, the boundary layer separates on these edges Schlichting et al. [6], and the separation bubbles are formed at the sides and at the top of the building. The main stream is deflected around it, and a large stele is at leeward. Figure 5.1 shows a partial scheme, which is a 2D CFD output, partially reproducing what happens (the stele is missing, since this is a 3D phenomenon).

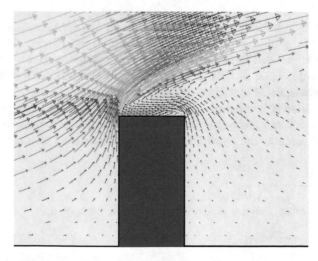

Fig. 5.1 Wind speed vectors produced in surroundings of buildings (2D model)

The results of this separation are a region with low velocities, a level of high turbulence, and the recirculation of flow in the roof and in the sides of the building. This recirculation region must be avoided for the location of the wind turbines. Therefore, it is important to know the size of the recirculation region.

To determine the best location for wind turbines on the roof of a building, we must know, among other things, the characteristic of the wind in that city and the size and shape of the building, and it is relevant to make use of analytical and experimental tools to solve the problem.

5.3   Climatology of Local Winds

As the buildings are fixed to the ground, and the wind can change direction, so the choice of type of wind turbine and its location will depend on the prevailing direction of the wind in the place.

Figure 5.2 shows the wind roses for several Argentine cities, where it can be observed that particularly in Patagonia the winds are very directional, and then the

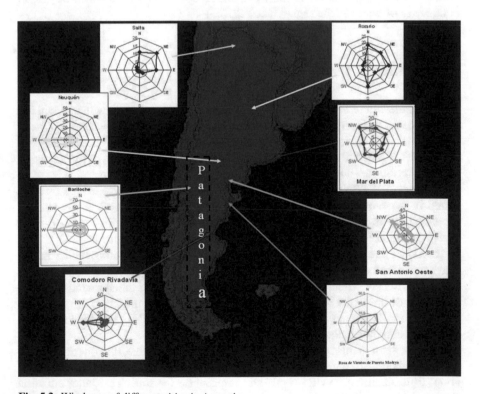

Fig. 5.2 Wind rose of different cities in Argentina

Fig. 5.3 West face of the Faculty of Engineering (FENUCO) building in Neuquén (left); aerial sight of FENUCO building (right)

buildings can be aligned with them, and this situation permits to offer the face that receives the wind to install the wind turbines.

This is the case of the building of the Faculty of Engineering of the National University of Comahue, located in the city of Neuquén, where the longitudinal axis of this building is aligned with the north-south direction, so there is a side face of about 80 m long that looks to the west, which coincides with the side of greater occurrence of winds in the region (see Fig. 5.3).

5.3.1 Wind Tunnel

It is very relevant to find the location of maximum wind and areas with and without turbulence for a given building, and then define where to install the wind turbines.

It is very relevant to find on a building points of maximum wind speed and places with and without turbulence on it. This data permits to locate wind turbines on building roofs. The wind tunnel is an adequate tool for this purpose.

In the case of FENUCO building, a model of it with scale 1:100 was made to be tested in the Wind Tunnel N°2 of the Laboratory of Dynamics of Environmental Flows. The tests on it are made at different preferred wind speed directions to obtain and optimize the best possible location of wind turbines on its roof (Fig. 5.4).

Reynolds number on the FENUCO building tests was around 300,000, taking into account the main length of the building.

These tests consisted, firstly, of determining the height of greater wind acceleration to windward of the building, for that smoke was used as marker of the trajectory of the air flow when passing over the building. Figure 5.5 illustrates the location of its points' maximum flow.

It can be seen in this figure that downstream of the roof edge, the flow accelerates and rises about 40°, and 2 m on the terrace, it reaches its maximum at a height of about 5 m. Then the flow descends and impacts on the roof. This latter situation produces a very turbulent flow.

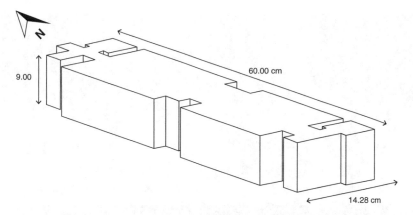

Fig. 5.4 Dimensions of scale model (1:100) of the FENUCO building used for wind tunnel tests

Fig. 5.5 Wind flux pattern developed on the roof

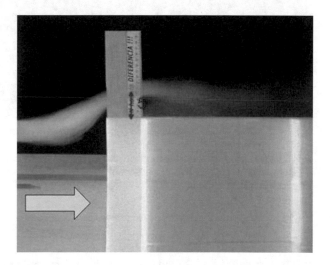

In order to qualify which part of this flow is more stable, and to determine the possible location of the wind turbines, we proceeded to install on a part of the roof of the building model, several tufts. Figure 5.6a shows the orientation of the markers, representing their location one meter above the roof with a westward wind (270°) that is perpendicular to the building. Figure 5.6b corresponds to the same wind direction (west) but the tufts were located representing 2.5 m, 5 m, and 7.5 m height above the building.

Comparing the two images, it is observed that one meter above the ceiling, the flow is very chaotic (indicated in the photograph by the multidirections that the markers and the "moved" strokes of them, indicating the presence of vortices).

On the other hand, in Fig. 5.6b where the tufts were located at higher heights of the roof, they are more stable at the front of the building, and particularly at the

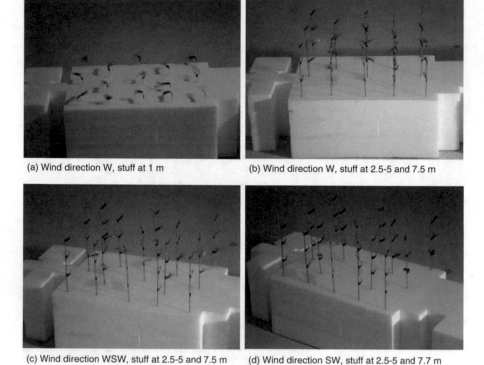

(a) Wind direction W, stuff at 1 m   (b) Wind direction W, stuff at 2.5-5 and 7.5 m

(c) Wind direction WSW, stuff at 2.5-5 and 7.5 m   (d) Wind direction SW, stuff at 2.5-5 and 7.7 m

Fig. 5.6 (**a**) Tufts 1 m above the roof with the west wind. (**b**) tufts at three heights above with west wind; (**c**) tufts at three heights above the roof with the wind from west-southwest; (**d**) tufts at three heights above the roof with wind from the south west

levels of 5 and 7.5 m. In the center of the ceiling and at 2.5 m height, the markers indicate the presence of strong vortices.

Figure 5.6c corresponds to the west-southwest wind (247.5°), and Fig. 5.6d shows the effect of the southwest wind (225°) above the building roof. In both cases, the observation is similar to that discussed for Fig. 5.6b, except for slight changes in wind direction. Therefore, the most stable area for the location of wind turbines is on the wall facing west, above 2.5 m. In the center of the building the most stable area is above 10 meters. In the center and leeward of the building the flow is very turbulent, and it is not recommended to install wind turbines there (Fig. 5.7).

Fig. 5.7 Roof zoning
which results as
conclusions of Fig. 5.6.
(Green) maximum and
stable winds, (yellow)
turbulent, (red) turbulent
and wake effect

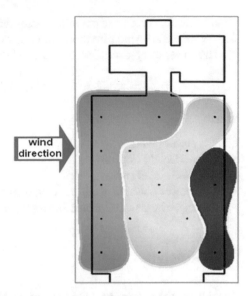

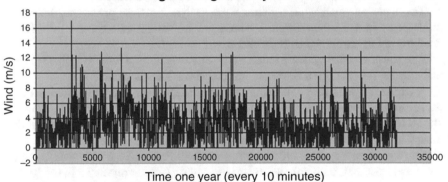

Fig. 5.8 Records of average wind speed on the FENUCO roof, 2010

5.3.2   Meteorological Data

FENUCO has for two years (2010–2011) an automatic meteorological station
(Anemos™) installed on the roof (north side) of the building with two anemometers
at two different levels that ensure to obtain records of the wind. Figure 5.8 shows
2010 record data. The average annual wind power is defined as:

$$P_m = \frac{1}{2} \cdot \rho \cdot \int_0^\infty f(v) \cdot V^3 dv$$

where $f(v)$ is the wind distribution function for the place of measurement.

As a result, the power potential measured for the west edge of the building was 122 W/m² for all wind directions and 105 W/m² for west direction.

The power to be extracted from a wind turbine is:

$$P = \frac{1}{2} \cdot \rho \cdot V^3 \cdot C_p \cdot A$$

where

- P is power (Watt)
- ρ is air density (kg/m3)
- V is wind speed (m/s)
- C_P is power coefficient of the wind turbine
- A is area of the wind turbine facing the wind

5.4 Results

To estimate the power potential to be extracted from the wind on the FENUCO roof, we must define the area to be used and the C_p of the wind turbine.

Two types of small wind turbines were analyzed, whose prototypes have been constructed in Comahue Region. These are of horizontal axis but of different designs, which are shown in Fig. 5.9a, b.

The power coefficient for a small horizontal axis turbine with two blades can be estimated at 0.4, and for a Savonius turbine with a low fastness of 0.3.

If we use the 80 m side of the roof that looks to the west, and we estimate a height of 1.5 m, this gives us about 120 m² of useful area for wind turbines. Table 5.1 shows the electric energy possible to obtain on the roof of the FENUCO building during a year taking into account the type of wind turbine.

The annual energy measured for the entire building is 316 MWh/year. If we use fixed wind turbines, with only westward orientation (green zone at 2.5 m in Fig. 5.7), it is possible to save 10–14% of the energy to be consumed with the proposed ones.

If the turbines location is above 7.5 m from the roof, then they could be used with free rotor orientation and the turbines could contribute to 12–16% of the energy consumed per year (considering winds from all directions).

5.5 Conclusions

Energy obtained by means of wind turbines which take advantage of the acceleration of wind on the sides and roofs of buildings is an actual possibility for energy efficiency in buildings. In fact it is possible to apply in several cities in Europe where there are examples of this.

A working methodology has been presented to evaluate the possible wind potential on the roof of a building, to perform a climatic analysis of the wind in the city.

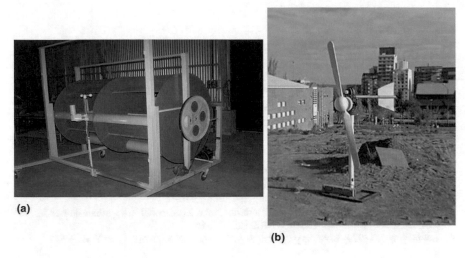

(a)

(b)

Fig. 5.9 Wind turbine prototypes made in North Patagonia (**a**) Savonious wind turbine of low solidity, (**b**) Horizontal axis wind turbine

Table 5.1 Electric energy possible to obtain during a year with a wind turbine on the FENUCO roof

According to wind data on FENUCO wind turbine developed in Neuquén	MWh/year $C_p = 0.3$	MWh/year $C_p = 0.4$
All directions	38	51
Only west wind directions	33	44

This methodology needs to make measurements "in situ" on the building roof and to use a wind tunnel to determine the most suitable place to locate the wind turbines.

In the case of the FENUCO building, the possible energy savings has been quantified, which is between 10% and 16% of the annual energy consumption, where this range include values found by other researchers mentioned above.

Many buildings, with many small wind turbines on its roofs in a city, could contribute to save a significant percentage of energy demand to the grid and also savings in installations.

Argentina, particularly provincial legislation during the last 5 years, permits domestic users to inject electricity to the electrical grid in LV and MV. But this advantage is not in all provinces.

The advantage of distributed generation is that it does not require any type of investment by the state and permits to reduce the investments in large power stations between 20 and 40% for local energy.

The cost and investment required are not analyzed in this paper, since this is the subject of other variables that escape wind engineering.

References

1. Jadranka Cace, RenCom; Emil ter Horst, HoriSun; Katerina Syngellakis, IT Power; Maíte Niel, Axenne; Patrick Clement, Axenne; Renate Heppener, ARC; Eric Peirano, Ademe (2007) Urban wind turbines – guidelines for small wind turbines in the built environment. Wind Energy Integration in the Urban Environment (WINEUR); http://www.urbanwind.net/ (10/Sep/2009)
2. Wang F, Baia L, Fletcher J, Whiteford J, Cullen D (2008) The methodology for aerodynamic study on a small domestic wind turbine with scoop. J Wind Eng Ind Aerodyn 96:1–24. Ed. Elsevier
3. Grant A, Johnstone C, Kell N (2008) Urban wind energy conversion: the potential of ducted turbines. Renew Energy 33:1157–1163. Ed. Elsevier
4. Lu L, Ip KY (2009) Investigation on the feasibility and enhancement methods of wind power utilization in high-rise buildings of Hong Kong. Renew Sust Energ Rev 13:450–461. Ed. Elsevier
5. Mithraratne N (2009) Roof-top wind turbines for microgeneration in urban houses in New Zealand. Energ Buildings 41:1013–1018. Ed. Elsevier
6. Schlichting H (1979) Boundary layer theory, 7th edn. McGraw-Hill, New York, p 817

Chapter 6
Natural Light in Architecture: Use Inspired by the Constructive Tradition

Fabio G. S. Giucastro

Abstract This paper is centred around the subject of natural light and its applications in architecture.

Solar light has always played an influential role among traditional construction techniques, to the extent that it is regarded as one of the invariant materials in the art of building.

As a case study, we put forward the analysis of the solar light in a typical farm of the Sicilian vernacular architecture. In this building, it has been possible to collect data on the natural light captured inside the individual rooms, each single day of the year.

Inspired by the nature, the current experimentation of "transparent building envelopes" has allowed to convey daylight, by adjusting the dimensions of both openings and obstructions, thus making these systems progressively more innovative and enabling them not only to capture natural light, but also to produce energy for men and for daily needs.

6.1 Introduction

This contribution – an integral part of the PhD thesis in Architecture Technology entitled "Conservation and Valorisation of Identity Characters. The Maddalena Peninsula", coordinated by Prof. C. Truppi – is developed around the theme of natural light and its applications in architecture.

Sunlight, in traditional construction techniques, has had a significant influence so much so that it is considered one of the invariant materials of the art of building.

Many scholars, analysing the relationship between natural and built environment, have identified in the study of traditional architecture the dynamics capable of suggesting the adequate use of natural resources and a perfect match of artifacts made by man with the surrounding environment. The aim of technology

F. G. S. Giucastro (✉)
SDS of Architecture of Siracusa, University of Catania, Catania, Italy

© Springer International Publishing AG, part of Springer Nature 2019
A. Sayigh (ed.), *Sustainable Building for a Cleaner Environment*,
Innovative Renewable Energy, https://doi.org/10.1007/978-3-319-94595-8_6

"becomes the appropriation of the elements object of building to modify, without destroying it, the existing environmental context. Relating to the nature system, it is crucial to determine the priority of a more natural adaptation of man to the environment by deepening the links with biology, bioclimate, natural resources, in a dialectical view of the cognitive moment and of the practice" [1].

So, in contrast to the modus operandi so far adopted by some architecture, to understand "the richness and complexity of the relationship between technology and the environment, it is necessary to gather all the message range that the environment itself transmits: history, materials, the culture of living, the lightness or gravity characters expressed by local architecture" [2], to make the constructive technique adapt to nature. The proof is the result of vernacular architectures built according to the place's respectful logic, "characterizing and conditioning the form: in an understanding that transcends the false belief that it can build any building in any environmental context by giving it the preeminent function that binds the environmental valences to architecture" [1] through knowledge transmitted for generations from man, a product of nature, created by nature in accordance with its laws.

From this, it is evident that it is not possible to ignore the boundary conditions in a particular project as well as the evolution that constructive systems have had over time in a particular geographic area.

6.2 Light as Material

At different latitudes, for different environmental conditions, solar rays and their incidence angles determine different types of constructions and architectural conformations, just like the materials of the site [3].

The interest in light vision problems has always been present in human history, both in the artistic and in the scientific sphere. The light natural or artificial, is used to illuminate objects, environments, and the presence is often denied obtaining shadow areas.

Capturing the sunlight, it is conveyed into the places by giving shape to the spaces in a similar way to what happens in the presence of any other material that influences the perception of man and "the solution of the internal-external relationship permits to collect air and light as building material" in a project "that reads the past in function of the present, aimed to a future which is conceivable, expressible in all its possibilities" [4].

A building material, that has always been a central theme in the art of building, from ancient times until today, aimed to shape the spaces, as men did in the past, while trying to survive in a hostile nature, they also thought of intercepting a ray of light so that it could illuminate a predetermined point and, at times, only in certain days of the year.

Originally, light control fuelled man's claim to dominion over nature, observing how astronomical conditions are invariant with respect to the moodiness of nature itself. Learning that the sun rises and sets always in the same position, on certain days of the year, and by harnessing sunlight rays at certain hours of the day, the building is simultaneously connoted of a temporality consisting of the formation and transformation of natural light, according to accurately calculated forecasts.

A presence that has ancient origins, but which continues, even today, to be the subject of study and interest, carefully measured and dosed, light has passed through the ages of man's history coming to us intact and arousing feelings and emotions despite the passage of time. A research that describes and investigates elements of reality, replacing "the usual constructive procedures (...) an authoritative principle derived from the intangible elements of nature: air, light, colours, giving visual and tactile consistency to the empty space that envelops the objects of everyday life. We are looking for a new root of things and man in nature, where today, instead, he is helpless to his destruction" [4], teaching us to coexist with it not only creatively, but also to perceive how the reading and the knowledge of its signs can become indispensable to explain scientific and design problems in today's architecture.

6.3   Case Study and Mediterranean Vernacular Architecture

This study investigated the logic that involved designing, as a function of the capture of direct sunlight, of a typical building localized in the Mediterranean area along the Sicilian east coast and "the ability to read the story in a design key", according to P. Valery, according to which "it is necessary to put the past not behind us, but between the present and the future, and read it according to the present for a perspective of the future".

The experimentation, in the field of transparent building envelopes and in the capture of direct solar input, cannot ignore the observation of the pre-existence and in particular of the development of the logic that led to the realization, affirming "the ariosity, the brightness, transparency as characters able to legitimize the use of new building materials and at the same time to restore consistency to those aged".

In a mild climate in winter and heat in summer, design choices are intended to maximize thermal input and disperse it to a minimum in winter, while shielding is needed to provide better environmental comfort in the summer. Among the elements that have made the interior comfortable, during the summer months, there are, on all, loggia, balconies, verandas and pergolas, which together gave birth to that characteristic image of architectural constructions in the Mediterranean area.

The sizing studied and not random of the openings and their orientation has allowed to develop typological solutions that in some cases have been trampled over time, while in others have remained unchanged since the realization of the vernacular architectures.

The constructive logics and invariants, which have been able to characterize the built environment landscape in the course of the centuries and to a different extent, are recalled to inspire and bring us back to the path of harmonization with the place.

Within the spatial scope of the study, there are some invariants inherent in the contribution of natural light:

Fig. 6.1 Adorno
farmhouse, type "C"
southeast of Sicily

- Continuous improvement regarding the size and exposure of the openings with regard to both thermal transmittance and luminous flux.
- The protection system both from direct sunlight and from external ventilation, for example, using the pergola or verandas typical of the Aeolian architecture. And yet the use of clear shades for façade plasters that reflect the incident light and reduce the percentage of absorption of solar radiation.
- Reduction of the *S/V* ratio of the building, thus reducing the direct solar incident on the lateral surface and covering of the building [5].

Depending on the orientation, the façade openings were adjusted by the greater or lesser exposure to external climatic conditions. For south-facing façades the use of small size openings and "a high mass, typical of high-density materials and common feature of traditional architectures in the Mediterranean area, greatly improved the energy performance of the building envelope giving it a thermal flywheel function and modulating element of heat flows between exterior and interior with consequent energy saving for summer air conditioning" [6] and an increased thermal inertia.

In the Sicilian areas, the windows were made of wood and of small size to have the double advantage of an air permeable surface, but refractory to the incidence of solar radiation on thc outer casing.

The typological study of traditional architectures has highlighted the presence and importance of the court or "bagghiu" not only from a formal point of view but also for the regulation of the internal microclimate. In fact, it was a huge air tank that cooled the surrounding environments during night time and gradually released fresh air during daylight hours. In addition, during the day, thanks to the arrangement of the openings on it and their formal composition, the sun rarely managed to act directly on the façades for a long time, allowing to have natural diffuse light within the environments [7] (Fig. 6.1).

6.4 Applying a Solar Analysis Method

These results were achieved through a method applied to a typical home of Syracuse vernacular architecture – "C" type with internal court – for which it was possible to determine all direct light data captured within individual environments.

With this method, we aim to obtain "data acquisition for the design of solar systems that are suitable for the recovery of solar energy (...) (trying) to obtain an architectural form capable of optimum sunlight," irrespective of 'use of sophisticated technologies' [8].

The method, which cannot always be valid for all boundary conditions, such as cloudiness or transparency in the atmosphere, is intended to provide a useful tool for producing a more conscious architecture, avoiding the correction, with a use deformed by artificial conditioning, or mistakes born in the project.

Polar diagrams were used on which the concentric circles represent constant solar (β) lines while the rays identify the constant (Ψ) solar azimuth lines [9].

The highest curve refers to the summer solstice (21 June), the lower curve at the winter solstice (21 December) (Figs. 6.2 and 6.3).

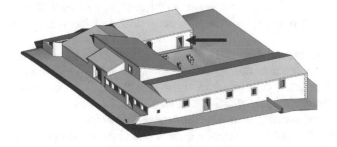

Fig. 6.2 Adorno farmhouse, type "C" 37 ° N. Picture by A. Alì

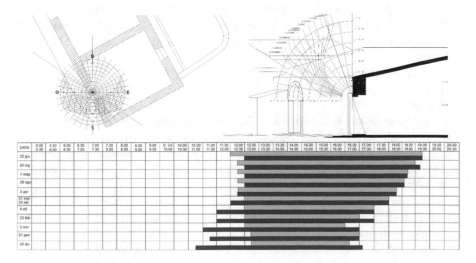

Fig. 6.3 Solar diagrams of a southwest opening

The results, summarized in the images shown, allow us to monitor the artefact for twelve days throughout the year, one for each month, and in particular provide the sunshine period, the direction of the solar rays – that is, their projection both horizontally and vertically – and drawing shadows.

6.5   Transparent Enclosure and Testing

As a result of adaptation with the place, the relationship between the size of buildings and living beings has changed; external and internal exchange roles, inside and out, meet in the transparencies. In fact, due to its external skin function, the diaphragm between the building body and surrounding environment, the envelope – considered energy-efficient if it is a dynamic interface, continuous and active interaction with external climatic factors – has undergone progressive modifications, following the evolution not only of the necessities of living, but also of the architectural techniques that allow the entire building industry to exploit the most important energy source for its own benefit, precisely because "through the transparency is created a different way of capturing light and the exterior," the air of nature, producing energy and allowing it to consume less and less.

Scientific studies have allowed to convey solar light by adjusting the dimensions of the openings, the materials of the windows, the exhibits and the obstructions by renewing "a classical view of architecture: experimentation, derived from technical innovation, integrates and updates a humanistic substance that Is an ineliminable and essential component of architectural design" [10].

In the recent past, solar radiation control was allowed through the use of shielding, which prevented access. Since the foundational core of the modern project lives in the constant tension of experimenting with a constructibility that is alternative to the present reality, today, these systems, and more and more innovations, allow not only to filter natural light but "exchange matter and energy in a programmed way to respond to the variations of environmental incentives and the needs" [11] of the inhabitants.

The International Energy Agency has drawn up a classification of the most important sunlight capture systems. These can be distinguished in active and passive systems.

The former modifies transmission characteristics according to external environmental variations, and among them, chromogenic systems, particularly electrochromic systems, which vary the transmission, reflection and absorption of sunlight within buildings, can be included, depending on the amount of electricity coming from outside.

The seconds change the amount of thermal and luminous energy following the angle of incidence of the solar ray by virtue of their formal composition. The latter systems are divided into the deflection and diffusion of solar radiation; among primes, prismatic panels and laser cut panels directing light to interior environments, some of the nanogel-composed devices, a granular and transparent material inserted into the intercommunication chamber systems.

Often the design of building enclosures did not take into account the amount of solar radiation incident which, if conveyed to a minimum, could produce a quantity of energy to meet the needs of those who live in them, reducing energy consumption and lowering CO_2 emissions. Solar technologies for the production of electricity are basically photovoltaic, fuelling an ever-expanding industry with new experiments that can provide continuous benefits in architectural applications and aim for the so-called next-generation photovoltaics with the introduction of novelties in cell materials not only silicon but also printed on thin aluminium films, with inorganic semiconductors, such as high-performance CIGS. The latest experiments have allowed the production of organic dye (DSSC) cells [12], whose nature-inspired operation has allowed to conceive an artificial process of photosynthesis obtained through the addition of organic molecules, following the principle of an architecture no longer excavated from nature, but energy derived from nature, producing force that transforms, through practical action, the same nature (or matter) by innovating spaces, places and times of living.

The use of these cells is advantageous also because they can be used with diffused light, whose intensity is largely inadequate for traditional systems. With new systems, it is expected to achieve efficiency levels of 20–25%, but with considerably lower costs and greater architectural integrity than any existing surface and structure.

6.6 Conclusions

Natural light, according to the latest sector analysis, still represents a low energy source. To overcome this lack, it would be useful to deepen the research on the relationship between the transparent part of the building and the matt part with the aim of obtaining high levels of natural illumination. The resulting benefit could be recorded not only by an energy-saving point of view with the proper use of solar radiation incident but also in terms of environmental comfort within the environments as well as ergonomic studies have largely demonstrated. On the other hand, but expressed in other words, Vitruvio also came to the same results, and in *De Architectura* he argued that the humour and character of man depends on the place where he lived on the basis of the amount of natural radiation present.

With the use of devices and systems that make changes to architectural artefacts to optimize thermal yield, it is preferable to identify a methodology that retrieves an appropriate approach to the environment. The building made up of thick masonry or equivalent materials brings benefits due to increased requirements related to the control of parameters defining the conditions of internal comfort, protection against cold and heat, sound insulation, light availability and natural ventilation, and energy saving.

The objective is to study and examine the formation of spontaneous architectures, to understand the constructive logic with which they were built, to reconstruct the design vocation of architecture within us by widening our attention and sensitivity to the problem of the link between pre-existence and innovation in all fields of knowledge, drawing useful considerations for the design of the new, not just pointing to one piece, but making the unique piece repeatable.

References

1. Truppi C (1984) Dall'apertura senza vetri all'infisso-facciata. Fratelli Fiorentino, Napoli
2. Vittoria E (1987) Le tecnologie devianti per la progettazione ambientale. In: Gangemi VE, Ranzo P (eds) Il governo del progetto. Edizioni Luigi Parma, Bologna
3. Cennamo MS (1976) Luci, colori, suoni: materiali da costruzione per l'architettura, in Costruire, 1976, n.97
4. Vittoria E (1994) Lo spazio vuoto dell'habitat. In: La Creta R, Truppi C (eds) L'architetto tra tecnologia e progetto. Franco Angeli, Milano
5. Fiorito F (2009) Involucro edilizio e risparmio energetico. Flaccovio Editore, Palermo
6. Rossi M (2009) Prodotti e sistemi di involucro innovativi per il progetto di edifici energeticamente efficienti Procedure, simulazioni termodinamiche e criteri progettuali per un'applicazione nel Sud Europa, Tesi di dottorato di ricerca di Tecnologia dell'Architettura XXI ciclo, Università degli Studi di Napoli "Federico II"
7. Giucastro F (2008) The environment built in Siracusa's countryside. Country houses, farmsteads and villas: memories from the past to exploit the territory's peculiarities. Maggioli, Santarcangelo di Romagna (Rn)
8. Truppi C (1980) Tecnologie bioclimatiche per il controllo dell'habitat. Edizioni della libreria, Napoli
9. Beccali GE, Butera F (1974) Determinazione grafica del soleggiamento su superfici verticali e orizzontali. Protezione delle facciate ed ombre portate, In Condizionamento dell'aria, riscaldamento, refrigerazione, 1974, 11
10. Truppi C (1994) Architettura Tecnologia Progetto, in Tecnologia e ambiente, Notiziario della Sezione Tecnologia e Ambiente del Dipartimento di Progettazione Urbana, Napoli, 1994, 1, 59
11. Peguiron G (2005) Prefazione. In: Altomonte S (ed) L'involucro architettonico come interfaccia dinamica. Strumenti e criteri per un'architettura sostenibile. Alinea Editrice, Firenze
12. Pagliaro M, Palmisano G, Ciriminna R (2009) Il nuovo fotovoltaico. Dal film sottile alle celle a colorante. Dario Flaccovio Editore, Palermo

Chapter 7
The Cost of Building to the nearly-Zero Energy Building Standard: A Financial Case Study

Shane Colclough, John Mernagh, Derek Sinnott, Neil J. Hewitt, and Philip Griffiths

Abstract The EU has mandated that all buildings are built to the nearly-Zero Energy Buildings (nZEB) standard from 2020. The *Passivhaus* standard has been in existence for over 25 years and potentially offers a tried and tested method of achieving nZEB, but can it be used as a cost-effective means of achieving nZEB?

This paper analyses the cost differential of building dwellings located in the south-east of Ireland to the nZEB standard using the passive house methodology, in comparison to building to the current prevailing minimum building regulations. A comparison of the two standards is also made to determine the suitability of using the passive house standard as a means of achieving nZEB compliance. In the analysis, the extra cost (compared with building to the minimum building regulations) include increased airtightness, insulation levels, a heat recovery and ventilation system and higher performing windows and doors. Cost reductions are achieved in the elimination of the traditional heating system, chimney stack and reduced site overheads. Costs are based on a designated date for the works of 1 January 2017, exclude VAT at the prevailing standard and reduced rates, exclude cost of site purchase, and exclude any design team or professional fees arising.

The costs are compared on an element-by-element basis using the National Standard Building Elements and Design Cost Control Procedures [1] National standard building elements and design cost control procedures. ISBN-13:978-1850531647 edn. Environmental Research Unit) format for comparison, the accepted industry standard in the Republic of Ireland for subdividing the overall cost of construction into logical and defined cost headings, and are assembled in order of the sequence of construction. The comparison shows that while differences exist in individual elements, the overall cost differential between constructing a residential dwelling to current building regs and that of *Passivhaus* standard is just

S. Colclough (✉) · N. J. Hewitt · P. Griffiths
Ulster University, Newtownabbey, County Antrim, Northern Ireland
e-mail: S.Colclough@Ulster.ac.uk

J. Mernagh · D. Sinnott
Waterford Institute of Technology, Waterford, Republic of Ireland

© Springer International Publishing AG, part of Springer Nature 2019   71
A. Sayigh (ed.), *Sustainable Building for a Cleaner Environment*,
Innovative Renewable Energy, https://doi.org/10.1007/978-3-319-94595-8_7

+€131 excluding VAT. It is noted that while this specific analysis has been carried out on the basis of a case study, it is proposed that the analysis will be of general applicability given the similarities in the large cost items between those mandated by the building regulations and those required in order to achieve the passive house standard (such as insulation levels).

7.1 Introduction

Given the planned 2020 implementation of the nearly-Zero Energy Building (nZEB) standard across the European Union, the well-established passive house (PH) standard is seen as a viable means of achieving the mandated high-energy efficiency standard. While a number of publications have been written to investigate the potential for the passive house standard in the Irish climate (e.g. [2, 3]) and a number have considered net zero energy buildings [4, 5] and also compared the PH with the newly defined nZEB standard for the Republic of Ireland [6], little has been written about the cost of achieving nZEB in comparison with the current building regulations for constructed dwellings. This paper takes the approach of analysing the costs in detail for a case study scheme of houses.

The nZEB standard in Ireland (to be finalised in 2019) requires that dwellings must consume less than 45 kWh/m^2/a [7]. Moran [8] carried out research into the life cycle cost, energy and global warming potential analysis of nZEB in a temperate oceanic climate, and found that for a residential semi-detached nZEB, focus should be placed on minimising the space heating requirements through high thermal and air tightness performance, and covering the remaining energy demand, through renewable sources such as a biomass boiler or heat pump. Colclough et al. [6] demonstrated that a case study certified passive house dwelling complies with the nZEB standard using this approach, i.e. a passive house in combination with heat pump and solar PV. Thus the financial case study considered here is seen to be highly relevant for future dwellings in the temperate maritime climate.

While other developers have reported that building to the passive house standard is cost neutral compared with the current building regulations [9], obtaining detailed information on the construction costs of dwellings is difficult given commercial sensitivity. The developer of the case study dwelling considered here has carried out cost analyses since 2010 of building to the passive house standard compared with the prevailing building regulations, and it is seen that costs have reduced in line with the increasing energy efficiency standards of the building regulations. Mullins [10] reported that the additional cost for a single dwelling was €18,010, while this cost was seen to reduce to €4000 in the case of a scheme of dwellings in 2015 [11].

This paper provides a detailed breakdown of costs for a scheme of houses which is described below, comparing the cost of constructing the dwellings to the current minimum building regulations – i.e. achieving a building energy rating (BER) of A3, with the cost of construction to comply with the nZEB – i.e. a BER of A1. The BER is the measure of energy performance and reflects the amount of energy required for space and water heating, ventilation and lighting, based on standard occupancy.

The nZEB dwellings were designed using the Passive House Planning Package (PHPP) and are independently certified as achieving the passive house standard. In meeting the passive house standard, the dwellings outperform the nZEB requirements with respect to air infiltration (achieving 0.6 air changes per hour rather than the mandated 7 m³/m²/h), and make use of a mechanical heat recovery ventilation system which is not required for nZEB compliance. The passive house might therefore be expected to be more expensive than a dwelling merely complying with the nZEB standard. However, this analysis will show that the elimination of the traditional heating system, along with the more streamlined construction process for the passive house, offsets the aforementioned additional costs.

The detailed base costs presented are produced independently of the developer and are based on market costs current to 1 January 2017 subject to the caveats listed in the second paragraph of this report. The cost differential associated with constructing to the minimum building regulations (i.e. to a BER of A3) and constructing to the passive house standard (to achieve a BER of A1) is analysed in consultation with the developer of the scheme of case study dwellings.

7.2 Description of the Case Study Dwelling

The case study dwellings comprise three bedrooms and two-storey semi-detached houses, each with a total floor area of 103 m² (see Figs. 7.1 and 7.2). The PH dwellings are completed within a 13-week construction period, while an extra week is

Fig. 7.1 Front, rear and side elevations of case study dwelling

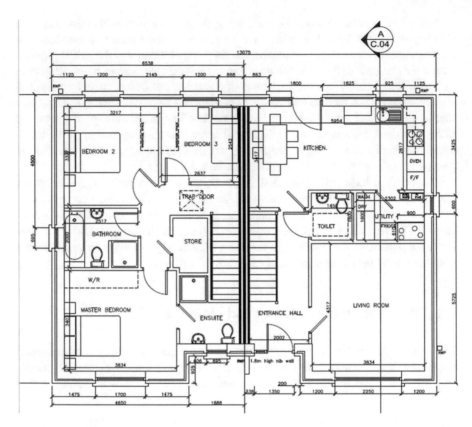

Fig. 7.2 Plans of case study dwelling

required in the case of the A3 dwelling, primarily for work associated with the wet heating system and associated chimney.

While the scheme was designed to achieve passive house certification and to meet the minimum renewable energy required by the current building regulations, analysis [6] has shown that such a dwelling is also compliant with the nZEB standard as it achieves a building energy rating (BER) of A1, i.e. less than 25 kWh/m²/a.

For both the A1 and the A3 dwellings, the construction method is that of 10 cm external rendered blockwork, 5 cm cavity and insulated internal timber frame leaf. Each dwelling has 6 m² of solar PV panels in order to meet the renewable requirement of the building regulations.

However, while the dwellings have many similarities, there are a number of significant differences. For example, in the case of the Passive house, certified PVC triple-glazed windows and doors are used, and an airtight attic hatch ensures that the required air infiltration standard of less than 0.6 changes per hour is achieved at 50 Pa. The heating system used is a Nilan Compact P heat recovery ventilation system which provides both space and DHW heating via an integrated electric heat pump. In comparison for the A3 dwelling, double-glazed windows, less external

Table 7.1 Parametric comparison: energy-influencing elements

Item			A3	A1
Thermal envelope			{W/m²K}	{W/m²K}
		Roof	0.16	0.07
		Floor	0.21	0.08
Insulation U value		Walls Windows & Doors	0.21 1.24	0.17 0.74
Ventillation system			Natural ventilation	Heat recovery Ventilatio
Heating system	Main backup		Oil fired central Htg & Multifuel stove	2 × 550 W electric rads 2 × 175 W towel rads & air post heater
DHW			Immersion	Heat pump

insulation and natural ventilation result in reduced construction costs, whereas extra costs are incurred in the provision of the traditional wet heating system with associated chimney. Table 7.1 gives an overview of the parametric comparison comprising the main energy influencing elements of the dwellings.

7.2.1 Cost Comparison

The approach taken in this analysis is to compare the cost of constructing the house depicted in Figs. 7.1 and 7.2 to achieve a BER of A3 with that required to achieve an A1 rating. The analysis is based on a scheme of houses (for both nZEB and A3 costs examined) and not a single house project and therefore includes the economy of scale and buying power a developer enjoys when constructing this number of concurrent buildings of a similar design.

Table 7.2 gives a breakdown of the estimated construction costs for both the same dwelling to comply with the current building regulations (i.e. achieve a BER of A3) and comply with the future mandated nZEB regulations (i.e. achieve a BER of A1).

In Element (19) (substructure), 80 mm less insulation is required in the foundations of the A3 dwelling compared with the nZEB, resulting in a cost reduction of −€725, while an additional cost of +€10 excluding VAT is required to reduce cold bridging associated with the chimney stack inherent in the design of the A3 dwelling, leading to an overall net reduction in cost of −€715 excluding VAT for the A3 dwelling in this element.

Examining Element (21) (external walls) in a similar fashion, the current building regulations require a less costly wall build-up and level of insulation to the cavity wall (saving −€1786) and also attract less cost for the detailing required to eliminate cold bridging in the building envelope (resulting in further savings of −€1384),

Table 7.2 Cost comparison (2015) and nZEB building regulations

Schedule of Areas (M2)	M2	M2
Gross Floor Area of New Build (GIFA M2)	102	102
Total Gross Floor Area (GFA M2)	**102**	**102**
Elemental Breakdown of Estimated Costs	**€**	**€**
Item	nZEB (A1)	Current regs (A3)
(19) Substructure	6923	6208
(21) External Walls	12,108	8938
(22) Internal Walls	7462	7462
(23) Suspended Floors	4233	4233
(24) Stairs/Ramps	1894	1894
(27) Roof	8114	8114
(28) Frames	–	–
(31) External Wall Completions	11,850	10,950
(32) Internal Wall Completions	7989	7989
(33) Suspended Floor Completions	–	–
(34) Stair completions	621	621
(37) Roof Completions	–	–
(41) External Wall Finishes	4554	4554
(42) Internal Wall Finishes	4905	4905
(43) Floor Finishes	1946	1946
(44) Stair Finishes	–	–
(45) Ceiling Finishes	5444	5444
(47) Roof Finishes	6665	6665
(52) Drainage/Wastes	704	704
(59) Mechanical Services (including associated builders works)	9,838	12,488
(66) Transport Services	–	–
(69) Electrical Installation (including associated builders works)	7890	7890
(74) Sanitary Fittings	2266	2266
(79) Building Fittings	3156	3156
(-) External Works	2500	2501
Subtotal 1	**109,131**	**106,997**
Preliminaries	3800	6065
Subtotal 2 EX VAT	**114,862**	**114,993**

which combined lead to an overall cost reduction of €3170 excluding VAT in this element compared with the nZEB standard.

External wall completions (Element (31)) compare the cost of the external windows and doors and show a cost saving of −€900 excluding VAT when constructing the current building regulation dwelling compared with the PH dwelling, as double-glazed windows (at a cost of €3300) are less costly than the triple-glazed nZEB windows (at a cost of €4200).

Element (59) (mechanical installation (plumbing and heating)) shows that there is an additional cost of +€2650 excluding VAT associated with constructing to the current A3 regulations over the nZEB dwelling (via the passive house route). All associated and ancillary costs associated with the plumbing and heating of each unit in this element are included. Cost savings are achieved in the A3 unit by not having to provide an HRV system to meet the low space heating demand of 10 W/m² of the *Passivhaus* standard (thus saving −€8748). Similarly, the two 550 watt heaters to the upstairs bedrooms (saving −€1090) can be omitted. However additional cost will be incurred for the A3 dwelling when compared with the PH standard in the following areas: (1) a chimney stack and associated plasterwork and capping will be required at +€2931; (2) mechanical ventilation will be required to the A3 unit with 5 fans included in this design at a cost of +€922; (3) the A3 unit will require a traditional heating system with associated radiators, oil burner and cylinder at a cost of +€6125; (4) an electrician will be required to wire the foregoing at +€350; (5) a carbon monoxide alarm will be needed in the A3 unit at a cost of +€85; (6) the builders' work associated with the foregoing, i.e. trenches, boxouts, oil line, cradle, opes and plinth, will carry a cost of +€775; and (7) stove and hearth +€1300.

Finally (06) preliminaries include the time-related costs for the construction of the units but not incorporated in the finished dwelling, e.g. the cost of site supervision, cost of scaffold, cost of insurances, cost of plant, etc. It is seen that there are additional costs for the construction of the A3 dwelling due to an increased length in the programme critical path of 5 working days (which arise in the mechanical and electrical first fixed and the construction of the chimney stack in the A3 unit not required of the PH). These cost differences can be summarised as follows: add cost of site overhead and preliminaries associated with additional 5-day programme (+€2750), add planning contribution discount not available to A3 unit +€225, omit cost of PHPP Fee −€540, and omit cost of blower door test −€170; therefore net additional cost to construct the A3 unit €2265 excluding VAT.

The total cost of constructing an nZEB dwelling (i.e. achieving a BER of A1 by building to the passive house standard) therefore amounts to a reduction of €131 excluding VAT when compared to an A3 rated dwelling at current building regulations.

7.3   Conclusion

This analysis has shown that the construction cost differential of ensuring compliance with the nZEB standard (i.e. achieving a BER of A1) compared with constructing to the minimum building regulations (i.e. achieving a BER of A3) have been reduced to €131 for the case study dwelling which was constructed to the passive house standard, with the nZEB dwelling being less expensive.

Acknowledgements The authors wish to acknowledge the support of the Interdisciplinary Centre for Storage, Transformation and Upgrading of Thermal Energy (i-STUTE) under EP/K011847/1 and Enterprise Ireland for funding this research. In addition, the authors are grateful for the assistance of Michael Bennett and Seamus Mullins, QS without whose assistance this paper could not have been completed.

References

1. Anon (1993) National standard building elements and design cost control procedures. ISBN-13:978-1850531647 edn. Environmental Research Unit.
2. Clarke J, Colclough S, Griffiths P, Mcleskey JT (2014) A passive house with seasonal solar energy store: in situ data and numerical modelling. Int J Ambient Energy 1:35–70
3. Colclough SM (2011) Thermal energy storage applied to the Passivhaus standard in the Irish climate. University of Ulster, Newtownabbey
4. Colclough S, Griffiths P (2016) Financial analysis of an installed small scale seasonal thermal energy store. Renew Energy 86:422–428
5. Hernandez P, Kenny P (2010) From net energy to zero energy buildings: defining life cycle zero energy buildings (LC-ZEB). Energ Buildings 42(6):815–821
6. Colclough S, O'Leary T, Griffiths P, Hewitt NJ (2017a) The near Zero Energy Building standard and the Passivhaus standard – a case study. In: Passive and low energy architecture conference, Edinburgh, 3–5 July 2017
7. Anon (2012) Towards nearly zero energy buildings in Ireland, planning for 2020 and beyond. Department of Environment, Community and Local Government, Dublin
8. Moran P, Goggins J, Hajdukiewicz M (2017) Super-insulate or use renewable technology? Life cycle cost, energy and global warming potential analysis of nearly zero energy buildings (NZEB) in a temperate oceanic climate. Energ Buildings 139:590–607
9. Colley J (2017) Ireland's largest passive house scheme shows way to nZEB. Passive House Plus pp 20. Available from: https://passivehouseplus.ie/magazine/new-build/ireland-s-largest-passive-house-scheme-shows-way-to-nzeb
10. Mullins S (2010) Last update, Rosslare case study – passive house cost analysis. Available: http://www.seai.ie/Renewables/REIO/SEAI_REIO_2010_Events/See_the_Light_Conference_9th_September_2010/Rosslare_Case_Study_-_Passive_House_Cost_Analysis.pdf. [July 2017]
11. Mullins S (2015) Last update, Madeira Oaks Affordable passive housing project – cost analysis (a case study). Available: http://www.passivehouseacademy.com/downloads/Cost_Analysis_of_Bennett_Wexford_Passive_House.pdf. [14 July, 2017]

Chapter 8
Policy of Intensification, Diversification, Conservation, and Indexation in Pursuing Sustainable Transport

Koesmawan

Abstract The Indonesian government has announced at least two regulations in energy policy, consisting of Law No. 30/2007 and Government Regulation No. 79/2014. Two provisions relate to national energy policy (NEP) and sustainable energy needs. The energy utilization paradigm has also changed. Formerly, energy was considered an export commodity as a part of state revenue, but now energy is crucial development capital to improve the people's welfare.

New and renewable energy (NER) has been outlined in the General Planning of Energy. From now to 2020 has been targeted at 135 gigawatts (GW), which is compulsory, and the figure of 45 GW should be covered by new and renewable energy (NRE). With a contribution of 45 GW, that means NRE has contributed 23% to the total national capacity; the rest is fueled by other energy, that is, oil, gas, and coal, with almost the same contributions at 25%, 22%, and 30%, respectively. Geothermal 7.2 GW, hydro 17.9 GW, micro-hydro 3 GW, solar 6.5 GW, wind 1.8 GW, and other sources at 3 GW are all government targets.

Of course, it is not easy to realize practically all the targets. There are many challenges to face. Some concerns are how to have the same thinking pattern in the development of new and renewable energy among institutions. Another problem is that intensive business schemes are not yet optimal. Also, technology is still dependent on other countries; consequently, the level of domestic or local content is still small. Therefore, at this time renewable energy is still expensive, but in the future this restriction can be overcome so that it will be cheaper than fossil fuel energy.

Furthermore, this chapter discusses another issue that requires the preparation of a large amount of sustainable energy for transportation, which of course must be in line with attention to the concept of eco-energy. Thus, the right choice for transportation problems for all fuel vehicles, especially cars and motorbikes, must be bio-energy; however, we have used bio-diesel and applied it long term. In addition, bio-gas is included in selecting sustainability transport with eco-energy orientation. The type of transport vehicles such as by land, sea, and air must be well considered.

Koesmawan (✉)
The Commisssioner of FID, Indonesian Energy Consultant and The Chairman of Quality Assurance, Ahmad Dahlan School of Economics, Jakarta, Indonesia

© Springer International Publishing AG, part of Springer Nature 2019   79
A. Sayigh (ed.), *Sustainable Building for a Cleaner Environment*,
Innovative Renewable Energy, https://doi.org/10.1007/978-3-319-94595-8_8

8.1 Introduction

8.1.1 Transportation for Mankind

Since there have been people in the world, transportation has been an important part of their lives. Of course, one of the first means of transportation was just one foot after another, from one location to another, so that a long distance traveled by foot could take as much as a day or many days. Then, faster transportation was developed by riding a fast-paced animal; the horse, a high-end animal, can run at speeds of up to 50 km/h. Thus, for a location that earlier took days to reach, on a journey by horse can be counted in hours. Further development by river or sea transportation was by raft, with human power; then we produced ships, from small boats to finally giant ships capable of moving 10,000 to 200,000 passengers, from a close destination to a very distant goal, even allowing living for months In the middle of the ocean.

With the invention of the machine, modern transportation tools were also found, in addition to the horse, which was originally a very simple choice, to a very sophisticated flying boat. The fuel that is used is still energy derived from fossil resources. This chapter discusses sustainable transportation by using alternative energy instead of fossil fuel, as the energy from fossil sources will be exhausted before long, so that raw materials for sustainable transport are required with respect to the environment. One result is to extend the average human lifespan to 65 years, because this kind of transport fuel will be less polluting and allow longer life. Even in the city of Yogjakarta, Indonesia, people on average can still live as long as 70 years.

8.1.2 Modus of Transportation

In general, from the past until today there have been only three modes of transportation, namely, transportation by land, sea, or air. Ground transportation generally occurs on a place on one island where there is already a road vehicle; if there is no access, then we search for sea transportation, even air transport.

Along with the development of the era, many people have been able to have land transport or a vehicle; thus, many people can have the opportunity to buy a car or motorcycle without any obstacles when using transportation. In Indonesia, if you have a permanent standing, the address is clear, and the interviews are with honest faces, then the vehicle purchase is made without a down payment, so that you can have a new car easily. As a result, of perhaps the total population of Indonesia, Jakarta numbers 14 million by day, but at night it remains at 8 million, as 6 million more have their living in border areas such as Jakarta, Bogor, Tangerang, Bekasi, and Depok. As a result, many big cities in Indonesia, ranging from Medan, Palembang, Merak, Jakarta, Bandung, Yogyakarta, and Madiun, to Surabaya and others, are filled with traffic, which then creates a traffic jam. For example, a trip that usually takes only 10 min during a normal situation could be 3 h or more during traffic jam conditions.

8.1.3 Current Transport Fuel

To run the wheels of various modes of transportation, we prepared raw materials in the grades of premium, premium plus, and other fuels. All transportation materials in the world are still dominated by fossil fuel. Abandonment of this material is beginning because the pollution level is very high, and the level of congestion is increasing, which then increases the calculated disadvantages. This congestion is one reason why the government intends to move the capital from Jakarta on the island of Java to Palangkaraya on the island of Borneo.

8.1.4 Disadvantages Using Fossil Fuel

A. Congestion
In this case, the traffic jam is generally caused by the wishes of people with good vehicles such as motorbikes or a private car. The government has asked us not to use these vehicles because the bus facility is well established. If half the workers are willing to obey these instructions, there will be a significant effect. When not using a car or motorcycle that is causing traffic, then one can easily ride the bus-way, which will definitely reduce the number of cars on the road used as a means of transport.

B. Pollution
Motor vehicle accidents are very bad. Almost all provincial capitals are heavily polluted because of the amount of carbon that is hard for people to process, even if it is ejected into the sky. It affects the ozone (O_3) holes. This action is in conflict with the program of "Environmentally Friendly:" we must pay more attention to pollution.

C. Run Out of Oil in 2050?
It is highly probable that by 2050, the petroleum reserves will have begun to decrease and may even be exhausted. For that, we as human beings, have to start with many people helping other people. One of the good ways for everyone to help is by not buying subsidized fuel and thus being willing to change to renewable energy if the conditions are possible. Do not burden the country by buying subsidized fuel.

8.2 Theoretical Framework

8.2.1 Alternative Energy for Transportation

8.2.1.1 Types of Energy

There are various forms of energy, but all types of energy must fulfill various conditions: can be converted into other forms of energy, follow the law of conservation of energy, and cause changes in body mass subjected to such energy. The common forms of energy are the kinetic energy of a moving object, the radiation energy of light and electromagnetic radiation, potential energy stored in an object by its position in the gravitational field, electric field, or magnetic field, and the heat energy consisting of potential energy and kinetic microscopic energy from the movements of irregular particles. As fossil fuels will soon be depleted, several countries have displaced their supplies of fossil energy (for example, The Netherlands) and replaced them with other, alternative energy forms, namely, wind and wave power. Soon, oil supplies will run out in this country. Indonesia's oil reserves are only about 3.5 billion barrels (MB). If the average production is 800,000 barrels per day, then the reserves are only enough for 11 years. As proof for this statement, Indonesia is no longer an oil-exporting country. Thus, currently Indonesia is not a member of OPEC (Organization of Petroleum Exporting Countries) [3].

8.2.1.2   Potential Conditions and New and Renewable Energy

(i) Potential
The fossil energy resource potentials versus production reserve ratio are 7.73 MB petroleum (18 years), 152.89 trillion cubic feet (TCF) natural gas (61 years), and coal 103 metric ton (147 years). Moreover, potential versus non-fossil energy resources with the percentage used in parentheses are marine power (ocean waves, tides, ocean currents, sea temperature differences), 60 GW (0.000458%), geothermal, 29 GW (4%), solar, 4.8 KWh/m^2 or 11 GW (0.00001%), wind, 9.9 GW (0.005%), micro-hydro, 0.769, 7 GW (0.1%), bio-fuel, 9 GW (ethanol, bio-diesel, bio-gas, biomass). The share of fossil energy consumption of 94.3% is 41.8% oil, 23.8% gas, and 29.7% coal) [3].

(ii) Condition of Energy Used in Indonesia
Energy is used for many purposes, such as for industry, transportation, and households. According to the latest data, energy use for industrial needs was 37.2%, for transportation 33.9%, for domestic 10.1%, for commercial 6.5%, and for raw materials 12.3%. Production is 777 thousand bpd (barrels per day) of oil and fuel with consumption of 1.6 million bpd, with average growth around 5% to 8%. In addition, fuel consumption in 2025 is estimated to be 2.5 million bpd [4].

(iii) New and Renewable Energy
According to RUEN (National Energy General Planning), the targets for primary energy supply by new and renewable energy by 2025 are a total of 400 million tonnes of oil equivalent (MTOE), including gas 22%, coal 30%, oil 25%, and new and renewable energy 23%; the target of new and renewable energy amounted to 92.2 MTOE, consisting of electricity 45 GW from new and renewable energy, amounting to 69.2 MTOE (geothermal, hydro, biofuels, solar, wind, etc.), and the remaining 23 MTOE of biofuel, biomass, biogas, and coal bed methane (CBM). Also, the implementation of the pilot project reactor nuclear power plant has a capacity of 10 MW and implementation of the pilot project power at least 1 MW sea current/waves [7].

To accelerate economic growth, in the future there are no more restrictions on electricity consumption for the community and the business community/industry in Indonesia. Therefore, it is necessary to prepare available electricity in sufficient quantities and inexpensively. Because it is no longer possible to rely on fossil-based electricity, the government has requested acceleration of the electrification program and sovereignty of renewable energy based on environmentally friendly concepts.

The Government will formulate a regulatory body-independent supply of energy so that energy policy no longer causes rivalry among the sectors. The Government maximizes the utilization of renewable energy. By involving the public and private sectors, the government's financial burden can be reduced. Provision of energy should provide benefits to all concerned, namely, the government, private companies, and the community.

Explanations about renewable energy that can be used for transportation and is environmentally friendly can be listed as follows.

(i) By definition, renewable energy is energy derived from "natural processes which are sustainable," such as solar power, wind power, current air-process biology, geothermal, and biomass. The concept of renewable energy came to fame in the 1970s, in an effort to offset the development of nuclear energy and fossil fuels. The most common definition is an energy source that can be quickly restored naturally, and that the process is ongoing. By this definition, nuclear and fossil fuels are not included. Eventually, such can be discarded. The Indonesian government faces very harsh pressure from protestors if announcing the use of nuclear energy.

(ii) All renewable energy is certainly also sustainable energy, because it is available in nature for a relatively very long time, so there is no need to worry it will outrun its source. The proponents of non-nuclear energy do not include nuclear power as part of sustainable energy because the supply of uranium 235 has natural limits, say, hundreds of years. However, nuclear activists have argued that nuclear energy is sustainable if it is used as fuel in a fast breeder reactor (FBR) because the nuclear fuel reserve could be renewed hundreds to thousands of times. Again in Indonesia, the nuclear reactor is the last option if it is not to say delete it forever.

(iii) Geothermal energy comes from radioactive decay in the center of the Earth, which produces heat from inside the Earth, as well as from the sun heating the surface of the Earth. There are three methods of geothermal utilization: for power plants and used in the form of electricity; as a heat source utilized directly from a pipe into the interior of the Earth; and as a heat pump, which pumps straight from the bowels of the Earth. This kind of energy should be used for generating electricity, then electricity for electrical train application.

(iv) Solar energy, such as a solar panel (photovoltaic array) on a small yacht at sea, which can charge 12 V to 9 ampere under conditions of full sunlight and directly. Because most renewable energy ultimately comes from "solar energy," this term is a little confusing. Solar power can be used for generating electricity using solar cells, generating electricity using a solar tower; then it is able to heat buildings directly, heating buildings through a heat pump, reheating food using a solar oven, and heating water with solar-powered water heaters. Of course, the sun does not provide constant energy to any one point on earth, so its use is limited. Solar cells are often used to charge the battery at noon, and the power of the battery is then used at night when sunlight is not available. Solar energy can be used for land, sea, and air transportation. Reference for Solar Spectral can be seen in [1]

(v) Wind power comes from the temperature difference at two locations that produces different air pressures, resulting in wind. Wind is the movement of matter (air) and has been known for many years as usable to drive turbines. The wind turbine is used to generate kinetic energy and electrical energy, which can be used for land and sea transportation. For further investigation can bee seen in [5].

(vi) Hydropower from water energy is used because water has mass and is able to flow. Water has a density 800 times that of air. Even a slow movement of water can be converted into other forms of energy. Water turbines are designed to get energy from different types of reservoirs, calculated from the amount of water mass, altitude, and speed of water. Water energy is used in several forms: (a) a dam power plant (the largest is the Three Gorges Dam in China); (b) micro-hydro built to generate electricity up to a scale of 100 KW; and (c) mini-hydro commonly used in remote areas that have water sources. The run of the river is built by harnessing the kinetic energy of the water flow without the need for a large water reservoir. This kind of energy can be used for generating electricity; then electrical power is available to run an electrical train.

(vii) Biomass energy has several advantages: (a) biomass is the world's oldest and most efficient solar battery; (b) biomass means healthy forests; (c) biomass is also abundant and one of the best available sources on earth; (d) biomass is net neutral carbon management as well as a greenhouse gas control mechanism; and (e) biomass is a new energy economy. In relationship to sustainable transport, this is the best choice.

In rural areas of Indonesia, because fuel from biomass or wood is used as a substitute for fossil fuels, forests become one of the important resources. The tree species selected for firewood are those which have these properties: fast-growing species, high increment, can be grown on marginal lands, are easily cultivated, quickly sprout after being cut, have a high calorific value, and do not produce much smoke or poison gas on burning. About 70 species of trees seem to have more potential for energy purposes wherein the calorific value is more than 100 GJ/ha/year, providing significant support.

Plants usually use photosynthesis to store solar energy, air, and CO_2. Biofuels (biofuels) are fuels derived from biomass, organisms, or products of animal metabolism, such as feces from cows. Biomass is also one of the sources of renewable energy. Typically biomass is burned to release the chemical energy stored in it, the exception being when bio-fuels charge the fuel cell (e.g., direct methanol fuel cells and direct ethanol fuel cells). Biomass can be used directly as fuel or to produce other types of fuels such as bio-diesel, bio-ethanol, or bio-gas, depending on its source. Three forms of bio-mass are used: solid, liquid, and gas. In general, there are two methods for producing biomass, that is, by growing the organism producing biomass and clicking[1].

(viii) Liquid biofuels usually shaped bio-alcohols such as methanol, ethanol, and bio-diesel. Bio-diesel can be used in modern diesel vehicles with little or no modification and can be obtained from waste vegetable and animal oils and fats. Depending on the potential of each area, corn, sugar beets, sugar cane, and several types of grass are cultured to produce bio-ethanol; bio-diesel is produced from crops that contain oil (palm oil, copra, castor seeds, algae) that has gone through various processes such as esterification.

 (ix) Solid biomass direct use is usually in the form of combustible solids, either wood or flammable plants. Plants can be grown specifically for combustion or may be used for other purposes, such as processing in specific industries, and waste processing results that can be burned as fuel. Briquetting biomass can also use solid biomass, wherein the raw material can be a piece or pieces of solid biomass raw or having gone through certain processes such as pyrolysis to increase the percentage of carbon and reduce the water content. Solid bio-mass can also be processed by means of gasification to produce gas.

  (x) Bio-gas comes from various organic materials, biologically by fermentation, or in physicochemical gasification, which can release flammable gases. Bio-gas can easily be produced from a variety of industrial wastes that exist today, such as paper production, sugar production, or animal husbandry. Various waste streams must be diluted with water and allowed to naturally ferment, producing methane gas. The residue from the fermentation activity is fertil-izer rich in nitrogen, carbon, and minerals.

 (xi) Small-scale energy sources and their relationship to transportation:

(a) Piezoelectric, an electric charge resulting from the application of mechanical stress on the solid object. This object converts mechanical energy into electrical energy.

(b) The automatic clock (automatic watch, self-winding watch) is a watch that is driven with stored mechanical energy that is generated from the movement of the hands of users. The mechanical energy is stored in the spring mechanism inside it.

(c) Platform electrokinetic (electrokinetic road ramp) is a method of generating electrical energy by harnessing the kinetic energy of a moving car on a foundation built on the street. A foundation has been installed in the parking lot of a supermarket, Sainsbury's in Gloucester, United Kingdom, where the electricity generated is used to drive the cash register.

(d) Catching the electromagnetic radiation that is not utilized and converting it into electrical energy using a rectifying antenna is one method of harvesting energy (energy harvesting). It is used also by the car driver.

---

[1] The meaning of clicking is the process whereby the residue from the biomass process being gathered together in the pellets and reused.

Therefore, any supply of energy no matter how small a contribution to a group of us will help in understanding more about energy.

8.2.1.3   Cases of Renewable Energy Development

(1) Design and Hazard
   (i)   The aesthetics of wind turbines or issues of conservation of nature are raised when the solar panels are installed in rural areas; these must be well designed.
   (ii) Wind turbines can be set up to avoid flying birds, whereas a hydroelectric dam power plant may create a barrier for migrating fish. Renewable energy should be safe for the life of these animals.
   (iii) The technologies must function in the way you wish; for example, take advantage of solar collectors as a noise barrier along the way, combine them as a shade for the sun, put it on a roof that is already available, or even replace the roof entirely.
   (iv) The carbon released into the atmosphere can be properly reabsorbed if the biomass-producing organisms are continuously cultivated.
   (v)   Another problem with many renewable energy sources, particularly biomass and bio-fuels, is the large amount of land needed for cultivation efforts. Fortunately, in most parts of Indonesia, there are still possibilities to develop land for wider energy production.

(2) Concentration
   (i)   Renewable energy is not merely concentrated in one particular area; there are setups in selected locations.
   (ii) Wind energy is the most difficult to focus; thus, the solution should be requiring large turbines to capture as much wind energy as possible. There will be no problem in Indonesia as many areas have stable wind conditions.
   (iii) Water energy utilization methods depend on the location and characteristics of the source water, so the water turbine should be well designed.
   (iv) Utilization of solar energy can be done in various ways, but to get this energy areas that can catch large amounts are necessary. For comparison, in the standard conditions of testing in the United States, energy received by 1 m^2 of solar cells with an efficiency of 20% will produce 200 W. Condition testing standards in question are air temperature of 20°C and irradiance of 1000 W/m^2.

(3) Distance to the Receiver of Electrical Energy
Renewable energy can solve the problems of geographic diversity because it can be distributed in several locations throughout the country (a national grid). This distribution can also be a significant problem, as some renewable energy sources such as (i) geothermal, hydro, and wind can be in a location some distance from the receiver

of electric energy; and (ii) geothermal in the mountains, water in upstream energy, and offshore wind energy or plateau. The solution in facing the problem of utilization of these resources on a large scale is likely to require considerable investment in transmission and distribution networks as well as the technology itself in the face of environmentally related concerns. Renewable energy is well established close to people in certain locations.

(4) Availability in Nature
Renewable energy is available throughout the country, and it must be fully intensified for the benefit of local or rural peoples. However, one significant shortcoming is the availability of renewable energy in the universe; some of the sources are available only occasionally and not at all times (intermittent). For example, (i) as sunlight is available only in daytime, the solution is preparing a reservoir of solar energy of which the power varies over time. (ii) The energy from water cannot be used when the river is dry, so that another reservoir must be prepared. (iii) Biomass can be affected by climate, pests, and other problems. Because these four problems, for the time being, mean these kinds of renewable energies are not yet feasible, we must make a feasibility study for better solutions. The foregoing explanation must be properly taken into account in the discussion of sustainable transportation.

8.2.2   The Role of the Indonesian Government

The government's policy has been formulated and needs to be realized: (i) maximizing renewable energy; (ii) minimizing the use of petroleum; (iii) optimizing the utilization of natural gas and new energy; (iv) using coal as the mainstay of the national energy supply; and (v) utilizing nuclear energy only as a last option.

Conservation strategies for energy efficiency that have been proposed will be continued and maximized, namely, the following: (i) a mandatory minimum energy performance standards program and labeling of household appliances; (ii) human resources development for the managers and energy auditors for industries, buildings, etc.; (iii) implementation of the Presidential Instruction No. 13 Year 2011 on Energy and Water Savings in Buildings and Buildings owned by the Government; (iv) the use of LEDs in public facilities such as street lighting; and (v) a campaign to cut 10% of the budget for energy efficiency by 10% (Ministry of Energy and Mineral Resource 2016).

To accelerate economic growth, in the future there are no more restrictions on electricity consumption for the community and the business community/industry. Therefore, it is necessary to provide the availability of electricity in sufficient quantities and inexpensively. Because it is no longer possible to rely on fossil-based electricity, the Government has requested acceleration of the electrification program and sovereignty of renewable energy based on environmentally friendly concepts.

The Government will formulate a regulatory body independent of the supply of energy so that energy policy is no longer creating rivalry among the sectors. Government can maximize the utilization of renewable energy. By involving the public and private sectors, the government's financial burden can be reduced. Provision of energy should provide benefits to all parties, namely, the government, private companies, and the community. The four main programs that have been introduced are as follows:

A. Intensification
B. Diversification
C. Conservation
D. Indexation

A. Intensification of Energy Resources Policy

The need to improve the survey and exploration of resources in an effort to determine the potential resources that could be used to meet the need of improving the welfare of the people, especially in using available energy (Dept. of Energy and Natural Resources).

B. Diversification of Energy Resources

A strategy is needed to reduce dependence on nonrenewable resources in an effort to meet the needs of domestic resources utilized and replace them with other resources [diversification (diversity of), the search for alternative energy beyond oil and gas]

C. Energy Conservation Resources

The need to use resources efficiently in an effort to preserve the resources through the use wisely to achieve balanced development and environmental conservation or effectively utilize (appropriate, effective, and efficient).

D. Government Indexation

Using the scientific way, for each sector of activity, it is necessary to determine the type of energy used where there is the most appropriate resource availability. Insights/sustainability/development/environment wisely/indexation (Decision of appropriate use of energy).

Type of Energy, Existing, Age, Potential, Notes

No	Energy type	Existing	Age	Measurement	Potential development	Notes
1	Gas					
2	Petroleum					
3	Carbon					
4	Solar					
5	Wind					
6	Hydro					
7	Bio-diesel					
10	Biomass					
11	Bio-gas					
12	Bio-ethanol					
13	Waves					
14	Tides					

8.2.3 Government Guidance for Developing Renewable Energy

Beside the four concepts of intensification, diversification, conservation, and indexation, the government of Indonesia announced the nine principles of renewable energy programs [6].

1. Achieve clean, accountable, effective, and efficient service for bureaucrats.
2. Complete the regulations.
3. Simplify licensing and non-licensing.
4. Provide incentives.
5. Provide subsidization of energy.
6. Improve coordination with ministries.
7. Promote energy-saving campaigns.
8. Renew new and renewable energy potential data.
9. Strengthen networking.

8.3 Implementation in Providing Energy for Sustainable Transportation

8.3.1 The Reason for Developing Sustainable Transport

A. Loss of Fossil Energy
As we know that sooner or later oil will be depleted from the surface of the Earth, human beings need to replace it with a rake or renewable energy because fossil energy once discharged cannot be renewed. Therefore, the discussion of new energy or renewable energy must be intensified. If the government plans economic growth at about 6%, then the growth of energy supply must be twofold, that is, 12%. Because its fossil energy will run out in approximately 11 years, Indonesia has immediately turned to provide renewable energy based on the list mentioned earlier on alternative energy.

B. Reduction of Pollution Level
Almost all cities in Indonesia, even in all the world, are exposed to air pollution caused by the high carbon content of exhaust gases from motor vehicles, especially motorbikes and cars. In fact, this air pollution causes ozone in the sky to leak; consequently, solar radiation to the earth cannot be absorbed by ozone when the atmosphere has leaks. Therefore, the utilization of bio-gas and bio-diesel derived from growing plants will reduce air pollution.

C. Human Mobility
Humans will continue to struggle in their lives in the presence or absence of fossil reserves for the provision of petroleum fuels. Humans need mobility in three areas:

land, sea, and air. Therefore, consideration of energy options with transportation should include attention to these three criteria.

D. Environmentally Friendly
The provision of fuel with bio-gas and bio-diesel is not only useful in reducing pollution, but also aids the Eco Energy program, which is an energy procurement program gaining environmental attention. Why? because environmental problems cannot be considered only by environmental experts but also require energy experts. In fact, energy has a very strong influence on environment conditions. Thus, sustainable transportation is closely related to the sustainable environment.

E. Sustainability
Sustainability is closely related to the provision of materials to make bio-gas, bio-diesel, and even biomass. Many vegetable materials in Indonesia that can be used for this purpose are spread over certain areas. The ingredients may include, for bio-ethanol as a substitute for gasoline, the species of crops (a) such as sugar cane (*Saccharum officinarum*), palm sugar (*Arenga pinnata*), cassava (*Manihot esculenta* Crantz), corn (*Zea mays*), and sorghum (*Sorghum* spp.); (b) bio-diesel as a substitute for diesel fuel, which is palm oil (*Elaeis guineensis*), coconut (*Cocos nucifera*), jarakpagar (*Jatropha curcas*), kemirisunan (*Reutealis trisperma*), and nyamplung (*Calophyll uminophyllum*); and (c) integrated power generation and bio-fuels, which can be provided by a micro-hydro generator.

8.3.2 Sustainable Air Transport

For air transportation, dependence on Avtur (Aviatur Turbo Oil) is still dominant, because it requires a strong turbo force to rotate the propeller. To reduce the usage of fossil energy, solar panels can be set on the wing of the aircraft so that the use of Avtur can be reduced. However, there is still very strong dependence on climate conditions. Avtur fuel is able to overcome the problem of climate, but solar energy still requires further examination.

8.3.3 Sustainable Land Transport

For land transportation, the options are still numerous, including these:

A. Electricity, for a while, can be produced by macro-hydro or mini-hydro, especially for electric trains. The turbines can be driven by water power, fossil diesel power, solar panels, and even windmills. For power with propeller, The Netherlands is still a pioneer and has long used this, so the Dutch may be close to using all means of land transport with renewable energy.

B. Solar panels can be directly set into the transport tool, especially as seen in many experiments, but commercially solar panels still do not attract investors. In the future, as fuel prices increase, along with fuel scarcity, the use of solar panels is attractive and will experience increased investment in this field.
C. Wind and magnetism: as wind capacity is still somewhat less feasible as a power tool, it needs to be combined with magnetic power. The wind rotates the positive and negative magnet, then saves the energy so that later when the wind is not available, the electricity is stored automatically to rotate the electric turbine. Materials are bio-diesel (Kemirisunan, Jatropha, Nyamplung) and there is also bio-ethanol from sorghum, sugarcane, or cassava. The exact usage for cars as well as the motorbike must be electricity, or a combination of wind and magnetism.

8.3.4 Sustainable Sea Transport

For sea transport there are only two options, namely, the combination of fossil fuels as a buffer if there is less solar and wind power available. Like a sailing ship, sea transport is based on combinations of diesel fossil fuels with solar panels or fossil diesel combined with wind and magnetic power.

Thus, if we conclude that the possible choices of fuel for transport are very great, then we have to take into account environmental friendliness, low pollution, and sustainability guarantees.

8.3.5 Some Consequences from the Change to Renewable Energy

According to technological dialectics, the following terms apply. If a technology develops, there will always be two aspects: technological benefits and technological losses. The benefits of technology continue to be enjoyed by humans. Just one example is the android, causing relationships to feel closer. The term "longing to meet" others will be lost because every day we are using Facebook, Instagram, etc. People have no need to see their fathers or mothers. Every day when grandparents miss their grandchildren, they just click the button, immediately search, or order something good. But none of the children or grandchildren comes to visit because contact is provided through the development of communication technology in the social media.

In another case in Jakarta, there was a famous and very successful taxi named "Bluebird," so that indeed it was not expected that its earnings would drop drastically because of the new Online Taxi that offers arrangements for passengers online at a fixed price and more cheaply. Here we see technological development with the dynamics that cause other businesses to go down. Forecasts of other consequences, with the development of this renewable energy, if not covered by error, are that public refueling stations or gas stations can be unfortunately run out of business in the future.

8.4 Conclusion and Discussion

The following statement can be further discussed in [2]

1. The need of transportation is a problem that persists so long as human life goes on. This need, along with needs for clothing, food, and housing, reflects the requirement of transportation to provide full support for moving human goods as well as personal mobility, that is, people moving from place to place.
2. Fuel for transportation at present is still fully dominated by fossil fuels, which cause severe environmental pollution, and as well its global reserves will soon be depleted. Therefore, the idea of converting fossil fuels to non-fossil fuels for transport or for renewable energy to increase availability energy for life needs must be included in community planning. Fossil fuels cannot be sunk (sunk cost effect) and will soon run out, in Indonesia by approximately 2030, and in the world possibly by 2050.
3. For air transportation, the fuel used is still Avtur (Avition Turbo Oil). If combined with photovoltaic (PV) solar panels mounted on the plane's wings, Avtur may be needed only for takeoff from the ground. Solar energy will support flight continuously through the air.
4. Transportation on land has many options, starting from the electric train. Although its running fuel is still diesel, it may be possible to place a solar panel at the top, as well as using combined wind and magnetic energy.
5. Transportation at sea can use a combination of a diesel fuel that is environmentally friendly or magnets combined with wind power.
6. Bio-diesel materials: *Kemirisunan, Jatropha, Nyamplung*. Bio-materials include ethanol: sorghum, sugarcane, cassava. Best usage for the motorbike must be electric, or a combination of wind and magnetism; for cars, better use includes bio-diesel as well as bio-ethanol.
7. Government policy must be supported by all parties. Further, the government itself must subsidize and follow the requirement that all transport shall transition from fossil to non-fossil fuels so that sustainable transportation can be provided properly.

References

1. ASTM G 173-03 (2013) Standard tables for reference solar spectral irradiances: direct normal and hemispherical on 37 tilted surface. ASTM International, New York
2. Demirbas A (2009) Political, economic and environmental impacts of biofuels: a review. Appl Energy (Lond) 86:S108–S117
3. Demirbas A (2007) Progress and recent trends in biofuels. Progress in Energy and Combustion Science 33(1), 1–18
4. Demirbas A (2009) Progress and recent trends in biodiesel fuels. Energy Conversion and Management 50(1), 14–34
5. European Wind Energy Association (2011) EWEA executive summary: "Analysis of wind energy in the EU-25." European Wind Energy Association, Brussels
6. Ministry of Energy and Mineral Resources (2015) Statistics of energy and mineral resources. ESDM, Jakarta
7. RUEN (2015), Rencana Umum Energy National (National Energy General Planning). Committee of National Enegy (KEN). Jakarta

Chapter 9
The Problem of Education in Developing Renewable Energy

Ellya Marliana Yudapraja

Abstract The Indonesian government's energy development program consists of nine policies, namely, (1) achieving clean, accountable, effective, efficient energy and serving the bureaucracy well, (2) completing the regulations, (3) simplifying licensing and non-licensing, (4) providing incentives, (5) providing energy subsidies, (6) improving coordination with ministries, (7) promoting energy-saving campaigns, (8) renewing new and renewable energy potential data, and (9) strengthening networking.

Of course, it is not easy to realize all these targets. There are many challenges to face. Some of them are the same pattern thinking in the development of new and renewable energy. Some of the buffers for intensive business schemes are not yet optimal and the technology is still dependent on sources abroad; consequently the domestic content is still small. Therefore, at this time renewable energy is still expensive, but in the future this barrier can be overcome. The next challenge is that some people are still resistant to the development of renewable energy. Therefore, here the importance of policy in education needs to be added specifically to socialize new energy for society. Thus, community education for the people is crucial.

This chapter used the following theoretical framework: first, the concept of Bloom's taxonomy; second, the four education principles of UNESCO; third, the four steps of maintaining energy; fourth, the nine programs of the Indonesian Government about energy.

In the fourth step, the policies are namely (a) implementing intensification, (b) diversification, (c) conservation, and (d) indexation of energy policies. The concept of conservation can make energy durable, to not run out quickly, although in practice this policy is more psychosocial because it involves awareness of community members in future energy needs. Indexation is the policy of selecting the right energy source for a region: for example, solar panels for housing and windmills to run factories. The nine programs were mentioned in the beginning, the Introduction (Chap. 1).

The main problem is whether the Indonesian people as a whole understand the concept of government energy policy in general and how the state of energy is provided for the future, both the existence of socialization or education of awareness

E. M. Yudapraja (✉)
Faculty of Teaching and Education Science, Pasundan University, Bandung, Indonesia

© Springer International Publishing AG, part of Springer Nature 2019
A. Sayigh (ed.), *Sustainable Building for a Cleaner Environment*,
Innovative Renewable Energy, https://doi.org/10.1007/978-3-319-94595-8_9

for future energy needs because we are no longer dependent on fossil fuels. Therefore, the author is certain that education of the people about energy must be starting now, or we will be too late to do so.

This education about renewable energy should be socialized in formal education or in informal education, then provided also according to the level of education, so it is necessary to arrange education, including the curriculum, according to education level (preliminary, secondary, even university) and formally or informally in daily living. In Indonesia, education and socialization about oil palm issues has already taken place and is completed with more education examples, especially for children. So the understanding of renewable energy must come from an early age. In implementing energy knowledge for rural or urban children, we purpose to include their mothers or their families, to reach more attendants.

9.1 Introduction

I believe that God Almighty is the One and the only one of the ultimate energy resources and it is Unlimited forever. One of God's creations that is the source of energy on Earth is the sun. The sun shines upon the whole world. Solar energy is absorbed by all living things, including plankton that absorb energy and then die and sink and, under heavy pressure, eventually become a petroleum resource: this is called fossil energy. Some of the other energy is absorbed by plants and animals, as well as by humans. This is where the human exceeds; in addition to energy absorbed by the physical, energy is also absorbed also by the brain, both physically and non-physically. The absorption of energy by the human brain makes people think, so I dare to conclude that the law of conservation of energy will last forever: "Energy will not be lost, but change shape." An illustration: the sun is shining on the grass, the sun's energy is absorbed by the grass, the grass is eaten by the cow, and the beef is eaten by humans. Man continues to live with the energy he consumes from this realm. The human brain that will think about energy is an educated brain. So the role of education is to make people smart. With this intelligence, humans seek and keep energy to support their life. Thus, to build the nation's economic growth of 6%, it must be supported by doubling energy growth by 12%: energy must increase. So, when fossil energy later runs out, new and renewable energy is needed. Here I dare to state that human beings think with the ability of brain energy . The human step in maintaining this energy, among others, through the first process, intensification, is to expand the existing state in the sense of the creation of the types of fuel: there is a high (low) premium octane value (premix), intensification also in the sense of expanding the region with search and exploration. Second, diversification, multiplying the choice of energy options in various forms, whether fossil and non-fossil, is called bio-energy. Third, conservation, which is to conserve usage, among other means by regulating electrical lighting, limiting the use to be effective and efficient, and recommending that energy is used only if really necessary. The four indexes are choosing and sorting for energy to be in accordance with their usage. For example, gas energy is more appropriate for household use. In subsequent chapters, we describe more fully how the process takes place. All processes of maintaining and supplying energy must be in an environmentally friendly context. The problem is, when humans understand all that.

So here is my chapter, inviting you to think that energy understanding should be an important task for education, where I am a high school teacher, elementary teacher, and even a lecturer at a college. This is my field every day. My advice is that energy understanding should start at an early age. Of course, it should not be discontinued at an early age, but should be taught according to the level of education. The consequence, of course, differs in the material taught for elementary school level compared with university level. So, it will be a different curriculum. If curriculum change is difficult, then it could be a renewable energy theme, only in a lecture course on a particular relevant subject.

9.2 Theory of Education and Energy Knowledge

The concepts of educational theories to be used in the analysis of this chapter include (a) Bloom's Taxonomy, (b) the UNESCO Concept, (c) the Four Concept of Energy Policy of the Indonesian Government, and (d) nine principles from the National Energy Committee of the Republic of Indonesia. According to Bloom et al. [1], in teaching a person we must divide this into three domains, namely, (1) cognitive, (2) affective, and (3) psychomotor. Cognitive is something related to science without being associated with reality and other variables beyond the education knowledge. Affective is associated with the interest of learners for the science being taught; the interest is because science can add to self-confidence in facing nature. Psychomotor is referring to what we can do by learning this subject after completion of the knowledge, and imagine what variables will be found in the real world. By creating research questions and then developing these to become a model that approaches real nature, then it can be concluded and continue to be analyzed [1].

Moreover, according to UNESCO the education process should cover these four stages: (1) learning to know, (2) learning to do, (3) learning to be, and (4) learning to live together. First, a learner must know something. Second, he must know how to implement, and third, he must how to be himself. Fourth, he lives with others, and he must do so for the best for all mankind.

The main program of the Indonesian government has been introduced as follows:

1. Intensification
2. Diversification
3. Conservation
4. Indexation

The questions asked are according to the type of the government program.

9.2.1   Intensification of Energy Resources Policy

The need is to improve the survey and exploration of resources in an effort to determine potential resources that could be used to meet the needs and improve the welfare of the people, especially in using available energy (Dept. of Energy and Natural Resources).

Types of Energy: Existing, Live, Potential, Notes

Number	Energy type	Existing	Age	Measurement	Potential developed	Notes
1	Gas					
2	Petroleum					
3	Carbon					
4	Solar					
5	Wind					
6	Hydro					
7	Bio-diesel					
10	Biomass					
11	Bio-gas					
12	Bio-ethanol					
13	Waves					
14	Tides					

9.2.2   Energy Conservation Resources

The need is to use resources efficiently in an effort to preserve resources through wise use to achieve balanced development and environmental conservation or effectively utilization (appropriate, effective, and efficient).

9.2.3   Government Indexation

Using the scientific way, for each sector of activity, it is necessary to determine the type of energy used where the most appropriate resource is available: that is, Insights/Sustainability/Development of environment wisely/Indexation (decision for the appropriate use of energy).

Beside the four concepts of intensification, diversification, conservation, and indexation, the government of Indonesia announced the nine principles of the renewable energy program (Ministry of Energy and Mineral Resources).

1. Achieve clean, accountable, effective, efficient energy while serving bureaucrats well.
2. Complete the regulations.
3. Simplify licensing and non-licensing.
4. Provide incentives.
5. Provide subsidized energy subsidies.
6. Improve coordination with ministries.
7. Promote energy-saving campaigns.
8. Renew new and renewable energy potential data.
9. Strengthen networking.

9.3   Implementation of Concepts in Practice

Just as an illustration, for the first 5 years we teach images that give the impression of the importance of energy. For example, we show pictures of trees and forests. Students are asked to choose whether the green or red color will beautify the forest image. If a child chooses a red image, this means the symbol of a forest fire; explain what it is to lose a forest by burning. Should it be a green color, it is an early invitation that the forest should be preserved. It is the human task that lives today, for the life of our children and future generations. In university education, students are taken to the laboratory, where they study mixtures of chemical compounds relevant to energy. They are required to create energy from growing plants or solar and wind energy as well as other energy options. If we follow the concept of Bloom's taxonomy, then make a formulation embodied in the concept of Bloom, namely, (1) cognitive domain, (2) affective domain, and (3) psychomotor domain.

1. Cognitive Domain. For this domain, we will teach more technical renewable energy, accomplished with the related fields of science being taught. For example, we are teachers/lecturers of mathematics, meaning we will teach students math problems related to the calculation of its use. Just for example, what size energy-efficient lighting is needed in a room of 5×5 m^2? This is a more technical domain about energy, a particular knowledge about energy.
2. Affective Domain. After students understand the calculation of the power of lighting in a room, then the lesson must be made interesting, for example, how to lay out the room: where to put the window, the ventilation, etc. All the concepts of the lessons are made so that learners are interested in using cognitive science to implement in the real world.
3. Psychomotor Domain. In the next step, the students must be taught to do, then do, the experiments that reflect how something that is taught actually happens in the real world. Of course, some variables should be considered, for example, aesthetics, strength, then the issue of the price of the goods we be using, and no less importantly, other requirements. Of course, because the energy equipment, whether electricity or fire, is concerned, there must be a specification, and other provisions that govern its use.

Furthermore, in the provisions of UNESCO that we will use in renewable energy education, then the stages follow the pattern (a) Learning To Know, (b) Learning To Do, (c) Learning To Be, and (d) Learning To Live Together.

(a) Learning To Know. Learners are taught the sciences they should know. For solar energy one must know how to bend solar panels into the roof, for example.
(b) Learning To Do. Learners understand and understand how to practice the science of solar panel installation and how can we operate it in daily living.
(c) Learning to Be. A learners must realize that now, he has become an educated man or woman who should use the knowledge well, what is one's position among communities, and how to be one's self.
(d) Learning to Live Together. Learners should be aware that all knowledge and practice that are mastered are for a wider human life. They must be aware that they will live with other people of different colors, religions, cultures, and behaviors. Therefore, it is not only the energy science that must be understood, but the social culture, tolerance, togetherness, diversity, and other factors that affect every decision taken by the central government as well as local government.

Moreover, UNESCO has also added what is called Education for Sustainable Development, including following the UNESCO concept of ESD (Education for Sustainable Development). I remind us of this concept because of its relationship to sustainable energy for human beings.

ESD encourages people, including these points:

1. Relevant education system. Education for sustainable development equips learners with the knowledge, skills, and values for the social life, environment, and economic challenges of the twenty-first century.
2. Educational transformation. Education for sustainable development uses innovative learning, student-centered instruction, and multiple learning styles. Empower students and make them agents in the process of education, from an early age to old age. It can improve learning beyond the limits of education.
3. Enhance the sense of justice and mutual respect. Education for sustainable development helps learners understand the situations, views, and needs of people living elsewhere or belonging to other (subsequent) generations.
4. Help overcome climate change: 175 million children will be affected by climate-related disasters that occur in the next decade. Sustainable development education prepares students to adapt from the impacts of climate change and empowers them to address the causes.
5. Build an environmentally friendly society. Education for sustainable development equips students with eco-friendly skills to help save or restore environmental quality, and to improve human well-being and social justice, motivating learners to choose a sustainable lifestyle (Source: ESD-UNESCO).

The four principles of maintaining energy consist of (a) intensification, (b) diversification, (c) conservation, and (d) indexation. The educator's job is, in accordance with his capacity, to spread to the community according to his education level, so that everyone understands it.

(a) Intensification. This concept allows for the expansion of energy consumption from existing ones in an area. For example, in a village already using solar panels, the usage is extended to adjacent neighboring villages, especially villages that have not been reached by such electrification.

(b) Diversification. This concept allows the expansion of the type or use of an energy product. For example, in a village that already has solar panels, then develop with wind energy with magnets or biomass. More options are better.

(c) Conservation allows the necessity of austerity measures in all community activities. In my city of Bandung, Indonesia, we have "Concept 17–22" that means we may only turn on electric appliances at 17:00 until 22:00; after this time, the electricity must be switched off.

(d) Indexation is the formulation of appropriate usage for the type of energy in the appropriate tool or place. For example, biomass is only for household needs, and bio-gas is for industrial and bio-diesel for transportation purposes. Finally, the implementation of renewable energy education should support one of the nine government programs:

(1) Achieve clean, accountable, effective, efficient energy and serve bureaucracy. (2) Complete the regulations. (3) Simplify licensing and non-licensing. (4) Provide incentives. (5) Provide energy subsidies. (6) Improve coordination with ministries. (7) Promote energy-saving campaigns. (8) Renew new and renewable energy potential data. (9) Strengthen networking.

Based on the existing capabilities and facilities owned by the education sector, then, the role of education, among others, contributed to the following programs:

1. Complete regulation program: then education can also disseminate the understanding of regulation. From here, the field of education can take grains of provisions that must be known to the public.
2. In the inter-ministerial coordination program, the education sector participates in drafting the concept of what can be coordinated with other ministries; for example, with Kominfo, the nature of the process of dissemination or dissemination of regulatory or austerity procedures.
3. Data potential dissemination program. Can help disseminate data to schools.
4. Network reinforcement program. The field of education is involved in this, so that there is a network of work, a inter-agency both inside and outside the country.

9.4   Conclusion

Important conclusions can be drawn.

1. Energy is the Grace of Almighty God, which must be nurtured and best utilized by mankind. Maintenance and utilization must be well programmed, and there is supervision of its implementation to get more benefits.

2. Education is a very crucial part of the whole life, especially in the dissemination of science on energy as well as forecasts and provision through the four programs of intensification, diversification, conservation, and indexation. Energy education must be done from the early age of children throughout the entire life (long life education).
3. The field of education can have a strategic role in filling four of the nine government programs by utilizing all existing facilitation in the field of education. Education has important roles to ascertain our awareness of energy.

Reference

1. Bloom BS et al (eds) (1956) Taxonomy of educational objectives: handbook I, cognitive domain. David McKay, New York

Further Readings

2. Gronlund NE (1978) Stating objectives for classroom instruction, 2nd edn. Macmillan, New York
3. Gendler ME (1992) Learning & instruction: theory into practice. Macmillan, New York
4. Krathwohl DR et al (eds) (1964) Taxonomy of educational objectives: handbook II, affective domain. David McKay, New York
5. Ministry of Energy and Natural Resources (2015) Annual report, 2015
6. UNESCO (2005) Annual report, 2005

Chapter 10
Winter Performance of Certified Passive Houses in a Temperate Maritime Climate: nZEB Compliant?

Shane Colclough, Philip Griffiths, and Neil J. Hewitt

Abstract The low-energy Passive House (PH) standard is over 25 years old and has a wealth of peer-reviewed academic publications relating to its performance in continental Europe. However, relatively little has been written about it's recorded performance in the Temperate Maritime Climate (TMC), such as that prevailing in Great Britain and Ireland. With the requirement to build to the near Zero Energy Buildings (nZEB) standard throughout EU member states by 2020, the PH standard offers a well proven methodology of building the required low energy dwellings. This paper reviews the recorded performance of certified passive houses and compares it with the recorded performance of contemporaneous houses built to the minimum building regulations in Northern Ireland. It then and considers their suitability for achieving the nZEB standard. The metrics being recorded at five-minute intervals for the houses include:

a. occupancy profile
b. indoor air temperature
c. indoor relative humidity
d. indoor carbon dioxide concentrations
e. outdoor temperature
f. outdoor relative humidity
g. wind speed
h. barometric pressure
i. energy consumption

S. Colclough (✉) · P. Griffiths · N. J. Hewitt
Ulster University, Newtownabbey, County Antrim, Northern Ireland
e-mail: S.Colclough@Ulster.ac.uk

© Springer International Publishing AG, part of Springer Nature 2019
A. Sayigh (ed.), *Sustainable Building for a Cleaner Environment*,
Innovative Renewable Energy, https://doi.org/10.1007/978-3-319-94595-8_10

10.1   Introduction

Nearly Zero Energy Buildings (nZEB) are mandated across Europe for all dwellings built after 2020. As efforts increase to reduce winter-heating demand and build low-energy and nZEB dwellings, e.g. using the well-established passive house (PH) standard [1], UK post occupancy analysis of low-energy dwellings (e.g. [2]) has focused on dwellings in mainland UK, with publications on Northern Ireland PH dwellings primarily focusing on social housing (e.g. [3, 4]). This paper presents recorded energy and Indoor environmental quality (IEQ) performance of owner occupied certified passive houses (PH) and contemporaneously constructed dwellings which comply with the minimum building regulations in the temperate maritime climate of Northern Ireland over the winter period. The potential of the PH in meeting nZEB is considered given that the cost of building to the nZEB standard has been shown to be comparable to meeting national building regulations [5].

10.2   Method

This paper presents an assessment of the monitored winter performance of four certified new build passive houses (PH) [6] and compares it to five houses built to comply with the minimum building regulations standard (B Regs), located in Northern Ireland (NI). Given the similarities in the cost of nZEB and PH standards [7], it is likely that the PH standard will be used to comply with nZEB requirement and therefore affords an opportunity to compare the recorded performance of the potential future nZEB building stock with that which is currently being constructed. The energy performance certificates (EPCs) of the constructed PHs are also used to determine if they are nZEB compliant. The paper represents a subset of the dwellings being monitored over a full year as part of a study of energy consumption and indoor environmental quality (IEQ) of 23 houses on the island of Ireland (see Fig. 10.1).

In common with the methodology employed in recent metadata studies [4], data is presented for the living rooms and master bedrooms. Data covers the period November and December 2016 and January 2017 for the NI dwellings (see Table 10.1) and supplements an analysis which has been carried out over the 2016 summer period [8].

The metrics being gathered at 5-min intervals for the nine houses include occupancy profile, indoor air temperature, indoor relative humidity, indoor carbon dioxide concentrations, outdoor temperature, outdoor relative humidity, barometric pressure and energy consumption.

Bands have been established for the key metrics being monitored, and the percentage of time individual metrics that exceed the thresholds is presented to assist the reader obtain insights. Passive houses are designed to have a uniform set temperature of 20 °C throughout. A temperature threshold has therefore been set at

Fig. 10.1 Locations of monitored passive houses and building regulations houses

Table 10.1 Overview of the monitored houses

House	Building type	Construction type	Year of construction	Size {m²}
PH 1	Two-storey house, detached	TF	2014	158
PH 2	Bungalow, detached	TF	2013	220
PH 3	Bungalow, detached	TF	2011	145
PH 4	Detached	TF	2016	247
PH 5	Under construction	n/a	2017	n/a
BRegs 1	Two-storey house, detached	Block	2010	329
BRegs 2	Two-storey house, detached	TF	2014	294
BRegs 3	Two-storey house, detached	Block	2013	230
BRegs 4	Two-storey house, detached	Block	2016	210
BRegs 5	Two-storey house, detached	Block	2015	246

TF: Timber Frame; n/a: Not Available

20 °C. Thresholds have been defined to reflect the set temperatures in SAP at 21 °C for the living room and 18 °C for the other parts of the dwelling. A set temperature of 24 °C is required in SAP in the case of air-conditioned buildings, and the final threshold temperature of 25 °C reflects the temperature that passive houses are allowed to exceed for no more than 10% of the time.

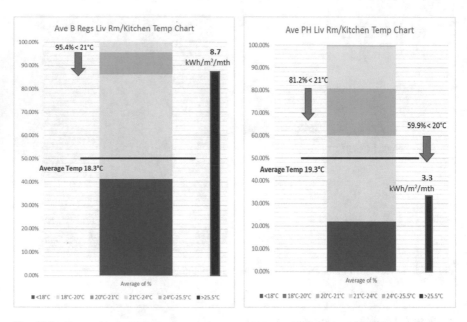

Fig. 10.2 Comparison of living room temperatures and building heating energy consumption

10.3 Results: Overview

Overall, it was found that the passive houses required only 38% of the heating energy per m² per month while at the same time experiencing a 1 °C higher average temperature than the houses built to the minimum building regulations (Fig. 10.2). The average temperature of the bedrooms is similar, with the passive house bedrooms being on average 0.2 °C cooler than houses built to the minimum building regulations.

The passive houses were also seen to exhibit lower and more uniform average concentrations of carbon dioxide (612 ppm for the living rooms and 690 ppm for the bedrooms) compared with the houses complying with the minimum building regulations (677 ppm in the living rooms and 824 ppm in the bedroom). Initial investigations indicate that the mechanical heat recovery and ventilation systems employed in the passive houses may be contributing to the improved indoor air quality.

10.4 Results: Temperature Profiles

Figures 10.3, 10.4, 10.5, and 10.6 give the proportion of time that temperatures were experienced in the building regulations and passive houses both individually and as groups over the period broken down into the distinct temperature bands previously defined in addition to the average temperatures.

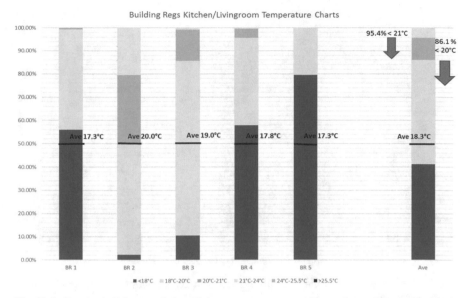

Fig. 10.3 Average building regulations living room temperatures November and December 2016, January 2017

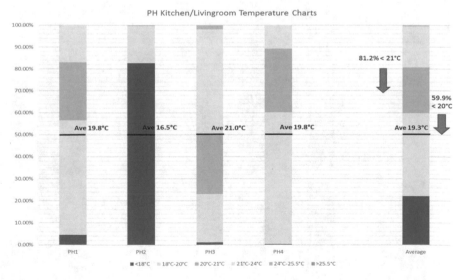

Fig. 10.4 Chart of the NI passive house living room temperatures, November and December 2016, January 2017

Figures 10.3 and 10.4 refer to the living room temperatures. The average temperature in the group of building regulations houses is seen to be 18.3 °C compared with 19.3 °C for the group of passive houses.

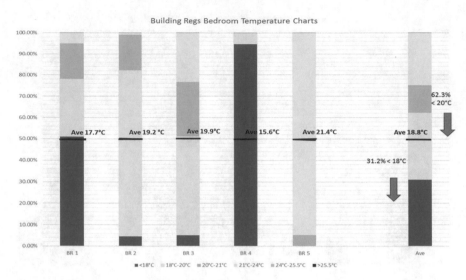

Fig. 10.5 Chart giving the building regulations bedroom temperatures November and December 2016, January 2017

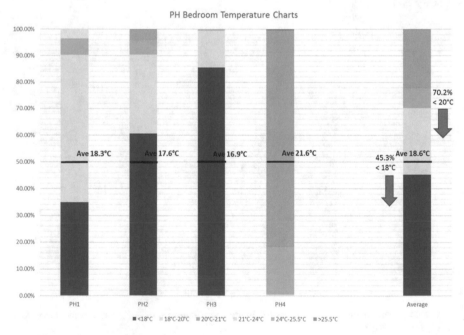

Fig. 10.6 Chart giving the passive house bedroom temperatures for November and December 2016, January 2017

Considering the set temperature for passive houses (20 °C), the passive house living rooms on average are seen to meet or exceed this threshold for 40.1% of the time, while the building regulations houses meet or exceed 20 °C for only 13.9% of

Table 10.2 Comparison of passive houses and building regulations houses – key parameters averages

Building Regulations Houses

Kit/Liv Rm	Temp	RH	CO2		Bedroom	Temp	RH	CO2
Ave	{°C}	{%}	{ppm}		Ave	{°C}	{%}	{ppm}
B Regs 1	17.3	60.7	581.6		B Regs 1	17.7	50.6	536.2
B Regs 2	20.0	60.5	437.8		B Regs 2	19.3	55.7	927.6
B Regs 3	19.0	63.4	960.1		B Regs 3	20.0	56.5	1242.5
B Regs 4	17.8	63.4	701.3		B Regs 4	15.6	60.4	621.2
B Regs 5	17.3	68.8	706.2		B Regs 5	21.4	50.8	792.7
Average	18.3	63.3	677.4		Average	18.8	54.8	824.0

Passive Houses

Kit/Liv Rm	Temp	RH	CO2		Bedroom	Temp	RH	CO2
Ave	{°C}	{%}	{ppm}		Ave	{°C}	{%}	{ppm}
PH 1	19.8	54.5	627.0		PH 1	18.4	50.7	721.2
PH 2	16.5	55.6	455.9		PH 2	17.6	48.0	596.8
PH 3	21.0	47.1	646.1		PH 3	16.9	50.9	493.7
PH 4	19.8	55.4	722.1		PH 4	21.6	48.2	946.9
Average	19.3	53.1	612.8		Average	18.6	49.5	689.7

the time and fall below the assumed building regulations set temperature of 21 °C for 95.4% of the time.

Considering the individual houses, PH 2 is seen to exhibit the lowest temperatures of all houses, with an average temperature of 16.5 °C. This reflects the fact that the house is unoccupied (and unheated) for large periods of time. In the same manner, B Regs 5 is seen to be an outlier – the house records the coolest living room and the warmest bedroom of all the building regulations houses monitored. It was found that the monitoring equipment had been moved from the standard positions, thereby significantly affecting the recorded temperatures.

Considering Figs. 10.5 and 10.6, it is seen that the trend of higher average temperatures in the passive houses does not continue in the monitored bedrooms, with the average temperature being on average 0.2 °C lower in the passive houses (18.6 °C compared with 18.8 °C, respectively). During the winter period, the temperatures are below the building regulations assumed set temperature of 18 °C for 45% of the time (passive houses) and 31.2% of the time (B Regs). Considering the set temperature of 20 °C of the passive house standard, the temperature is below the threshold for 70% of the time in the passive houses and 62% time in houses complying with the minimum building regulations.

10.5   Results: Review of Key IEQ Parameters

Table 10.2 shows the key parameter averages for the monitored living rooms and bedrooms of the passive houses and houses complying with the minimum building regulations. It is seen that the average carbon dioxide concentrations appear lower and more uniform in the passive houses compared with the houses complying with the minimum building regulations. While this could be due to the forced ventilation system of passive houses, further investigation is needed, as there appears to be an error in CO_2 readings in the B Regs 2 for 2 months, and readings are high for B Regs 3, despite the presence of a positive input ventilation system and more widespread monitoring that is required to validate the findings. The possibility of high CO_2 readings due to insufficient ventilation in non-PH does need further investigation.

It is noted that the relative humidity recorded in the living rooms is approximately 4–8% higher than that recorded in the bedrooms of both the passive houses and building regulations houses. This may be a characteristic of the monitoring equipment, as during calibration, the units located in the living rooms have been found to read higher relative humidities than the module type used in the bedroom. Despite this potential instrumentational error which may be influencing the results, it does appear that the relative humidities in the B Regs houses exceed those recorded in the passive houses in both the living rooms and bedrooms.

10.6   Results: Energy Consumption

The details of the electrical energy consumption and space heating and DHW energy consumption for the passive houses and building regulations houses for the months commencing 1 June 2016 to 1 February 2017 are given in Fig. 10.7.

10.6.1   DHW and Space Heating Energy Consumption

As can be seen from the first group of columns, the average energy consumption of the houses built to the minimum building regulations is 8.7 kWh/m²/mth, almost three times that of the passive houses at 3.3 kWh/m²/mth.

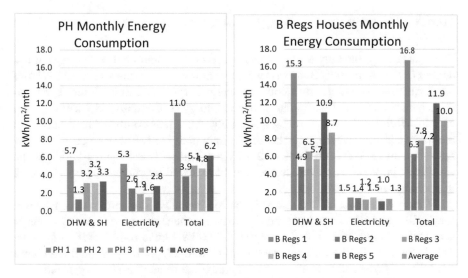

Fig. 10.7   Energy consumption of PH and B Regs dwellings June 2016 to Jan 2017

There is considerable spread in the energy consumption figures among the dwellings; with PH 2 having the lowest energy consumption figure at 1.3 kWh/m²/mth, PH 2 recorded a low occupancy level and lower than average interior temperatures (living room temperature averaged 16.5 °C compared with an overall average of 19.3 °C for the group of passive houses). Equally PH 1 had a higher than average interior temperature, with the living room temperature recording 19.8 °C on average.

Equally, looking at the DHW and space heating energy consumption figures for the building regulations houses, there is seen to be a significant spread, especially between B Regs 1 and B Regs 2, with a spread of three times the energy consumption per metre squared per month between the two dwellings. B Regs 1 is the oldest house in the sample, having been built in 2010. In addition, as can be seen from Fig. 4.5, the indoor air quality is very good, indicating a good ventilation rate, which may be impacting on the energy consumption figures.

10.6.2   Electrical Energy Consumption

The electrical energy consumption can be seen in the second group of columns in Fig. 10.7 for the houses. On average the passive houses are seen to have an electrical energy consumption twice that of the building regulations houses (2.8 kWh/m²/mth versus 1.3 kWh/m²/mth). A portion of the difference is to be expected given that passive houses use a heat recovery and ventilation system which also provides electrical space heating to the dwelling. The high electrical energy consumption of PH 1 is due to the use of electricity for space heating, as it uses an electrical Aga.

10.6.3 Discussion on Energy Consumption

The total average monthly energy consumption of the minimum building regulations houses (10 kWh/m²/mth) is almost twice that of the passive houses (at 6.2 kWh/m²/mth). Almost seven times the amount of energy is spent on heating building regulations houses compared with their electrical energy consumption (8.7 kWh/m²/mth vs 1.3 kWh/m²/mth, as can be seen from Fig. 4.6). This compares markedly with the passive houses where space heating is only marginally above electrical energy consumption (3.3 kWh/m²/mth vs 2.8 kWh/m²/mth). These figures highlight the significant impact that passive houses can have on the overall energy consumption of dwellings in Northern Ireland.

While there is uncertainty caused by Brexit, the nZEB standard has been defined for the UK (and NI) in a recent report (zero carbon hub) as requiring primary energy for regulated loads of less than or equal to 43.6 kWh/m²/a (or 44 kWh/m²/a to the nearest integer). Of the four passive houses monitored, publicly available EPCs are available for three and indicate that the "approximate energy use" unregulated loads are 40 kWh/m²/a (PH 2), −39 kWh/m²/a (PH 3) and 44 kWh/m²/a (PH 4). While further more detailed analysis will be required once the nZEB standard for the UK is finalised, based on the primary energy use metric, the three passive houses are seen to comply with nZEB requirements.

10.7 Conclusions

Monitoring has been carried out of four NI passive houses and five "standard" houses (built to the minimum building regulations) over the winter months 2016/2017.

A key finding is the difference in terms of the interior temperatures with the group of passive houses recording on average a 1.0 °C higher temperature in the living rooms despite requiring only 38% of the space heating energy consumption of the houses constructed to the minimum building regulations. The available EPCs of the monitored passive houses also indicate that the passive houses are currently nZEB compliant, in advance of the 2021 requirement.

Acknowledgements The authors wish to acknowledge the support of the interdisciplinary centre for Storage, Transformation and Upgrading of Thermal Energy (i-STUTE) under EP/K011847/1 and Invest NI for this research.

References

1. PHI (I), 28 November 2011-last update, what is a passive house? Available: http://www. passipedia.org/passipedia_en/basics/what_is_a_passive_house. May 22 2013

2. Gupta R, Dantsiou D (2013) Understanding the gap between 'as designed' and 'as built' performance of a new low carbon housing development in UK. In: SEB. Springer, Berlin/Heidelberg, pp 567–580
3. McGill G, Oyedele LO, McAllister K (2015) Case study investigation of indoor air quality in mechanically ventilated and naturally ventilated UK social housing. Int J Sustain Built Environ 4(1):58–77
4. McGill G, Sharpe T, Robertson L, Gupta R, Mawditt I (2017) Meta-analysis of indoor temperatures in new-build housing. Build Res Inf 45(1–2):19–39
5. Colclough S, O'leary T, Griffiths P, Hewitt NJ (2017). The near Zero Energy Building standard and the Passivhaus standard – a case study. In: Passive and low energy architecture conference, Edinburgh, 2017, 3rd to 5th July 2017
6. PHI, 2017-last update, passive house database – certified passive houses in Northern Ireland. Available: http://www.passivhausprojekte.de/index.php?lang=en#s_bf725c60608e8a 13788f7d294d22c2bb
7. Colclough SM, Hewitt NJ, Griffiths PW (2017) Financial analysis of achieving the nZEB standard through the passive house standard. In: Mediterranean green buildings and renewable energy foum 2017, 30 July to 2nd August 2017
8. Colclough SM, Griffiths P, Hewitt NJ (2017) Summer performance of certified passive houses. In: Temperate maritime climates, passive and low energy architecture conference, Edinburgh, 2017, 3rd to 5th July 2017, pp 340–341–347

Chapter 11
A Parametric Tool for Assessing Optimal Location of Buildings According to Environmental Criteria

Giacomo Chiesa and Mario Grosso

Abstract Site-climate design is the first phase to be carried out for a sustainable design of a building. In some local building codes and environmental rating tools, specific requirements related to building location on the plot are included. Generally, in site design, solar access and wind protection/exposition are the two most considered parameters, together with local temperature conditions.

An effective tool to analyse the environmental quality for localising a specific function and activity on a plot is the "site microclimate matrix". This technique, firstly developed in the US, includes solar and wind access analyses for specific days of the year and hours of the day (able to represent seasonal limits). Sunny and shaded areas in the plot are overlapped to areas protected and exposed to wind in order to classify the entire plot in four classes (sunny-calm, sunny-windy, shaded-calm, shaded-windy). These classes can help in applying a score of bioclimatic suitability for the location of several indoor/outdoor functions and activities. Solar access is generally determined by using instruments such as solar charts and shading mask protractors, while few science-based tools have been applied so far for assessing wind access and protection. For overtaking this gap, a method is presented here, including a parametric definition of wind wake core areas.

The paper will introduce a parametric tool able to calculate, for parallelepiped solids, wind wake cores and shadow profiles to calculate a site microclimate matrix. This calculation can be performed not only on the plot plane but also in three dimensions. By using this matrix, it can be possible to classify the volume of the plot for a sample of activities and suggesting their more suitable localisation. Considering the algorithmic nature of the proposed tool, able to directly interact with a CAD-based environment, it will be possible to define indicators for assessing the optimal environmental quality of an urban design project.

G. Chiesa (✉) · M. Grosso
Politecnico di Torino, Dipartimento di Architettura e Design, Torino, Italy
e-mail: giacomo.chiesa@polito.it

© Springer International Publishing AG, part of Springer Nature 2019   115
A. Sayigh (ed.), *Sustainable Building for a Cleaner Environment*,
Innovative Renewable Energy, https://doi.org/10.1007/978-3-319-94595-8_11

11.1 Introduction

One of the main aspects of sustainable urban and building design is the conception of the design object as an integrated part of its physical context. This context can be seen as both a constraint and a resource: a constraint since it limits the freedom of the designer and builder to conceive any form and volume they want in order to avoid a negative impact to the environment and the public perception, a resource in the sense that the climate variables of a location, specifically solar radiation and wind, can act as environmental forces, which can be used to reduce the consumption of fossil fuel in technical energy systems and, hence, the related greenhouse gas emissions causing global warming.

To allow for this use, a simple awareness is not enough; scientific and technical knowledge is necessary. And, a correct approach, based on analytical methods and supporting tools, needs to be applied from the earliest design phases, i.e. site analysis and building programming.

In the present paper, a method and a tool that can be used for analysing the microclimate characteristics of a site as well as defining environment-friendly location criteria of buildings and activities are described. This approach can be applied to both didactical activity and professional work.

11.2 Environmental Criteria and Optimal Localisation of Buildings

11.2.1 Indicators to Assess Environmental and Sustainable Design Issues

Criteria to assess environmental sustainability of site design choices are listed in Table 11.1, with a focus on sun and wind exposition as main climatic variables affecting the site design. In fact, differently from other climate parameters, e.g. dry bulb temperature and RH%, solar exposition and wind access may be modified adding shading systems, wind protection barriers or wind catchers.

11.2.2 Site Microclimate Matrix: A Tool for Environmental Assessment in Site Design

In the environmental design process, the characteristics of a building site affect not only technical and regulation constraints to building construction (e.g. type of soil and related foundation typology or the specific constraints related to urban planning) but also the local microclimate. The main microclimate parameters influenced by a specific plot and its surroundings are the following:

Table 11.1 Criteria to assess environmental sustainability of site design choices – see also Grosso and Chiesa [1]

Morphological aspects	Climatic variables	Environmental aspects	Sustainability criteria
Plan shape (rectangular, squared, fragmented, curved, expanded, in line)	Sun exposure	Spatial characteristics Visual comfort	Access to diffuse solar radiation in principal spatial units (average daylight factor)
		Spatial characteristics Thermal comfort	Winter access to direct solar radiation in principal spatial units Level of control of direct solar radiation in summer in principal spatial units
	Wind exposure	Spatial characteristics Indoor air quality	Yearly wind access in principal spatial units
		Spatial characteristics Thermal comfort	Summer wind access in principal spatial units
Height	Sun exposure	Impact on the environment	Winter solstice shadow depth on surrounding buildings and on vertical facades (negative effects on the building heating demand)
	Wind exposure	Impact on the environment	Wind wake core depth considering surrounding buildings in winter (positive effects on the building heating demand) and in summer (negative effects on the building cooling demand)
		Thermal comfort Indoor air quality	Number of spatial units that overpass the average surrounding buildings' height (increase in the pressure difference between windward and leeward facades)
Orientation of the main facades with transparent surfaces and openings (N, NE, E, SE, S, SW, O, NW)	Sun exposure	Visual comfort	Average daylight factors in boundary spatial units and glaring control
		Thermal comfort	Winter exposure to solar radiation of transparent elements localised in the SE-S-SW quadrant Level of control of the incident solar radiation reaching transparent surfaces localised in the SW-W-NW quadrant
	Wind exposure	Indoor air quality	Opening exposure in relation to the prevalent wind direction (entire year)
		Thermal comfort	Opening exposure in relation to the prevalent wind direction in summer Opening protection from the prevalent wind direction in winter

(continued)

Table 11.1 (continued)

Morphological aspects	Climatic variables	Environmental aspects	Sustainability criteria
Urban plan density (distance between buildings)	Sun exposure	Impact on the environment	Increasing in shading depth on the surrounding buildings both in vertical and horizontal in winter (negative effect on winter heating need and on outdoor comfort)
	Wind exposure	Impact on the environment	Increasing in the wind wake core depth on the surrounding buildings both in vertical and horizontal plans in winter (positive effect on the building heating demand) and in summer (negative effect on the cooling demand and on outdoor comfort)
Surface/volume ratio	Air temperature	Thermal comfort	Thermal losses through the building envelope according to the climatic zone
Roof type (flat roof, pitched roof, pitched inclinations)	Sun exposure	Renewable sources (solar panel)	Amount of direct solar radiation reaching solar thermal panels (winter optimization of the slope of the pitched roof)
		Renewable sources (PV panels)	Amount of direct solar radiation reaching PV panels (optimization of the slope of the pitched roof)
	Wind exposure	Impact on the environment	Increasing in the wind wake core depth according to roof types (positive winter effect and negative summer ones)
		Thermal comfort	Increasing in the difference in the wind pressure between roof pitches and consequent effect on the wind-driven ventilation potential for inhabited attic

– Solar radiation intensity, especially the direct component, influencing heat transfer through the building envelope and glaring, both affecting indoor and outdoor comfort (thermal and visual) and the life spam of exposed materials
– Exposure/protection to wind, influencing thermal comfort, heat transfer and dispersion of air pollutants, i.e. volatile organic compound (VOC), CO, CO_2 and particulates (PM10, PM2.5).

Hence, these parameters have to be assessed from the early design phases (e.g. site analysis and building programming [2, 3] in order to define the optimal localisation of users' activities [4–6] both indoor and outdoor, based on a performance-driven approach [7, 8]. For example, incidence of solar radiation causes solar gains, reducing the amount of required energy for heating but increasing thermal load in summer. On the contrary, wind exposure induces heat gain dissipation, hence decreasing the energy demand for cooling in summer, but increasing the energy need for heating in winter. In addition, the presence of wind, depending on its direction and force (velocity), influences the potential for wind-driven ventilation (cross ventilation) to passively assure both indoor air quality and comfort cooling [9–11]. The combination of the above-mentioned microclimate parameters can be evaluated according to local climate conditions, types of activities, time of day and season, in order to analyse positive and negative environmental localisation aspects for comfort and reduction in energy consumption, using a "Site Environmental Quality"(SEQ) indicator [1].

Amid several possible methods for SEQ assessment, only a few consider solar shadowing and wind access together as it is done by the approach here described based on the site microclimate matrix. It allows for early evaluating, in the building programming phase, of an environmentally optimal location of buildings and outdoor activities [12]. This method was firstly developed by Brown and Dekay [4], who described a graphical technique to draw a microclimate matrix by overlapping the shadowed/sunny areas with windy/lee zones. Based on this approach, a simplified method was developed by Grosso [11] and further described in Chiesa and Grosso [12], to calculate the extension of wind exposed and calm zones around solid obstacles, based on the definition of wind wake core and reinterpreting results from wind tunnel tests and analyses carried out by Boutet [10] and Evans [13]. The wind wake core represents a calm zone where wind velocity is reduced by at least 50% with respect to the undisturbed flow upwind the considered obstacle. Considering the limitations of this approach in comparison to more complex computational fluid dynamic (CFD) simulations, only the effect on the leeward side of the obstacle is considered. An example of microclimate matrix representation is shown in Fig. 11.1.

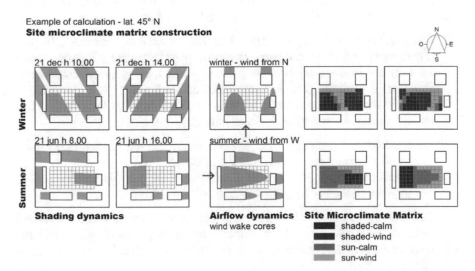

Fig. 11.1 Example representation of the process to calculate a microclimate matrix on a plot for four reference periods (winter: Dec 21 at 10:00 and 14:00; summer: Jun 21 at 8:00 and 16:00) [12]

11.3   A Parametric Tool to Calculate Wind Wake Cores and Shadow Profiles

At present, several tools, some of which CAD-integrated, are available to calculate solar shadow dynamics in a given location considering geopositioning and timing, while no instruments are available to automatically calculate a wind wake core without using CFD software. In addition, these tools are not coupled with the most diffused CAD software and are far to be of common use in architectural practice due to high costs, long elaboration time and required engineering skills. The basic method proposed by the authors in Grosso [11] and Chiesa and Grosso [12] allows for drawing the extension of wind wake cores downwind parallelepiped obstacles (e.g. building, urban furniture and barrier), by using a hand-calculation method related to the geometrical properties of the considered obstacle. An example application of this method to define the optimal localisation of different outdoor activities on a plot is described in Sect. 11.3.1. Even if this procedure is simple, it is a time-consuming procedure. In order to overcome this limitation, an automatic tool, called "STEMMA" (SiTE Microclimate Matrix model), was developed in a parametric scripting environment connected with the commercial CAD platform Rhinoceros®. This model is described in Sect. 11.3.2.

11.3.1   Hand-Made Methodology

The proposed hand-calculation procedure for optimising the localisation of outdoor activities using the microclimate matrix is based on the following steps:

I. Characterise the building site related to both geometries (obstacles, shapes, dimensions) and climate aspects (seasonal prevalent wind directions, solar intensity and outdoor dry bulb temperature profiles, geographical coordinates to define the local sun path diagram) – see Sect. 11.3.3.
II. Calculate the shading profiles of obstacles and buildings on the plot in reference periods of the year (e.g. seasonal, monthly…) such as

   (a) Northern hemisphere – Winter December 21 at 10:00 and 14:00 (plus 12:00), summer June 21 at 8:00 and 16:00 (plus 12:00)
   (b) Southern hemisphere – Winter June 21 at 10:00 and 14:00 (plus 12:00), summer December 21 at 8:00 and 16:00 (plus 12:00)

III. Calculate the seasonal wind calm/exposed areas on the plot defining the wind wake cores – see Sect. 11.2.1 – of obstacles facing the prevalent seasonal winds.
IV. Overlap the graphical results of steps II and III to define four (or six) site microclimate matrixes.
V. Attribute a score to each condition (sun/calm, sun/exposed, shaded/calm, shaded/exposed) according to each season and activity/type of buildings.

Possible outdoor activities can be classified in relation to their high, medium and low metabolic rates. Buildings' localisation can be assessed according to climate conditions (hot, cold, humid, dry – e.g. by following the Koppen-Geiger classification [14]) and building envelope (level of insulation, internal mass).

VI. Sum seasonal and yearly scores to obtain a global map of optimal positioning for each outdoor activities/buildings.

Note: Microclimate matrixes are developed on the plot before and after having located the designed buildings: the former matrixes are used for localising the buildings, the latter for localising the outdoor activities, taking the effects that the new buildings have on the microclimate context into account.

This paper is focussed on steps II and III.

Step II: Shadow Analysis
Local sun path diagram or shadowing profiles are used to draw the shadowed areas on the site (see Fig. 11.2).

Step III: Wind Analysis
The hand-made calculation method uses fitting correlations of wind tunnel tests to calculate the depth of the wind wake core downwind each considered obstacle, as described in Sect. 11.2.2. Accepted geometries are parallelepipeds with rectangular (and square) base, T-shape, L-shape and U-shape obstacles as shown in Fig. 11.3 [10, 11].

11.3.2   Definition of the STEMMA Model

The new proposed model STEMMA aims at overcoming some weaknesses of the above described hand-made calculation procedure (steps II and III) by translating it into a software environment. In particular, two main problems were solved: the need

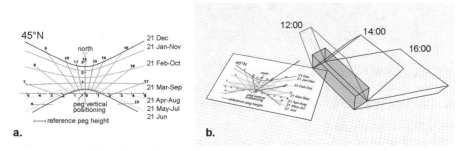

Fig. 11.2 Graphical methods to define the shadow profile of an obstacle given the day of the year and the calculation hour, (**a**) sun path diagram, (**b**) shading profile – the examples refer to the city of Turin (Italy, 45° N)

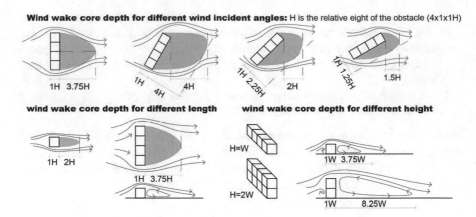

Fig. 11.3 Geometrical relations between the chosen obstacle shape and the expected wind wake core depth for different wind incident angles. (Re-elaboration from Boutet [10])

to interrelate CAD geometries with the proposed procedure (e.g. chose the right coordinate system, translate geometries into usable inputs) and the difficulties in translating implicit aspects into explicit and calculable ones (e.g. passing from graphs to expressions, with consequent definition of interpolating points).

The flowchart of the proposed model within the entire procedure as described in Sect. 11.3.1 is shown in Fig. 11.4.

Step II: Shadow Analysis

The shadow analysis is based on the definition of a sun vector able to represent the sun position at the given day of the year and hour of the day. This vector includes the sun elevation angle, α [°], and the solar azimuth, β [°], that are calculated according to the following expressions – see also Grosso [15]:

$$\sin\alpha = \sin\phi\sin\delta + \cos\phi\cos\delta\cos\omega \qquad (11.1)$$

$$\sin\beta = \frac{\cos\delta\sin\omega}{\cos\alpha} \qquad (11.2)$$

Where

φ is the latitude of the considered plot.
δ is the solar declination for the considered day of the year.
ω is the hourly angle considering the local solar time.

Step III: Wind Analysis

The wind wake core calculation, which is graphically synthetized in Fig. 11.5 and whose construction is described in Fig. 11.4, is a function of the relative dimensions of the considered obstacle, calculated dividing each absolute dimension by the absolute height of the solid "h" ($H = h/h = 1$, $W = w/h$, $L = l/h$, where capital letters represent the absolute dimensions). Based on these ratios, a correlation is used to

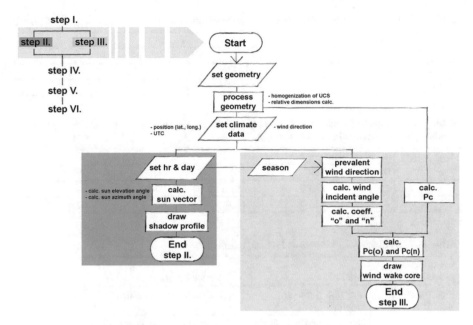

Fig. 11.4 The proposed flowchart to calculate steps II (shading profile) and III (wind wake core) in a software environment. On the left, it is also reported the full methodology to evaluate the optimal localisation of activities in a plot according to Sect. 11.3.1

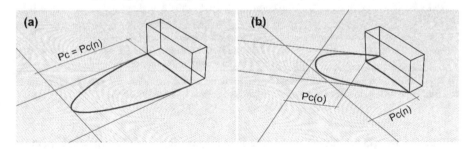

Fig. 11.5 Example calculations of a wind wake core with a wind direction perpendicular to the main façade (**a**) or tilted, e.g. 50°, (**b**)

calculate a depth reference for the wind wake core – Pc. Table 11.2 shows the correlation for $L = 1$ and $L = 2$ considering different values of W, where L is the relative length of the obstacle and W is the relative width – see also Grosso [11] and Chiesa and Grosso [12].

Furthermore, the wind wake core geometry is also function of wind incident angle on the main (longest) windward façade. According to this angle, the Pc value previously calculated is used to determine the two distances defining the wind wake core boundary conditions parallel to the main leeward façade, Pc(n), and opposite to the leeward edge of the obstacle on the wind direction, Pc(o). Table 11.3 shows the

Table 11.2 Pc values for different correlations between obstacle relative dimensions. $L = 1$; $L = 2$

	W					
	0.125	0.5	1	2	4	10
$L = 1$	2.67	2.4	2.0	1.6	1.5	1.48
$L = 2$	3.33	3.0	2.5	2.0	1.82	1.8

Table 11.3 Weights to calculate the Pc(n) and Pc(o) for different wind incident angles – parallelepiped building with rectangular base

	Wind incident angle				
	0°	30°	45°	60°	90°
n	1	1.07	0.53	0.4	–
o	–	1.07	0.6	0.33	0.4

Table 11.4 STEMMA tool inputs and related calculation

Calculation procedure	Input	Related calculations
Shadow profile Step II	Location (latitude, longitude)	Sun elevation and azimuth angles
	Day and month → day of the year [1–365]	Solar declination → sun elevation and azimuth
	Hour of the day [1–24]	Hourly angle[a] → sun elevation and azimuth
	Geometry	Shadow profile according to the solar vector
Wind wake core Step III	Prevalent wind direction	Wind incident angle → correction factors n & o → Pc(n) & Pc(o)
	Geometry	Wind incident angle; Relative dimensions ($H = h/h = 1$; $W = w/h$; $L = l/h$) → Pc calculation

[a]It is possible to include the calculation of the local solar time including the local time equation, the longitude difference from Greenwich expressed in time and the time zone

calculated weight to define Pc(n) and Pc(o) according to different wind incident angles – Pc(n) = Pc * n, Pc(o) = Pc * o. When these values are known, it is possible to draw the wind wake core of an obstacle for the considered wind direction.

11.3.3   Tool Development and I/O Definition

In order to automatically perform the two calculation procedures foreseen in steps II and III considering obstacles that may be differently oriented and designed, the STEMMA model reorients the obstacle's relative UCS (user coordinate system) according to a coherent framework: the x and y axes of the base plan of each parallelepiped have to be fixed as parallel, respectively, to the length and to the width dimensions. The list of inputs of the STEMMA model is shown in Table 11.4, while

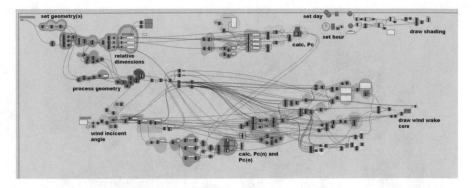

Fig. 11.6 The elaborated script to perform the calculation of the shadow profile and wind wake core geometry of a parallelepiped obstacle. The used scripting environment is Grasshopper™ and related plug-ins connected to Rhinoceros®

Table 11.5 Confidential internal of dimensions and wind angle of incidence for model definition

Variable	Confidential interval	Variable	Confidential interval
Relative obstacle length L (L = length/height)	[1, 8]	Relative obstacle width W (W = width/height)	[0.125, 10]
Wind angle of incidence relative to the perpendicular of the obstacle axis[a]	[0°,87°)[b]; [90°,90°]; (93°,267°); [270°,270°]; (273°;360°]		

[a]See the sketch in the table
[b]In the ranges (0°, 30°), (−30°,0°), (150°, 180°), (180°, 210°), the "*n*" and "*o*" weights for respective Pc calculation – see Sect. 11.3.2– do not follow a linear trend. Geometrical evaluations were performed to define a wind wake core shape considering the reference points at 0° and 30°. The wind wake core for angles in the confident intervals (87°, 90°), (90°, 93°), (267°, 270°), (270°, 273°) can to be drawn manually using the boundary lines automatically defined by the STEMMA tool

the procedure schemes translated in the used software environment is shown in Fig. 11.6. The developed STEMMA tool is based on the Rhinoceros® CAD platform [16] and scripted in Grasshopper™ [17], a graphical algorithm editor that allows to correlate different typologies of inputs and plug-ins, from geometries to calculation; on the theory behind the use of algorithms in the design process, see also Chiesa [18], Woodbury [19], Braham and Hale [20], and Terzidis [21].

At present, the "STEMMA" model is able to calculate the wind wake core of a parallelepiped building in the intervals of confidence reported in Table 11.5.

Fig. 11.7 The sample building plot (in yellow) and surroundings – axonometric view

11.3.4   Example Application

In this section, the STEMMA tool is applied to an example context to calculate its site microclimate matrix and optimise the localisation of a building. The chosen building plot is localised in Turin (lat. 45° N, long. 7.7° E) and is described in Fig. 11.7.

According to the nearest typical meteorological year (TMY) data – Torino Caselle TMY from Meteonorm 7.1 elaborated using Climate Consultant 6.0 – the prevalent wind directions are respectively NNW (−30° from North in the West direction) for winter season and NE (60° from North to the East direction) for summer. The shadow analysis was performed at the two solstices, which are able to represent the opposite extreme points, respectively for winter on December 21 at 10:00 and at 14:00 and for summer on June 21 at 8:00 and 16:00. Figure 11.8a shows the 4 + 2 considered situations after the automatic calculation (see Sect. 11.3.2) of steps II and III of the procedure described in Sect. 11.3.1. Furthermore, Fig. 11.8b illustrates the four correspondent site microclimate matrixes. Finally, these four matrixes were analysed according to the scoring system reported in Table 11.6 for building localisation considering a temperate climate condition and a low-internal gain building [4]. The suggested optimum localisation positioning on the plot is reported in Fig. 11.9.

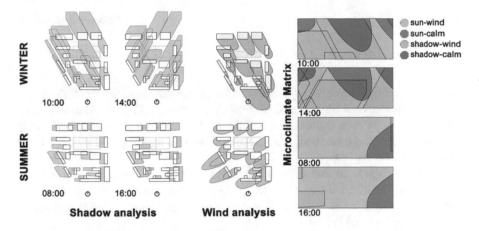

Fig. 11.8 Calculation of the site microclimate matrix of the considered plot by using the STEMMA tool. Fig. 11.8 (**a**) on the left shows the Shadows and wind wake cores, while Fig. 11.8 (**b**) on the right shows the four calculated matrixes

Table 11.6 Used scoring system for optimal building localisation – temperate climate low internal gain [4]

Season	Windy-sunny	Windy-shaded	Calm-sunny	Calm-shaded
Winter	4	1	5	2
Summer	4	5	1	2

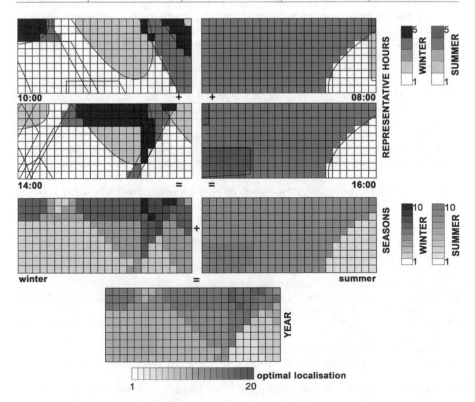

Fig. 11.9 Calculation of the optimal localisation for buildings according to the site microclimate matrixes reported in Fig. 11.8 and the scoring system of Table 11.6

11.4 Conclusions

A simple method to evaluate optimum environmental localisation of buildings and outdoor activities on a site is here described. A new model (STEMMA) to automatically calculate the cast shadow area and the downwind wake core of an obstacle is also introduced and described. Early results show that the proposed model can be easily applied to real cases.

A more complete version of the STEMMA tool is under development in order to automatically perform the site microclimate matrix calculation, even at different levels (3D analysis), hence suggesting optimum localisation for different types of outdoor activities and buildings.

Acknowledgements This research was co-funded by the RTD grant 59_ATEN_RSG16CHG.

References

1. Grosso M, Chiesa G (2015) Environmental indicators for evaluating properties. Territorio Italia 15(2):63–74
2. Chiesa G, Grosso M (2017) An environmental technological approach to architectural programming for school facilities. In: Sayigh A (ed) Mediterranean green buildings & renewable energy. Springer, Cham, pp 707–716
3. Grosso M (2005) Valutazione dei caratteri energetici ambientali nel metaprogetto. In: Progettazione ecocompatibile dell'architettura. Esselibri, Napoli, pp 307–336
4. Brown GZ, Dekay M (2001) Sun, Wind & Light. Wiley, New York
5. Dessì V (2007) Progettare il comfort urbano. Sistemi Editorialie, GIugliano. n.d. s.l.:s.n
6. Grosso M, Chiesa G (2017) Progettazione climatica di sito. In: Grosso M (ed) Il raffrescamento passivo degli edifici in zone a clima temperato, 4th edn. Maggioli, Sant'Arcangelo di Romagna, pp 155–184
7. Cavaglià G et al (1975) Industrializzazione per programmi. RDB, Piacenza
8. Chiesa G (2016) Model, digital technologies and datization. In: Pagani R, Chiesa G (eds) Urban data. Tools and methods towards the algorithmic city. FrancoAngeli, Milano, pp 48–81
9. Allard F (ed) (1998) Natural ventilation in buildings: a design handbook. James & James, London
10. Boutet T (1987) Controlling air movement. A manual for architects and builders. McGraw-Hill, New York
11. Grosso M (2017) Il raffrescamento passivo degli edifici in zone a clima temperato, 4th edn. Sant'Arcangelo di Romagna, Maggioli
12. Chiesa G, Grosso M (2015) Accessibilità e qualità ambientale del paesaggio urbano. La matrice microclimatica di sito come strumento di progetto. Ri-Vista 13(1):78–91
13. Evans BH (1957) Natural Air Flow Around Buildings - Research report no. 59, Texas: Texas Engineering Experiment Station, College Station
14. Kottek M, et al. (2006) World Map of theKöppen-Geiger climate classification updated. Meteorol Z 15:259–263
15. Grosso M (1986) Dinamica delle ombre, 2nd edn. Celid, Torino
16. Robert McNeel and Associates (2017) https://www.rhino3d.com/en/. [Online]. Accessed 06 2017
17. Davidson S (2017) http://www.grasshopper3d.com/. [Online]. Accessed 06 2017

18. Chiesa G (2015) Paradigmi ed ere digitali. Accademia University Press, Torino
19. Woodbury R (2010) Elements of parametric design. Routledge, London/New York
20. Braham WW, Hale JA (2007) Rethinking technology. A reader in architectural theory. Routledge, London-New York
21. Terzidis K (2006) Algorithmic architecture. Elsevier, Architectural Press, Oxford

Chapter 12
Methodology of Solar Project Managing Through All Stages of Development

Andrejs Snegirjovs, Peteris Shipkovs, Kristina Lebedeva, Galina Kashkarova, and Dimitry Sergeev

Abstract The unique methodology for managing solar projects through all of their development stages has been created. This paper provides detailed description of the specifics of this methodology and describes its benefits and necessity. The lack of a unified methodology is caused by the fact that there are many service providers, which offer various separate tools and services for solar energy companies. Most of the available separate services and tools focus on supporting a specific stage of solar project development. Overall, the lack of a unified methodology leads to unnecessary resource spending, making the final costs of solar projects higher, and lengthening payback time, thus making solar energy less attractive. Developed methodology is the result of cooperation between solar installers, solar software developers, and researchers. In the process of methodology development, the unique knowledge transfer space was created. This allowed to pinpoint the real needs of a solar energy project development process and discover the solutions to the burning issues. The developed methodology is represented as the service platform PVStream, and its services cover the full scope of the needs of solar energy project development. The tools and services united by the methodology include a lead generation and proposal tool, financial feasibility analysis, energy efficiency analysis, project design and engineering tool, performance modeling and simulation suite, and monitoring and maintenance tool. The usage of such methodology significantly reduces human resources time by automating design, performance modeling, documentation creation, and overall project management. Further development and dissemination of the united solar energy project development methodology will make solar energy more affordable and attractive, by cutting the costs of the project development and cutting overhead costs of solar energy companies.

A. Snegirjovs (✉) · P. Shipkovs · K. Lebedeva · G. Kashkarova
Energy Resources Laboratory, Institute of Physical Energetics, Riga, Latvia
e-mail: shipkovs@edi.lv

D. Sergeev
PVStream, MZ Consulting, Riga, Latvia
e-mail: info@pvstream.com

© Springer International Publishing AG, part of Springer Nature 2019
A. Sayigh (ed.), *Sustainable Building for a Cleaner Environment*,
Innovative Renewable Energy, https://doi.org/10.1007/978-3-319-94595-8_12

12.1   Introduction

The solar photovoltaic (PV) energy share in the overall European energy production amounted for 12% [1] of all renewable electricity in Europe. The considerable growth of this sector has been driven by the technological advances that led to the reduction of costs [2–4]. Since the technology costs have been dropping [5–7], it is quite important to provide the needed infrastructure to empower solar industry companies to reduce administrative costs.

Photovoltaic (PV), solar heating, and cooling projects require significant initial investments to set up a solar energy system [8] and substantial time for this investment to pay off. Precise calculations and simulations are essential for the solar energy system development process at each stage to provide an optimal system for the needs of the end user.

It is important to keep the administrative costs low to keep raising the attractiveness of solar energy for the consumer [9]. The easy-to-use infrastructure that facilitates development of links between potential customers and existing service providers is needed for further development of the solar energy industry. Such networking space will help to bring solar energy closer to the general public and facilitate the process of the transition from fossil fuels to the renewable energy sources.

Lack of a unified format and unified tool set for photovoltaic, solar heating, and cooling projects management hinders the development of cross-border cooperation [10]. Having a unified methodology for calculations and simulations of the performance of solar energy projects that is flexible enough to include all the necessary parameters into consideration would significantly simplify the financing and investment decisions for general public, financial institutions, and political institutions. PVStream methodology reduces the amounts of time and resources necessary for solar energy project development and support.

12.2   Methodology

PVStream methodology development required input from groups and people of various backgrounds. The methodology was developed as a collaboration between researchers (Latvian Institute of physical energetics, Latvia Energy Institute), practitioners in the industry (among them), developers of solar energy simulation algorithms (Sandia, NREL SAM) and various mathematical models (wind load, shading), and experienced software developers for the solar energy industry (S Fabrika).

Development of the methodology included the following: examination of the results of multiple research projects in the field of solar energy and renewable energy, study of the current solar energy project development process, discussion boards regarding the most common pain points from the companies in the industry, surveys and polls of the solar energy company representatives and solar energy users, study of the tools and services that are available for the solar energy companies on the global market, development and testing of the mathematical models, and algorithms for calculations and simulations.

The information received from universities, researchers, and industry representatives become the basis of this methodology. The result of the analysis was a set of specific rules and guidelines and a set of various calculation models and algorithms. These results already were critically examined and reviewed during common meetings with partners.

12.3   Results and Discussions

The major idea we put in the PVStream methodology is to organize and streamline the photovoltaic, solar heating, and cooling project development process by providing guidelines, algorithms, and calculation models on every stage of development and support process. These guidelines, models, and algorithms together with an overall development and support process have been implemented practically in PVStream service platform.

Over the duration of its development, every solar energy project goes through specific development stages: proposal, design, installation, and support (Fig. 12.1).

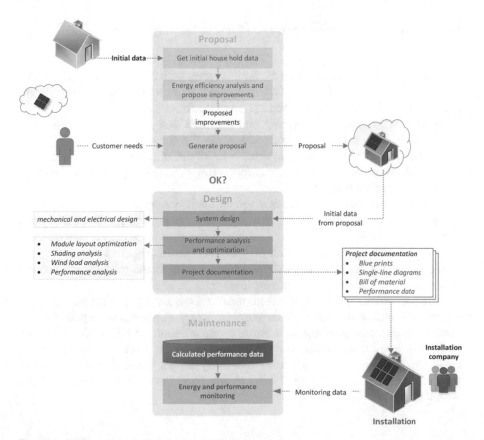

Fig. 12.1 PVStream methodology project life cycle

Fig. 12.2 Optimization possibilities and its influence on system results

The proposal stage begins with the initial contact with the potential customer and ends up with the defined goals of a solar project, chosen financing and investment model, and preliminary engineering design of the future project.

At the design stage, engineering and electrical design take place. This stage results in the finalized full project documentation for the installation team.

During Installation stage, physical installation of solar energy system is performed. Over the maintenance and support stages, further support of the installed solar energy system happens.

The developed solar project management methodology defines overall project management process and specific rules, regulations, and practices which apply to a particular life cycle stage. As practical implementation to proposed methodology software service platform PVStream has been developed (Fig. 12.2).

General rules apply to the overall project management process and are not focusing on a particular project life cycle stage.

It is beneficial to use a minimal number of tools for the project management process. Whenever multiple tools are used, the integration needs to be performed to transfer results and information from one tool to another; this forces spending additional time (up to 2 weeks per project) and resources and raises administrative costs. One of the ways to avoid these unnecessary costs is to use all-in-one solutions that lead the solar energy project development throughout all development stages within the scope of one tool.

Fig. 12.3 Capture roof data from satellite image

It is beneficial to avoid initial on-site visits. Elimination of the necessity of initial on-site visits allows to decrease administrative costs and to focus on the solar energy project development. PVStream methodology allows to utilize various sources for initial data capturing, starting from specifically dedicated services, such as Eagleview, to satellite images, including Google Maps (see Fig. 12.3).

Methodology allows to run project optimization, according to the customer's needs at the earliest stages of solar project development (Fig. 12.4). PVStream provides rules and guidelines on calculations for the optimal solar energy project for the instances when the customer aims to decrease the energy bill to a specific amount, when the customer wants to receive 100% of consumed energy from solar energy, or when the customer wants to invest specific sum to the renewable energy development. PVStream methodology contains a set of algorithms and actions that allow to meet the actual customer's goals during project development and keep the customer at the center of the project. Figure 12.5 shows part of financial analysis results presented to the customer.

Exploration of available financial support options, such as government incentives, rebates, feed in savings tariffs, etc., into the initial financial analysis of the planned solar energy project is necessary. This makes the benefits of transition to solar energy obvious to the end user. Services that provide information for a specific limited set of regions exist at the moment. In order to be able to provide Europeans with the accurate data, PVStream collects and aggregates data about the available support for solar energy projects, according to the region where the planned project is to be installed.

According to PVStream methodology, the engineering and design team continues project development, utilizing as much data as possible from the proposal stage. This includes roof and obstruction drawings, information about nearby objects that might influence solar system performance like trees and other roofs, energy

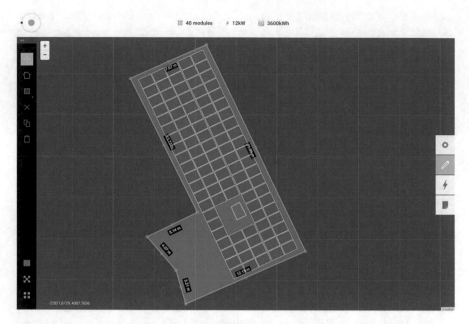

Fig. 12.4 Designer view – mechanical design stage

consumption before solar project installation expenses, and the customer's aims. This significantly reduces the time to set up a design project. Using the initial data collected at the proposal stage for the automated design process can reduce the costs for the solar energy project development up to 75%.

The design of a solar energy project must include a set of specific analyses. In order to avoid large discrepancies between the real performance of the system and the forecasted performance, it is essential to run full-scale shading analysis. PVStream methodology provides several shading algorithms, such as shadow pie to calculate the area in which specific objects cast shadows, ray tracing to run simulation for every hour within the entire year, and PV module self-shading calculation algorithms. These algorithms allow to estimate energy production for each individual PV module accurately.

Wind load and roof load algorithms are essential parts of the design process. Since the results of these calculations define the amount of ballast, inaccurate calculations lead to an inaccurate and nonoptimal bill of material which results in extra costs. The developed algorithms and methods for wind load calculation include various parameters, among them roof data and shape, objects located on the roof, the location of every PV module, type of roof cover, and of course climate data. This makes calculations much more accurate and results in the projection of a safe system with a more optimal bill of materials.

In the cases when any changes have been made in the project blueprints or calculations during the installation stage, the project should return to the design stage. This is needed in order to incorporate all the amendments, run additional

Energy Production

Annual energy production: 10.254 kWh

Energy offset

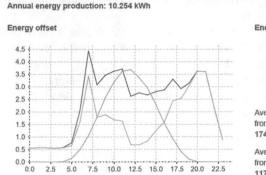

Energy consumption from utility vs solar

- 32% from solar
- 68% from utility

Average monthly energy consumption from utility before solar:
1740.70 kWh

Average monthly energy consumption from utility after solar:
1178.54 kWh

Average monthly feed-in energy sold back to the grid:
292.35 kWh

Post-Solar energy bill

Monthly energy bill before and after solar

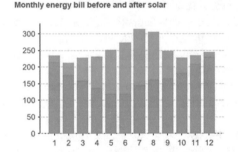

Average monthly bill before solar:
€250.00

Average monthly bill after solar:
€169.26

Average monthly bill savings:
€80.74 (32 %)

Feed-in tariff:
0.10 EUR/kWh

Average monthly feed-in earning:
€29.24

Fig. 12.5 Example of results of cost-efficient analysis

simulations, and ensure that the impact of the amendments to the initial design on the overall system performance is reflected and recorded in the project documentation. This allows to avoid any extra costs that might be caused by the amendments to the initial design.

Include monitoring of the actual system performance into the support of the solar energy process to provide continuous data acquisition and aggregation from the installed system on each individual system node. Using continuous analysis of actual data and comparing it with simulated data, it is possible to predict malfunctioning on different system nodes. Detecting a fault before it really occurs allows to fix the issue before it has an impact on the overall system performance, thus increasing the reliability of the system. PVStream methodology includes smart prediction logic which can detect faults by comparing actual system performance with forecast.

The web-based platform PVStream has been developed that incorporates most of the above-mentioned rules, recommendations, algorithms, and calculations. This set of tools covers all stages of the solar energy project's development and is flexible enough for further developments.

Development of PVStream methodology has both direct and indirect benefits. It widens the set of available tools for all the parties involved in the development of the solar energy industry: researchers, end users, solar energy companies, and regulatory institutions. The methodology also facilitates the development of cross-border and cross-sector collaboration and contributes to the development of a unified regional standard for solar energy project management.

12.4 Conclusions

The practical usage of the methodology using PVStream web-based platform showed that accepting the above-mentioned rules saves time and resources.

Minimizing the number of different tools used saves up to 2 weeks spent on data integration, and the usage of web-based software tools minimizes the number of possible errors and saves additional amounts of time.

Ability to focus most of the efforts on the actual client allows solar industry companies to cut the resources and time spent on the initial on-site visits by up to 95%.

By using PVStream financial analysis models, the sales team can save up to 80% of time on financial analysis and proposals generation.

Streamlining and automating design process, by following PVStream guidelines and using proposed simulation and calculation models, decreases the overall design cost by an additional 75%.

According to PVStream methodology, continuous monitoring of system performance on the individual node level is a must-have option. Together with continuous comparison with predicted data, this allows to detect failures earlier, thus avoiding costly repairs and extending the installed system life time.

Overall, accepting the rules developed within the scope of the methodology and using appropriate tools allow to save up to 70% of time and resources spent within the solar energy project development. This can help keep administrative costs low and contribute to the transition from fossil fuel to renewable energy sources.

References

1. European Commission (2017) Report from the Commission to the European parliament, the Council, the European economic and social committee and the Committee of the regions, p 18
2. Directive of the European parliament and of the Council on the promotion of the use of energy from renewable sources (2017) p 116
3. Masson G, Orlandi S, Rekinger M (2014) Global market outlook for photovoltaics 2014–2018. European Photovoltaic Industry Association, p 60

4. Trends in Photovoltaic Applications (2016) 21st edn (2016) Report IEA PVPS T1-30, p 72
5. Yenneti K, Day R, Golubchikov O (2016) Spatial justice and the land politics of renewables: dispossessing vulnerable communities through solar energy mega-projects. Geoforum 76:90–99
6. Cheng C, Wang Z, Liu M (2017) Defer option valuation and optimal investment timing of solar photovoltaic projects under different electricity market systems and support schemes. Energy 127:594–610
7. Turlough F, Guerin (2017) Evaluating expected and comparing with observed risks on a large-scale solar photovoltaic construction project: a case for reducing the regulatory burden. Renew Sust Energ Rev 74:333–348
8. Montagnino FM (2017) Solar cooling technologies. Design, application and performance of existing projects. Sol Energy 154:14
9. Jeffrey RS, Brownson (2013) Solar energy conversion systems. Elsevier-Academic Press, p 468
10. Pacudan R (2016) Implications of applying solar industry best practice resource estimation on project financing. Energy Policy 95:489–497

Chapter 13
Computational BIPV Design: An Energy Optimization Tool for Solar Façades

Omid Bakhshaei, Giuseppe Ridolfi, and Arman Saberi

Abstract In contemporary buildings, façades are generally the largest borders between the inside and the outside which determine the proportion of energy consumption of the buildings. With today's technology, they could also offer the opportunity of producing energy by adding photovoltaics into their systems, to cover a portion of the building's need for electricity and reduce its dependency on fossil fuels, especially where there is a high amount of global horizontal irradiation (GHI) and a high potential for generating electricity from photovoltaics. In the new concepts, building-integrated photovoltaics (BIPV) is even being used in the transparent sections of the façades which should be a cautious decision as they can highly affect the total energy demand of the building due to the change in the proportion of daylight and heat that can pass through. Thus, they should all be taken into consideration during the first stages of design to get the best result possible. While there have been some studies on this subject, we are still facing a shortage of tools and methods for BIPV design in the preliminary design phases.

This research aimed to provide a design tool for BIPV systems by making use of the integration of energy simulation programs with visual programming tools to spot the best façade solutions for any specific project. The optimization of these solar façades by this tool is discussed and compared to the nonoptimized alternatives. To put it briefly, the tests were done on a common vertical two-section façade with windows to provide natural light and solar heat to a certain amount that would be beneficial energy-wise, with crystalline silicon-based photovoltaics in the remaining parts of the façades.

The simulation results illustrated how a great quantity of inputs could affect the performance to a great extent. For instance, glazing material was put on a test and the results with four different alternatives in the south façade of Cairo (Egypt) showed that a wrong decision on glazing material alone could result in an increase of 28% in lower window-to-wall ratios (WWRs) and 51% in higher WWRs in the energy

O. Bakhshaei (✉)
Novin Tarh Studio, Tehran, Iran

G. Ridolfi · A. Saberi
Department of Architecture (DIDA), Università degli Studi di Firenze, Florence, Italy
e-mail: giuseppe.ridolfi@unifi.it; arman.saberi@unifi.it

© Springer International Publishing AG, part of Springer Nature 2019   141
A. Sayigh (ed.), *Sustainable Building for a Cleaner Environment*,
Innovative Renewable Energy, https://doi.org/10.1007/978-3-319-94595-8_13

consumption of an office room. Therefore, by choosing the optimal solution for each input, we could reduce the energy use of a building extensively which highlights the need for tools to come to the aid of the decision makers to find the best options and avoid choosing the inferior alternatives during the first stages of design.

13.1 Introduction

Throughout history, architects have always aimed to define some methods which were gained from their experience to achieve designs that lead to higher-performance buildings, but today with the emergence of energy simulation programs which can simulate the performance of buildings in terms of heating and cooling electricity consumption within 3% of mean absolute error [1] and with the high degree of complexity in projects, practicing rules of thumb would be a failure in catching up with the developments in the field.

There have been previous studies on using simulation tools for the optimization of solar façades with different methods. This was done by providing a tool to optimize the geometry of the building to achieve maximum insolation [2] and focusing on maximizing the amount of energy production rather than minimizing the net energy need or by trying to optimize energy production by finding the façade parts that have the highest insolation for BIPV positioning [3]. There have also been some studies on the PV area ratio on glazing to provide light and energy production at the same time and reach the lowest net energy consumption [4, 5]. In this research, minimizing also the dependency on fossil fuels was considered as the goal rather than maximizing the production of renewable energy; thus, the solutions which had the lowest net energy need were considered optimum.

In producing an optimizing tool for solar façade design, the first task would be to find the variables that could change a façade's performance for better or worse. One of the main effective parameters in changing the amount of energy loads of buildings is the size of windows. Nowadays, façades with vast glazing are more common to make the best use of daylight, which could result in higher loads due to the lower capabilities of glass to act as a heat insulator in comparison with walls in general. Therefore, they should have high-performance materials or preferably a reasonable size to provide the required daylight level while keeping the amount of heat transfer at its best rate.

There are other variables that could change the amount of solar gain and as a result the optimum window-to-wall ratio (WWR). An adjustment in the material properties of the walls or the windows themselves such as solar heat gain coefficient (SHGC) or U-value or whether the windows have external blinds or not could change the amount of energy loads and the ideal WWR to a great extent. Other factors like geographical location, orientation, the required light level inside of the building, etc., would also affect the optimum window-to-wall ratio for façades. In designing these solar façades where building-integrated photovoltaics (BIPV) are a part of their system, the amount of energy that photovoltaics produce would also be added as a factor that plays a role in determining the optimized WWR.

Table 13.1 Required input

Geometry		Material properties (wall)	Material properties (glazing)
Length	Function	U-value	U-value
Width	Lighting set point	PV-to-wall ratio	SHGC (solar heat gain coefficient)
Height	Shading scenario	PV efficiency	VT (visible transmittance)
Orientation		PV temperature coefficient	PV efficiency
Location			PV temperature coefficient

13.2   Methodology

It was decided to study the effectiveness of the variables on the optimal solar façade solution by testing them on different WWRs to monitor the changes. Consequently, the enumerative method was used for the tests. The principle of this method is simple. Within a finite search space, or a discretized infinite search space, the algorithm assesses the fitness function at every point in the space, one at a time [6]. In this way, a better understanding of the impact of the variables could be reached.

13.2.1   Inputs

A list of all effective parameters on the ideal façade solution was gathered as changing variables for the simulations. A detailed list of the provided inputs is demonstrated in Table 13.1.

Function By choosing the function input, visual properties of the surfaces, equipment and lighting power density, or thermal loads conditions like ventilation and occupancy schedules would be set automatically by predefined values based on the given function.

Lighting set point This input sets the needed amount of light. When there is a lower light level than the set point input, it means there is a shortage of daylight, and the artificial lighting system will fill the gap with its automatic dimming controls.

Shading scenario By selecting true for the shading scenario, an optimized horizontal blind system would be calculated for the windows (based on the other inputs selected) at the starting point of the calculations, and it would be considered in the evaluations.

Each input could take more than one value that would be tested in the later rounds of the calculations. For instance, in the second round, all the second inputs would be taken into account altogether, and for the third round, the third inputs and so on. Where there is no value for the second test for an input, the tool will automatically use the same last value for the new round. The existence of any multiple values for one input would result in multiple calculations.

13.2.2 Predetermined Inputs for the Evaluations

In the calculations of this article, there are some fixed and some changing values as inputs and not all inputs would be tested and considered as variable. Three different cities with different climates in the Mediterranean region were selected as locations for a typical office room with an area of 35 m² and geometry inputs of 7 m width, 5 m depth, and 4 m clear height, which was the case study for all the evaluations (Fig. 13.1).

This single office is included in an entire building; thus, only one wall would be fully exposed to the outdoor space (Fig. 13.2). This external wall consists of a single window placed at its center, as the energy load of the office would be at its lowest when the windows are located in middle height [7]. The WWR test range is from 1% to 100%.

Fig. 13.1 Geometry of the case study

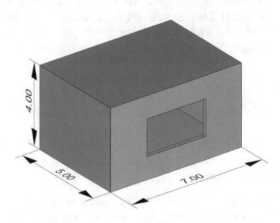

Fig. 13.2 Context's scenario

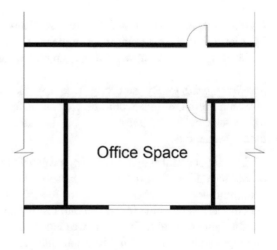

Fig. 13.3 Arrangement of daylight sensors

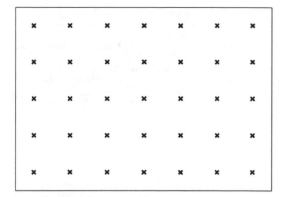

For the artificial lighting, a set point of 500 Lux, which is the minimum level for comfort in offices according to EN 1246-1 [8], was given as input to the auto-dimming system. The reference plane of the daylight simulations was placed at a working plane of 0.8 m height with a sensor dedicated to each square meter which would be a default action in the provided tool. In Fig. 13.3, the arrangement of these sensors in this office space is demonstrated.

13.2.3   Evaluation Tools

All the simulations are carried out in Grasshopper, Ladybug, and Honeybee which are all Rhinoceros 3D plug-ins, and they are all used to interface EnergyPlus [9] and Radiance [10] for the annual energy and illuminance computations. EnergyPlus is a free and open-source building energy simulation program. Its development is funded by the US Department of Energy that is open source. Radiance is also an open-source program, which is the most generally useful software package for architectural lighting simulation [11]. Even though Radiance is not a common tool for architects due to the lack of a graphic user interface, and as it needs an accurate model for simulation purposes, Grasshopper which is a visual programming tool was used as the modeling software.

13.3   Results and Discussions

13.3.1   Adding Photovoltaics to the Wall's System

For the first evaluation, the effectiveness of different WWRs on energy consumption of the office with a simple wall with no photovoltaics was compared with a BIPV wall system. In the second wall system, where BIPV was used, the whole area of the façade had photovoltaics added to its system except for the transparent section; thus,

Table 13.2 Inputs considered for the first test

Geometry	Material properties (wall)	Material properties (glazing)
Orientation: South	U-value: 0.34	U-value: 2.56
Location: Cairo	PV efficiency: None, 14%	SHGC: 70
Shading scenario: True, true	PV temperature coefficient: None, 0.45	VT: 80

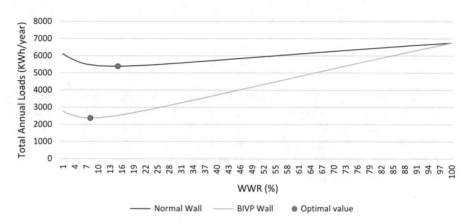

Fig. 13.4 Energy consumption based on window-to-wall ratio (WWR) for different wall materials. (Location: Cairo, Egypt; Orientation: South)

the PV-to-wall ratio was equal to one minus the WWR value. The other inputs that were used for this simulation are illustrated in Table 13.2.

The results in Fig. 13.4 show how the ideal WWR value changes by adding photovoltaics to the façade system. While the best WWR for a normal façade with the mentioned properties was 15%, this ideal ratio stepped down to 8% after adding the photovoltaics. This change is due to an added factor which is the energy production by the photovoltaics. The higher they produce energy, the lower the optimal WWR becomes.

13.3.2 Adding Photovoltaics to the Glazing's System

After photovoltaics was added to the wall's system in the first simulation, the effectiveness of different materials for glazing on the total energy consumption and the ideal WWR was put to the test. Three photovoltaic glazing types with different properties were chosen for a better investigation, as adding photovoltaics to the transparent part of the façade would be more complex in comparison with adding them to the opaque parts. The properties of these glazing materials are demonstrated

Table 13.3   Inputs considered for the glazing in the second test

Properties	Normal glass with optimized blinds	BIPV 1	BIPV 2	BIPV 3
U-value (W/m²K)	2.56	1.2	1.2	1.2
SHGC (%)	70	30	20	10
VT (%)	80	50	30	10
PV efficiency (%)	–	2.8	3.4	4
PV temperature coefficient (%/°C)	–	−0.13	−0.13	−0.13

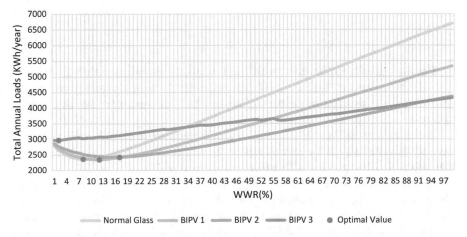

Fig. 13.5 Energy consumption based on window-to-wall ratio (WWR) for different glazing materials. (Location: Cairo, Egypt; Orientation: South)

in Table 13.3 as inputs along with the others. Other inputs remained the same as the previous test for the BIPV wall.

The results in Fig. 13.5 show how the optimal value of WWR changes for each material, from 8%, which belongs to the previous test results with only the wall as a source of energy production, to 12% for the first BIPV alternative, 17% for the second BIPV, and 2% for the third. The results illustrate how a wrong choice for the glazing material could result in a higher energy load. For instance, by choosing BIPV 3 instead of the normal window at a WWR of 8%, 655.64 kWh would be added to the annual energy consumption of the office which would be a 28% increase.

The graph also shows how each material has its own best WWR range which would be beneficial while it would be a wrong pick for other situations. For example, choosing the same BIPV 3 instead of normal glass in a WWR of 93% where BIPV 3 would be the best choice, energy consumption of the office drops from 6407.36 kWh per year to 4210.96 kWh, which would be a decrease of 34%. Figure 13.6 shows the most beneficial glazing material for different window-to-wall ratios.

Fig. 13.6 Best glazing
material based on energy
consumption for different
WWRs. (Location: Cairo,
Egypt; Orientation: South)

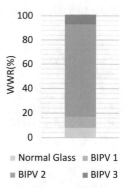

For the first 8%, normal glass with blinds would be the best solution, from 9% to 17% BIPV1, from 18% to 92% BIPV2, and from 93% to a whole glass façade; BIPV3 had the best results in comparison to other materials provided as inputs in this test.

It can be seen in Figs. 13.5 and 13.6 that highest range of WWR for being the best material belongs to the second alternative of BIPVs, while the optimized value belongs to the first (at 12% with a total energy load of 2345 kWh).

It is worth mentioning that the energy-wise optimal value would not always be the proper solution. By having access to this valuable data, decision makers could decide what would be the optimal value according to their own requirements. For instance, the optimal value of WWR for normal glass is 8% with a total energy load of 2368.5 kWh. Generally, the preference would be bigger windows and more natural light inside. The decision maker could expand the WWR without sacrificing net energy consumption. By choosing the BIPV1 for instance, the WWR could rise to 14% with a total energy load of 2361 kWh which is even less than the consumption of normal glass at 8%, while the amount of lighting energy would have a decrease of 26% due to having a higher amount of daylight (Table 13.4).

13.3.3 Orientation

For the third test, the same previously used materials for the wall and glazing were put to the test with different orientations of the façade. The given inputs to the tool were the same as the former simulation; only the other three orientations were added to the orientation input.

Figure 13.7 shows how the proper material for each WWR varies in different orientations which is due to the change of the amount of direct solar radiation on the façade, which reduces the effectiveness of photovoltaics on WWR optimal value and the amount of heat gain from the transparent part of the façade which would make higher WWRs and more transparent materials suitable to a greater extent as

Table 13.4 Comparison of two different alternatives for glazing

Material	Cooling consumption	Heating consumption	Lighting consumption	PV production	Total
Normal glass with blinds WWR = 8%	4573.24	712.87	362.64	3098.54	2368.53
BIPV 1 WWR = 14%	4578.31	694.61	267.23	2997.13	2361.34

Location: Cairo, Egypt; Orientation: South

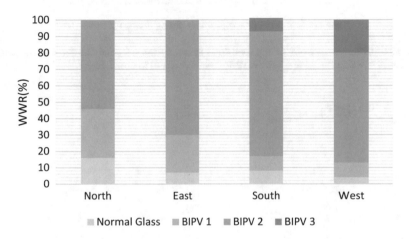

Fig. 13.7 Best glazing material based on energy consumption for different WWRs. (Location: Cairo, Egypt)

they would lead to a lower energy need for artificial lighting. Figure 13.8 clarifies the fluctuation of different sections of energy in different orientations while it also illustrates how the energy consumption of the same office cell could change up to 44% which would be 1913 kWh per year only by a change in the cardinal direction in which it is oriented.

13.3.4   Geographical Location

For the last test, the same simulations were done for all the selected location inputs. The cities of Palermo and Marseille were also added to Cairo to study the location input's impact on the façade solution. BIPV 1 was selected as it had the highest capability of reducing the energy loads in the previous tests.

Figure 13.9 visualizes how the ideal amount of WWR varies by the change of geographical location. Marseille has the highest recommended sizes for windows, and from Fig. 13.10, it could be seen that it also has the uppermost energy consumption among the selected cities due to its heating-dominated climate. Cairo

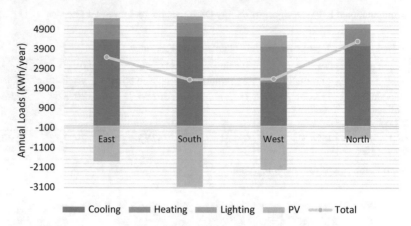

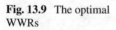

Fig. 13.8 Different sections of the total load at the optimized solutions for each orientation (East = 19% − South = 12% − West = 12% − North = 30%) in Cairo, Egypt

Fig. 13.9 The optimal WWRs

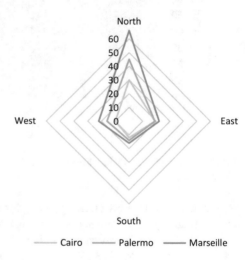

Fig. 13.10 The total energy loads at the optimal WWRs

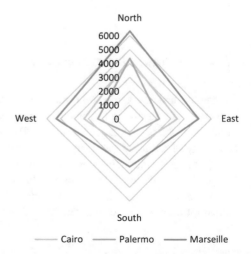

on the other hand is the most cooling-dominated climate which resulted in lower WWRs as optimum. The changes were small in east and south orientations, and the best result would be derived from almost the same WWRs as in Palermo. In general, west and north orientations had the most contrast between the three locations, while east and south had the least.

13.4 Conclusion

The test results demonstrated how a change in each of the material properties, like the efficiency of the photovoltaics, visible transmittance (VT), and solar heat gain coefficient (SHGC) of windows, window-to-wall ratio, the orientation of the façade, and the location of the building could result in a completely different ideal solution. They clarified that not only BIPV glazing materials would not be beneficial at all times, but also they could considerably increase the total energy consumption in some cases. This underlines the decision makers' need for a precise set of data from computational simulations to allow them to choose the best active and passive design options and avoid making energy-wise costly mistakes. It also helps them to make their decisions based on their own predefined requirements; if they need to meet some *regulations* for *daylighting or a demand for having larger windows to comply with visual needs of the occupants, the proper solution would be different. It is also worth mentioning that* the illustrated results are based on the specifications of today's existing building materials on the market. With the developments on material properties every day like the efficiency of BIPV or U-value of glazing or walls, the ideal solution would be different through the time due to these ever-changing values. By taking into consideration the valuable data from computational simulations, designers could catch up with these developments and have the highest performance designs possible with the latest materials.

References

1. Yezioro A, Dong B, Leite F (2008) An applied artificial intelligence approach towards assessing building performance simulation tools. Energ Buildings 40(4):612–620
2. Lobaccaro G, Fiorito F, Masera G, Prasad D (2012) Urban solar district: a case study of geometric optimization of solar facades for a residential building in Milan. In: International conference "Solar 2012", Melbourne. December 2012
3. Lovati M (2014) A BiPV design optimization method. In: International conference SOLARTR 2014, Izmir, November 2014
4. Robinson L, Athienitis A (2009) Design methodology for optimization of electricity generation and daylight utilization for façade with semitransparent photovoltaics. In: Proceedings of building simulation 2009: 11th international IBPSA conference Glasgow, Scotland, pp 811–818
5. Sandra M, Frontini F, Wienold J (2011) Comfort and building performance analysis of transparent building integrated silicon photovoltaics. In: Proceedings of building simulation 2011: 12th conference of international building performance simulation association, Sydney, pp 2080–2087

6. Goldberg DE (1989) Genetic algorithms in search optimization and machine learning. University of Alabama, Addison Wesley
7. Kim S, Zadeh PA, Staub-French S, Froese T, Cavka B (2016) Assessment of the impact of window size, position and orientation on building energy load using BIM. Proc Eng 145:1424–1431
8. EN 12464-1:2011. Light and lighting – lighting of work places – Part 1: Indoorwork places. European Committee for Standardization (CEN), 2011
9. EnergyPlus. Version 8.1. 2013. Engineering Reference The Reference to EnergyPlus Calculations. Lawrence Berkeley, National Laboratory, 2013
10. RADIANCE, the radiance 4.0 synthetic imaging system (2010) Lawrence Berkeley, National Laboratory, Berkeley,
11. Roy GG (2000) A Comparative Study of Lighting Simulation Packages Suitable for use in Architectural Design. Murdoch University. Perth

Chapter 14
Feasibility Study of a Low Carbon House in the United Kingdom

Timothy Aird and Hossein Mirzaii

Abstract In 2008 the United Kingdom passed a climate change act that set a target to reduce their carbon emissions by at least 80% in 2050 from the 1990 levels, and given that 27.8% of these emissions are from the housing and building sector, having high energy efficient buildings can contribute greatly to reaching this target. The primary aim of this project is to conduct a feasibility study of a low carbon emitting home in the United Kingdom to determine the possibility of reducing carbon emissions from newly designed homes and if it is financially viable to build and live in such a home. The objectives were in four parts: the first part being to design the house using 3D modelling software, the second part is the analysis of the passive cooling concept that was implemented using computational fluid dynamics (CFD) software, the third part is the design and analysis of the active system using Polysun software, and lastly the economic analysis of the active system that was designed. The house was designed using materials with a low embodied energy such as wood for the structure and walls of the house and natural hemp material for the insulation. A complete 3D model was completed using the exact dimensions and type of materials used to design the house in order to achieve most accurate depiction of the design as possible. The passive cooling concept using natural ventilation that was implemented and virtually tested using CFD software simulation and an average temperature of 23.1 °C was established inside the house, while the outside temperature was set at 30 °C. The active system which comprised of 14 solar panels and 2 evacuated tube collectors obtained a solar fraction of 54.9% and an electricity self-consumption of 47.3%. The annual space heating for the house was found to be 9.1 kWh/m^2 while the total primary energy demand was found to be 34.62 kWh/m^2. The LCOE of the PV system was found to be 11.17 p/kWh which makes the production of electricity economically feasible. The profit obtained from the active system was found to be £44,026.33 after 25 years with a profit being made after 4 years.

T. Aird · H. Mirzaii (✉)
School of Aerospace & Aircraft Engineering, Kingston University London, London, UK
e-mail: H.Mirzaii@kingston.ac.uk

© Springer International Publishing AG, part of Springer Nature 2019
A. Sayigh (ed.), *Sustainable Building for a Cleaner Environment*,
Innovative Renewable Energy, https://doi.org/10.1007/978-3-319-94595-8_14

14.1   Introduction

Across the European Union, a standard has been set across the European Union that all newly designed buildings are to achieve close to zero energy by the end of 2020 [5]. The passive house standard has therefore been seen as a viable means to meet the target set and to achieve the high-efficiency building standard. The passive house standard has four main requirements which are as followed: space heating demand to not exceed 15 kWh/m^2 of net living space per year, the total energy to be used by all domestic applications to not exceed 60 kWh/m^2 of treated floor area per year, an airtight house with a maximum of 0.6 air changes per hour at 50 pascals pressure and thermal comfort being met for all living areas during winter as well as in summer with no more than 10% of hours in a given year being over 25 °C [7]. Passive heating and cooling have been used in different variations over the years and have showed that the highest level of energy efficiency within households can be produced when they are combined together effectively in a way that reduces the energy demand of the house. These concepts were used in this project along with an active system to meet the energy demand.

14.2   Literature Review

The literature review that was done focused on three main parts being: the materials used to build the house, the passive heating and cooling concepts and the active system.

14.2.1   Structure and Insulation

The structural material was a major factor that needed to be considered, and the low carbon aspect of this thesis was at the forefront in choosing the type of material that was to be used for the structure of the house. It was important to choose a material with a low embodied energy, and timber was ideal for this criteria. Timber has high structural qualities and has been used by societies around the world for thousands of years. When compared to concrete, another material commonly used for construction, timber is more efficient using around only half of the energy needed for production as seen in Figs. 14.1 and 14.2 [8].

Insulation is another factor that plays a very important role in the design of a passive house as the heat transfer of heat needs to be contained as best as possible. Many different types of materials are currently used for insulation of walls, but a lot of these materials have high impact carbon footprints, and therefore it was necessary to find materials that would not be carbon neutral. Plant-like materials for insulation are considered very beneficial as it stores carbon during its usable life throughout

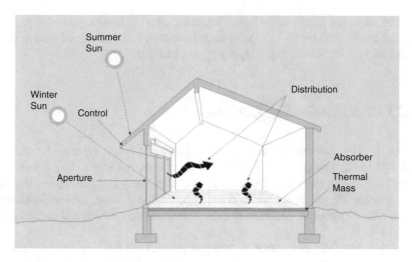

Fig. 14.1 Direct solar gain (Energy.gov 2015)

Fig. 14.2 Solar cooling
system [9]

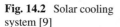
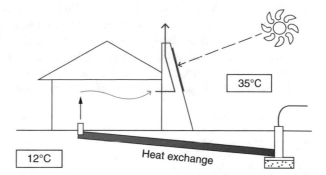

the plant growth and remains stored during its use as a building material [2]. Carbon is released only at the end of its life through natural decomposition which returns other nutrients to the soil or being burnt as fuel to generate heat/energy [2]. In 2011 a study was done by Korjenic et al. that focused on testing the insulation properties of natural thermal insulation materials, and results obtained found that hemp with a thickness of 40.2 mm produced the same thermal conductivity properties as other synthetic materials [6].

14.2.2   Passive Building Design Concepts

The orientation of the house is also very important to maximise heating from the sun. A study was carried out in 1984 for 25 climates in the United States with different building orientations using a developed version of a building analysis energy computer programme. The results obtained showed that for all climates, the total loads are significantly lower when the more extensive glazed exposure is oriented south when compared to the same buildings oriented east or west and with a north orientation having significantly higher loads [1].

14.2.3 Passive Cooling Concept

Due to the high-insulated walls and air tightness of passive design houses, it is necessary to have proper ventilation to remove the excess heat during the summer. Passive cooling of the house can be done through many mechanisms of transport such as naturally occurring wind, conduction to the ground, airflow induced by temperature differences, radiation to clear skies, and convection and conduction to clear skies [4].

For this thesis a concept that is currently used in hot dry climates in the world, such as the Middle East and Australia, would be implemented [9]. This design makes use of the ground as a heat sink to draw the heat from the air before it enters into the room, and a chimney with a clear glass is used to allow for the air inside to be heated up. This heated air would then create an induced flow system to help with the cooling effect. Even though the design is for a cold climate being in the United Kingdom, there are still times of the year that the temperature goes above 30 °C, and therefore having some type of cooling effect in high-insulated passive homes would be detrimental to the all year comfort of the home owners.

14.3 House Design

14.3.1 Wall Structure

A breakdown of the wall structure used for this design can be seen in Table 14.1. This wall structure had an equivalent U-value of 0.146 W/m^2.

Table 14.1 Wall structure

Exterior side		
Material	Thickness (mm)	Thermal conductivity (W/mK)
Wall cladding	11.5	0.12
Gypsum wall board	12.7	0.65
Hemp insulation	139.7	0.038
Gypsum wall board	12.7	0.65
Hemp insulation	110.3	0.038
Gypsum wall board	12.7	0.65
Interior side		
Total	299.6	

14.3.2   Windows

The windows are all designed to be triple glazed, argon filled and have a U-value of 0.7 W/m^2 K straight from the manufacturer.

14.3.3   Floor

The flooring of the house is made up of a mixture of concrete and recycled materials such as crushed glass, bits of recycled plastics and marble chips. This is covered by a SoyCrete Architectural Concrete Stain finish that allows for various colours of flooring to be chosen. Concrete flooring is chosen for this design due to its high thermal mass and ability to retain heat.

14.3.4   Passive Heating and Cooling Additions

A Trombe wall and a solar chimney are designed and added to the house to assist with the heating and cooling load, respectively, by natural means.

14.3.5   Passive Cooling Computational Fluid Dynamics (CFD) Simulation

Due to overheating during the summer being a problem for passively designed homes, a passive cooling concept was designed for this house. Warm air enters an inlet at the base of the house and flows underground where the cool ground removes some of the heat from the air allowing for cool air to then flow into the building. This airflow is increased due to the solar chimney that is placed at the top of the house which creates a forced convection of air which is possible due to the air tightness of the home. This air cools down the house to below the standard comfort level without the need for additional mechanical cooling equipment.

In order to test the operation of the design implemented into the house, a simulation was run to determine the temperature of the airflow through the house. The simulation was set to be in London during summer with the inlet mass flow rate of air being 4.26 kg/s. Figure 14.3 shows the results obtained from final design with average temperature of the indoor air being 23.1 °C.

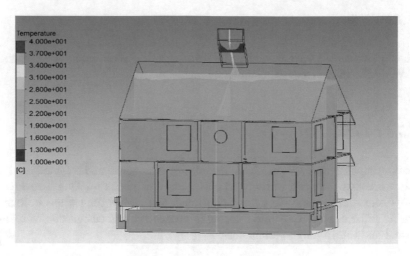

Fig. 14.3 Temperature profile of final concept with room vent

14.4   Active System

The house was designed for four individuals, and therefore based on their requirements and the size of the house, the active system was designed with respect to the heating and electricity demand. The house was designed to have 14 solar panels and 2 evacuated tube collectors. It was also designed to have a heat pump providing the excess heat energy and a heat recovery ventilation unit to ensure a high indoor air quality due to the air tightness of the home. Polysun was used to analyse the annual operation of the system and can be seen in Fig. 14.4.

14.5   Results and Discussion

14.5.1   CFD Simulation

Three different simulations were run with all having the same parameters but with different designs for the airflow. The final concept with air vents added to the ceiling proved to give the best results as seen by the grey bar in the graph in Fig. 14.5. Each room found to have lower temperatures with the average temperature in the house being 23.1 °C. This was due to the design allowing for the air to be cooled by a large ground area along with the cool air being evenly distributed into the rooms before it exited through the solar chimney.

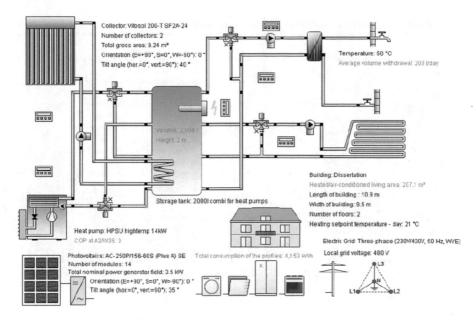

Fig. 14.4 Polysun simulation

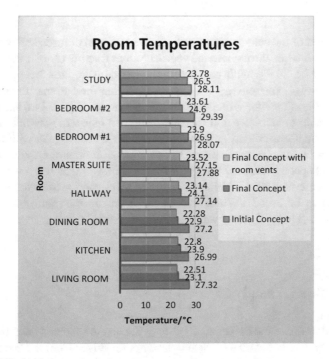

Fig. 14.5 CFD simulation room temperatures

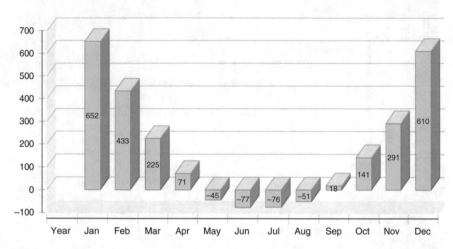

Fig. 14.6 Polysun simulation

14.5.2   Polysun Simulation

The simulation of the active system provided a great deal of information on how the system would operate throughout the year. The solar fraction of solar thermal energy was found to be 54.9% with the electricity self-consumption being 47.3%. The highest amount of solar energy was produced during the summer time which was expected as this is the time with the great sunlight hours. Figure 14.6 shows just this as for combining electricity and thermal energy the system was able to produce more power than it used during the months from May to August, while during the winter time when the heat demand was high, a large amount of energy was used.

An annual space heating of 9.1 kWh/m² was achieved which fell well below the standard requirement of 15 kWh/m². Also the total energy used for all domestic applications to be 34.62 kWh/m² which far from the minimum requirement of 60 kWh/m²6. These figures were calculated based on the results from the simulation.

14.6   Financial Analysis

In order to determine the financial viability of the active system, it was important to analyse the system based on its cost and the output received from it. Determining the savings from energy production along with any incentives that would be given for the location, it was possible to determine the payback period of the PV and thermal system. The economic analysis for the passive house design was not done as all the designs used are currently implemented in many passive homes around the United Kingdom and have already been found to be economically feasible, and therefore focus for the economics was put on the active side of the design instead. This is can be also seen in studies that were done, for example, Colclough et al. conducted a study this year [3] in Northern Ireland which solely focused on determining the cost of building to the near zero energy building standard (nZEB) [3].

14.6.1   Incentives

The incentives for the PV system included the general feed in tariff (FIT) and the export FIT which was totaled £203.07 for the first year and reduced for the consecutive years due to degradation of the solar panels. A renewable heat incentive is given for the heat produced by the thermal collectors and was found to be £765.29 each year.

14.6.2   Loan

In order to be as realistic as possible for the economic analysis, a loan payment was added to the analysis which had a great effect on the payback period and profit made after the 25-year life.

14.6.2.1   Present Worth Factor

In order to accurately calculate the payback period and profit of the system, the periodic payment for the loan was determined. This was done by first calculating the present worth factor as follows:

The sum of the present worth values over a total N number of years gives a present worth factor (PWF).

$$\mathrm{PWF}\left(N,i,d\right)=\sum_{j=1}^{N}\frac{\left(1+i\right)^{j-1}}{\left(1+d\right)^{j}}=\frac{1}{d-i}\left[1-\left(\frac{1+i}{1+d}\right)^{N}\right], \text{if } d \neq i$$

PWF = 14.09

14.6.2.2   PV System

The periodic payment for the loan is found from the following equation.

Yearly payment of loan:

$$\text{Periodic payment} = \frac{M}{\text{PWF}\left(N_i, i, d\right)}$$

M = Loan amount = £4550 × 0.8 = £3640
Periodic payment = £258.34

14.6.2.3 Thermal System

The periodic payment for the loan is found from the following equation.
  Yearly payment of loan:

M = mortgage = £13,011.92 × 0.8 = £10,409.54
Periodic payment = £738.79

14.6.3 Levelised Cost of Energy (LCOE)

$$\text{LCOE} = \frac{I + \sum_{n=0}^{n} \dfrac{\text{AO}}{\left(1+d\right)^n} + \text{LP} - \sum_{n=0}^{n} \dfrac{\text{CBI}}{\left(1+d\right)^n}}{\sum_{n=1}^{n} \dfrac{Q_n \left(1 - \text{System degradation rate}\right)^n}{\left(1+d\right)^n}}$$

I, initial payment; AO, annual operations and maintenance; LP, loan payment; Q_n, energy generation in the year; CBI, cost-based incentive; d, discount rate; and n, life of the system

14.6.3.1 PV System

$$I = £910 \ \ \text{AO} = £75 \ \ Q_n = 3073.3 \ \text{kWh} \ \text{LP} = £5166.80 \ \text{CBI}$$
$$= \left[\left(3073.3 \times 0.9952^{(n-1)}\right) \times 0.404\right] \ d = 3.5\% \ \ n = 25$$

System degradation = 0.48%/year
  From the calculation the LCOE was found to be 11.17 p/kWh.
  For the PV system, due to the 20-year loan that was assumed, resulted in a payback period being only after the 24th year with a small profit of £97.41. These poor economic values were received due to the loan that was assumed to be taken at the beginning of the project along with the low solar radiation available in the set location. The LCOE of the PV system was found to be 11.17 p/kWh which is lower than

the current electricity cost of 14.3 p/kWh, and therefore the production of electricity from the PV system was found to be economically feasible.

The economics of the thermal system involved the cost of the entire heating system which included the thermal collectors, the heating tank, the control system, the heat pump and the heat recovery ventilation system. This system proved to be much more valuable economically and would receive profits from the 3rd year onward with a total of £43,928.92 of profit after 25 years. Due to this high profit margin when being obtained, the combination of the two systems was found to be economically feasible with profit being made from the 4th year onward and a total profit of £44,026.33 at the end of the 25 years.

14.7 Conclusion

The house was designed using material with a low embodied energy, and all the UK passive house standards were considered and met. An annual space heating demand of 9.1 kWh/m^2 was achieved which fell well below the standard requirement of 15 kWh/m^2. Also this value can be considered very low when compared to the heating demand of new conventional households having an annual space heating demand of approximately 100 kWh/m^2. Also the total energy used for all domestic applications to be 34.62 kWh/m^2 which far from the minimum requirement of 60 kWh/m^2. These figures that were achieved were mainly due to the high levels of insulation used and the production of energy from the solar collectors and PV system. All the materials used to build the house and for the passive heating and cooling concepts fall within the norm for passive house designs currently in the United Kingdom, and therefore the total cost to build the home would not exceed the cost for an average new home.

The passive cooling concept implemented into the house design was able to reduce the average temperature inside the house down to 23.2 °C, while the ambient temperature was set at 30 °C. The active system was able to achieve a solar fraction of 54.9% for space heating and domestic hot water, while the electricity self-consumption from the PV panels was found to be 47.3%. The LCOE for the electricity produced by the PV panels was found to be relatively high at 11.17 p/kWh which fell below the current cost of electricity and therefore is economically feasible. The entire active system had an initial cost of £16,461.92, and after taking into account all the financial aspects of the system, a total profit after 25 years was found to be £44,026.33 with profit being made from the 4th year onward.

The house was able to achieve a comfortable temperature for the entire year while also reducing large amounts of CO^2 from the atmosphere by the use of low carbon materials and the methods of energy production for use by the house. This was done while also having a low overall cost which therefore labels this project a success as from this project it can be said that it is economically feasible to build and live in a low carbon house in the United Kingdom.

14.8   Improvement and Further Work

Different types of material can be considered for both the construction and insulation materials as there are many natural materials that may be deemed better to be used at the specific location such as mud for insulation.

Due to limitations of the solid works software, the effect of that the Trombe wall had on reducing the heating load was not recognised. To get a much more accurate representation of the system, the Trombe wall should be tested using CFD simulation software and the results from this simulation analysed and the entire system placed in another computer software for the analysis of the active system.

A miniature model of the design can be built in the future to test the airflow rather than just trusting the use of simulation software. This can provide a grounds to compare the actual tests with those done on the software.

A detailed economic analysis can be done in the future of the actual building of the house to determine the exact cost required to build the designed home.

References

1. Andersson B, Place W, Kammerud R, Scofield M (1985) The impact of building orientation on residential heating and cooling. Energ Buildings 8(3):205–224
2. Binici H, Eken M, Dolaz M, Aksogan O, Kara M (2014) An environmentally friendly thermal insulation material from sunflower stalk, textile waste and stubble fibres. Constr Build Mater 51:24–33
3. Colclough S, Mernagh J, Sinnott D, Hewitt NJ, Griffiths P (2017) The cost of building to near zero energy building standard – a financial case study. In: Sayigh A (ed) Sustainable Energy for All. Springer, Cham
4. Duffie J, Beckman W (2013) Solar engineering of thermal processes, 4th edn. Wiley, Hoboken, p 5
5. European Commission (2015) Nearly zero-energy buildings – Energy – European Commission. [online] Available at: http://ec.europa.eu/energy/en/topics/energy-efficiency/buildings/nearly-zero-energy-buildings. Accessed 21 Aug 2017
6. Korjenic A, Petránek V, Zach J, Hroudová J (2011) Development and performance evaluation of natural thermal-insulation materials composed of renewable resources. Energ Buildings 43(9):2518–2523
7. Passiv.de (2015) Passivhaus Institut. [online] Available at: http://www.passiv.de/en/02_informations/02_passive-house-requirements/02_passive-house-requirements.htm. Accessed 21 Aug 2017
8. Timber as a structural material – an introduction (2014) [ebook] Structural Timber Association, p 3. Available at: http://www.cti-timber.org/sites/default/files/STA_Timber_as_structural_material.pdf. Accessed 23 Aug 2017
9. Yourhome.gov.au (2014) Passive cooling | YourHome. [online] Available at: http://www.yourhome.gov.au/passive-design/passive-cooling. Accessed 23 Aug 2017

Chapter 15
Why Do We Need to Reduce the Carbon Footprint in UAE?

Riadh AL-Dabbagh

Abstract Today, the term "carbon footprint" is often used as shorthand for carbon (usually in tones) being emitted by an activity or organization. The carbon component of the ecological footprint, which we call the carbon footprint, takes a slightly differing approach. Our carbon footprint measurement translates the amount of carbon dioxide emissions into the amount of productive land and sea area required to sequester those carbon dioxide emissions. UAE enjoys a high level of human development but at the cost of a large ecological footprint. A quick action to overcome the raise of carbon footprint is therefore required. The paper explores all issues that affect the environment in UAE and the associated consequences. CO_2 emission trends are analyzed and showed the country has a high percentages compared with other worlds' emissions. Carbon cycle is presented to explain the real causes for these emissions. UAE enabling strategies, policies, and activities are surveyed to show how the country is acting and what are the existing practices followed in order to reduce the carbon footprint. The paper introduces extra practices that if followed will have a large impact on the carbon footprint. Green buildings and the deployment of renewable energy systems are some of the solutions that could insure a considerable amount of carbon emission. Energy efficiency assessment for buildings is an important tool that controls the right implication of the proposed practices. It has been concluded that UAE is developing a good policies and implementing the right practices to achieve their 2030 target. However, more is needed as emphasized by this work.

15.1 Introduction

This tells us the demand on the planet that results from burning fossil fuels. Measuring it in this way enables us to address the climate change challenge in a holistic way that does not simply shift the burden from one natural system to

R. AL-Dabbagh (✉)
University of Ajman, Ajman, UAE

© Springer International Publishing AG, part of Springer Nature 2019   165
A. Sayigh (ed.), *Sustainable Building for a Cleaner Environment*,
Innovative Renewable Energy, https://doi.org/10.1007/978-3-319-94595-8_15

Estimated size of major stores of carbon on the Earth.	
Sink	**Amount in Billions of Metric Tons**
Atmosphere	578 (as of 1700) - 766 (as of 1999)
Soil Organic Matter	1500 to 1600
Ocean	38,000 to 40,000
Marine Sediments and Sedimentary Rocks	66,000,000 to 100,000,000
Terrestrial Plants	540 to 610
Fossil Fuel Deposits	4000

Fig. 15.1 The estimated size of major stores of carbon on the earth

another. The carbon footprint is currently 60 percent of humanity's overall ecological footprint and its most rapidly growing component. Humanity's carbon footprint has increased 11-fold since 1961. Reducing humanity's carbon footprint is the most essential step we can take to end overshoot and live within the means of our planet. An average person in the UAE consumes ten times more energy than world average, emits nine times more CO_2 emission, and consumes six times more electricity than world average. UAE energy demand is expected to double by 2020; water demand is expected to grow by 44% by 2025. The desalination fuel requirements alone will take up 20% of the total fuel production by 2030.

Building sector today has an oversized footprint. The emirate believes it is well placed to invest its knowledge and financial resources in the world's future energy markets – renewable energy. So in April 2006, the Abu Dhabi government established Masdar. In this way, Masdar plays a key role in the development of Abu Dhabi's renewable energy sector, driven continual innovation, and commercialization of clean and sustainable energy technologies.

We live in one of the world's most arid environments, so it's no surprise that we're one of the world's biggest consumers of energy.

Within carbon cycle, there are various sinks or stores of carbon (Fig.15.1).

The study of the carbon cycle basically means that we want to know how carbon is exchanged between these reservoirs and at what rate. The approximate rates of carbon exchange are shown in the picture below. The amounts indicated are in gigatons of carbon per year (Fig. 15.2).

CO_2 makes up approximately 85 percent of total greenhouse gas (GHG) emissions (Fig. 15.3).

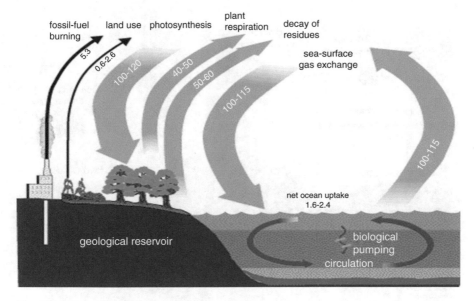

Fig. 15.2 The carbon cycle

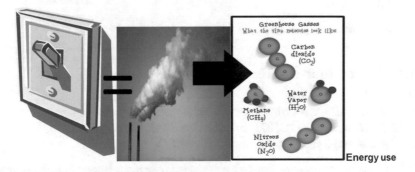

Fig. 15.3 The percentage of CO_2 out of the greenhouse gases

15.2   What Is a Carbon Footprint?

Carbon footprint (FP) is "the total set of GHG (greenhouse gas) emissions caused directly and indirectly by an individual, organization, event or product".
Everyone in this room has a FP.
The amount of GHGs which are emitted by production (including extraction of raw materials), use, and disposal of a product/service in CO_2 equivalents is called carbon footprint.
Not only products are assigned a carbon footprint, but also services or even kilometers have a carbon footprint or even breakfast (Fig. 15.4).

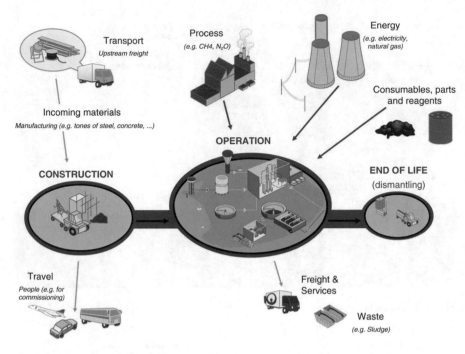

Fig. 15.4 The amount of greenhouse gas from different sources

15.3   Context for Numbers

The average UAE has a carbon footprint of 19 tons CO_2e/year/person.
World greenhouse gases (GHG) emissions are 34 Gt CO_2e/year.
The global average carbon footprint is ~6 tons CO_2e/year/person.

UAE had an ecological footprint of 9.5 global hectares per capita.
UAE consumes more food, fiber, energy, goods, and services than the planet can naturally regenerate.
One gram of CO_2e would be produced if you burned a pea-sized blob of gasoline.

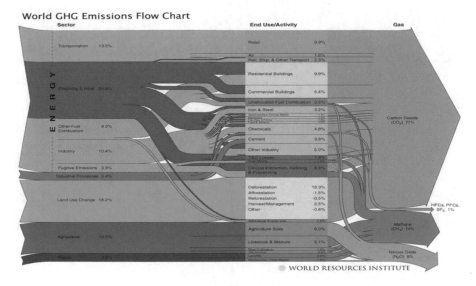

Fig. 15.5 The flowchart of the world greenhouse gas emission

One kilogram (2 lbs) of CO_2e would be produced if you burned two cups of gasoline.

One ton of CO_2e would be produced if you burned 60 gallons of gasoline.

Emissions of carbon dioxide, methane, and nitrous oxide in equivalent carbon dioxide units:

Carbon dioxide = 1 CO_2 equivalent.

Methane, CH4 = 21 CO_2 equivalents.

Nitrous oxide, N2O = 310 CO_2 equivalents.

Other gases, HFCs, PFCs, and SF6 = range 600 – 23,900x CO_2 equivalents.

Best estimate of full climate change impact of something.

Carbon – Usually talking about CO_2e (CO_2-eq) not actually just carbon.

Footprint – Total impact Fig. 15.5.

15.4   Ecological Footprints

A carbon footprint is only one component of the broader ecological footprint. An ecological footprint compares the population's consumption of resources and land with the planet's ability to regenerate. The earth's ecological footprint is currently 23 percent over capacity. It takes about 1 year and 2 months to regenerate what we consume in a year [source: Footprint Network] Fig. 15.6.

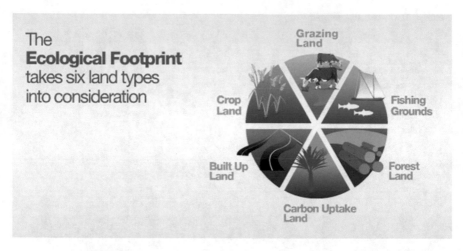

Fig. 15.6 Ecological footprint

15.5 Development of the Methodology Guidelines for National GHG Inventories.

Methodology guidelines for national GHG inventories have undergone several revisions and updates.

*Good Practice Guidance focuses more on how uncertainties and errors may be estimated and minimized.

The Revised IPCC Guidelines is a very good starting point for people who want to learn how national GHG emissions and removals are estimated and the national.

OECD Report on Estimation of GHG Emissions and Sinks (1991).

IPCC Guidelines for National GHG Inventories (1995).

Revised IPCC Guidelines for National GHG Inventories (1996).

*Good Practice Guidance and Uncertainty Management (2000) (Figs. 15.7 and 15.8).

Building sector today has an oversized footprint:

Building sector is called the industry of "thirds."

- Over *1/3 of all CO2 emissions* come from building construction and operations.
- Over *1/3 of all energy and material resources* are used to build and operate buildings.
- Over *1/3 of total waste* results from construction and demolition activities (Fig. 15.9).

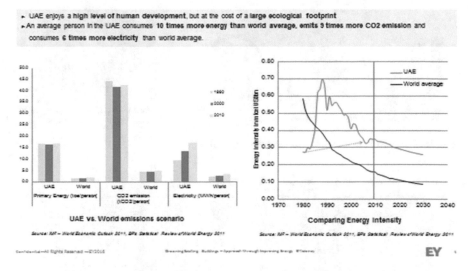

Fig. 15.7 UAE emission of CO_2 in comparison to the world emission

UAE Enabling Strategies, Policies, Activities

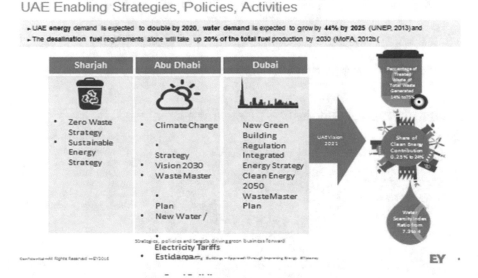

Fig. 15.8 UAE enabling strategies, policies, and activities

Sectorial projections for CO_2 mitigation potential in 2030

- Building sector holds the *greatest potential* to reduce GHG emissions
- A 29" reduction in projected baseline emissions by 2020 is achievable at zero cost
- About *3.5 GtCO₂ of emissions could be achieved through increased energy efficiency* (Figs. 15.10, 15.11, 15.12 and 15.13)

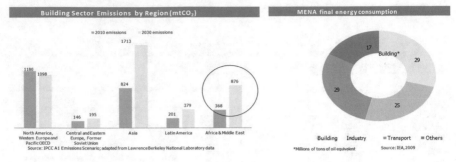

Fig. 15.9 Emission of CO_2 from building sector in UAE

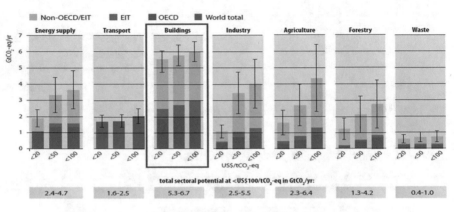

Fig. 15.10 Sectorial projections for CO_2 mitigation potential in 2030

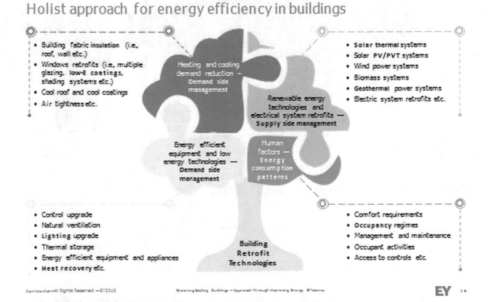

Fig. 15.11 Energy efficiency in buildings

Fig. 15.12 Sources of CO_2 from different usable energy technologies

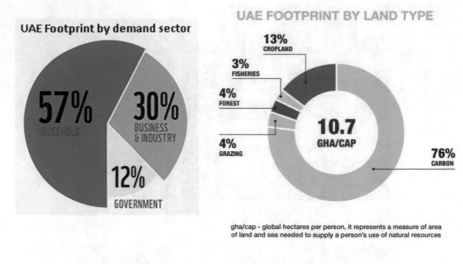

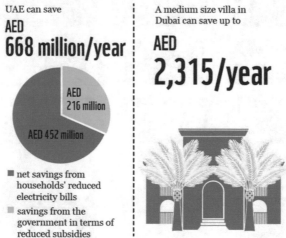

Fig. 15.13 UAE footprint by different sectors

15.6 UAE Tops World on Per Capita Carbon Footprint

UAE residents are still consuming more natural resources than anybody else on earth, ahead of the United States and Kuwait, according to the 1NVVF Living Planet 2008 report which is based on statistics gathered in 2005.

The 2008 report shows the UAE's per capita footprint was found to be 9.5 global hectares. Its overall demand on global resources in 2005 was less than half of one percent. However only 2.1 global hectares were available per person.

There has been a lot of change in the calculations and reviewing of the data, so we cannot just look at the number and see that it is high and say we are not doing well.

There is a very high demand on natural resources, and the report reflects this. The ecological footprint is 80 percent from fossil fuels for power and transport.

The carbon emissions data apparently comes from *"World Development Indicators 2012."*

By 2009 standards, the UAE and Qatar were over three and seven times worse, respectively, in per capita CO_2 emissions than an enormous emerging country like China. The good news is the numbers are slowly sloping downward; however it would be nice to have data from 2009 and 2013 to confirm.

The UAE and Qatar have been having massive development projects really only over the last two decades. Infrastructure in developed countries such as the United States and the United Kingdom is comparatively older and more mature. So the question really is whether this data just reflects the normal development patterns of countries modernizing themselves over the time period for which we do have data.

The negative aspects of country transformation patterns maybe seen in another set of ecological health checks called ecological footprint and bio-capacity.

Ecological footprint, measured in global hectares, is defined by the Footprint Network as a measure of how much area of biologically productive land and water an individual, population, or activity requires to produce all the resources it consumes and to absorb the waste it generates, using prevailing technology and resource management practices. Because trade is global, an individual or country's footprint includes land or sea from all over the world.

Bio-capacity, also measured in global hectares, is the capacity of ecosystems to produce useful biological materials and to absorb waste materials generated by humans, using current management schemes and extraction technologies.

Ecological footprint and bio-capacity vs time plots for some countries. It is not surprising to see by the green line that the available land per capita required to replenish consumed resources and absorb wastes for all these countries is on a decreasing trend. The point where the two cross each other is called overshoot beyond which there is a bio-capacity deficit to meet a country's footprint. Most, if not all countries, are under visible ecological deficits.

The UAE, Qatar, and Kuwait have a wide disparity between available resources and per capita resource consumption. The image that these points to region-specific issues such as high immigrant influx, high population growth (UAE's is 12% average!), massive urbanization, expansion and associated development projects, and the demands from a difficult desert climate for high amounts of energy for comfortable living. Significant amounts of natural resources are hence imported from outside to drive growth.

It's a good reminder that the monitoring site on a mountain in Hawaii that sets the world's benchmark for CO_2 emissions told us in March 2013 that the earth has passed the 400 ppm milestone. The ugliness here is that CO_2 could stay up for a long time and nations cannot really stop developing. It is necessary to keep pumping a lot of gas into the atmosphere, and no one really knows what the real consequences are going to be. But we can always manage risk, if not reverse consequences.

Fortunately, the UAE is very well acquainted about this issue. There are national initiatives to try and better understand consumption patterns. All this shows that this country is still a big beacon of hope and prosperity for many individuals in the Middle East. In that regard, trade-offs of expansion, of trade, and of integrating foreign people who come here to live and work will inevitably crop up.

In Dubai, an average person uses a mammoth 20,000 kilowatt-hours of energy and 130 gallons of water per year, according to the Dubai Electricity and Water Authority (DEWA), while water consumption across the UAE was estimated to be around 500 liters per day in 2013, according to the Federal Electricity and Water Authority. This was 82 percent above the global average and three times higher than average per capita consumption in the European Union. Waste levels per capita are also among the highest in the world, with Abu Dhabi alone producing between 1.8 kg and 2.4 kg per person per day – almost double that of the United Kingdom.

But the UAE remains resolute, even going so far as to pledge that the site of the World Expo 2020 event Dubai will host in 5 years' time will be the "most sustainable ever." At least 50 percent of the Expo site is to be constructed from recyclable materials and powered with renewable energy, claimed Reem Al Hashimy, UAE minister of state and managing director of the Dubai Expo 2020 Higher Committee, at a conference in March, as she insisted Dubai would play a big role in promoting global environmental sustainability in years to come (Figs. 15.14 and 15.15).

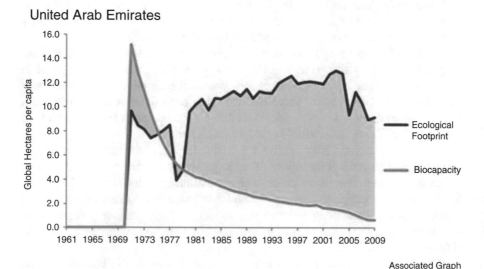

Fig. 15.14 Tracks the per-person resource demand *Ecological Footprint* and *biocapacity* in United Arab Emirates since 1961. Biocapacity varies each year with ecosystem management, agricultural practices (such as fertilizer use and irrigation), ecosystem degradation, and weather, and population size. Footprint varies with consumption and production efficiency

Fig. 15.15 The new face of Dubai after Expo 2020

15.7   Conclusion

The UAE has one of the largest carbon footprints in the world. Sources of carbon emissions are explored, and the main cause for this high amount of footprint is explained in this paper. The UAE produced almost 20 tons of CO_2 emissions per person in 2010, a 63 percent increase from 2000. Water and electricity generation account for 33 percent of greenhouse gas emissions, followed by transportation, responsible for 22 percent. Practices and policies that are introduced by UAE have been revised, and further recommendations to deploy green building regulations and renewable energy system were discussed. Green buildings and the deployment of renewable energy systems are some of the solutions that could insure a considerable amount of carbon emission. Energy efficiency assessment for buildings is an important tool that controls the right implication of the proposed practices. It has been concluded that UAE is developing a good policies and implementing the right practices to achieve their 2030 target. However, more is needed as emphasized by this work.

Further Readings

1. Prof. Dr. Riadh Al-Dabbagh: AL-Dabbagh RH (2012) Potential role of UAE in promoting environmental technology. European Affords of Social Sciences, vol. 2, 2012
2. United Nations Environmental Program: UNEP (2013) Annual report
3. Ministry of Foreign Affairs: MOFA, act amendment (2012)
4. Intergovernmental panel on climate change: (IPCC) (2007)
5. Super council of energy report (2008)
6. United Nations Environmental Program: Path towards sustainable development, UNEP (2011)
7. Dubai City Guide (2016) www.dubaicityguide.com

Chapter 16
Urban Farming in the Era of Crisis in Greece: The Case Study of the Urban Garden of Ag. Anargiri-Kamatero and Fili

Konstadinos Abeliotis and Konstadinos Doudoumopoulos

Abstract Over the past couple of decades, urban agriculture has increasingly gained recognition as a viable intervention strategy for the urban poor to earn extra income throughout the world. Greece is now entering the seventh year of an economic crisis, and urban agriculture in abandoned municipal urban spaces is seen as a means for fighting the poverty of certain social groups.

The aim of this paper is to present the main characteristics of the operation of a municipal allotment garden located in the western part of the greater Athens urban area, namely, in the border of the municipalities of Ag. Anargiri-Kamatero and Fili. The aforementioned allotment garden was initially operated as a poverty relief action under the supervision of the central Greek government and the financial assistance of a European Union-funded programme for poverty alleviation. The initial beneficiaries/participants in the allotment garden were selected by the municipalities based on their sociodemographic characteristics. In a second phase, after the end of the funded programme, the garden continued to operate until today with the support of the two municipalities and the personal initiative of beneficiaries, some of them being from the original group while others are newcomers.

Moreover, the paper aims to report the key attitudes and experiences of people that operate the garden. The research on the attitudes and experiences of the participants took place in 2016 via a structured questionnaire. Thirty-nine participants answered questions regarding their attitudes and experiences regarding the economic, social and environmental benefits resulting from their participation in the activities of the allotment garden. The participants, most of them being men, were originating from different educational, family and economic backgrounds.

K. Abeliotis (✉) · K. Doudoumopoulos
Harokopio University, Department of Home Economics and Ecology, Athens, Greece
e-mail: kabeli@hua.gr

© Springer International Publishing AG, part of Springer Nature 2019   179
A. Sayigh (ed.), *Sustainable Building for a Cleaner Environment*,
Innovative Renewable Energy, https://doi.org/10.1007/978-3-319-94595-8_16

The results of the research indicate that the main benefits resulting from the participation in the municipal allotment garden is the socialisation of the participants. Eighty-five percent of them spent more than two hrs daily within the garden trying to exchange services, vegetables and knowledge. In terms of the economic outcomes, the main benefits result from the savings that the participants have by not spending their money elsewhere, since they spent their free time in the garden. In addition, the participants share the products produced in the garden by establishing a network of exchange economy. Finally, in terms of the environmental sustainability, 92% of the participants reported that they never used chemical pesticides or fertilisers during their gardening activities.

## 16.1	Introduction

In 2008 the world's urban population outnumbered its rural population [1]. According to the United Nations Population Fund, the world's urban population is expected to double from 3.3 billion in 2007 to 6.4 billion by 2050. It is predicted that 60% of the world's population will live in cities by 2030 [1]. Therefore, urban agglomerations and their resource uses are becoming the dominant feature of the human presence on earth, profoundly changing humanity's relationship to its host planet and its ecosystems.

In the same time, today's cities are not sustainable; they consume too many natural resources and produce too much waste. Moreover, cities require vast areas of land for their sustenance and have come to depend on food systems that require large amounts of food to be brought in from outside the land area that cities actually occupy.

In addition, urbanisation also influences all aspects of food production and consumption, i.e., the food system. A *food system* includes all biological processes as well as the physical infrastructure involved in feeding a population: growing, harvesting, processing, packaging, transporting, marketing, consumption and disposal of food and food-related items. It also includes the inputs needed and outputs generated at each of these steps. A food system operates within and is influenced by social, political, economic and environmental contexts. It also requires human resources that provide labour, research and education. A food system is derived from and interacts with the ecosystem in which it is located [1]. Specific aspects of food security applicable to the urban context include (i) the necessity to purchase most of the food needed by the household and (ii) greater dependence on the market system and on commercially processed food. The diverse array of present "food systems" is changing rapidly on a global scale and will be transformed even more rapidly in the near future, due to the rising introduction of urban agriculture.

16.2   Definition of Urban Agriculture

Urban and peri-urban agriculture (UPA) can be defined as the growing of plants and the raising of animals within and around cities. Urban and peri-urban agriculture provides food products from different types of crops (grains, root crops, vegetables, mushrooms, fruits), animals (poultry, rabbits, goats, sheep, cattle, pigs, guinea pigs, fish, etc.) as well as non-food products (e.g. aromatic and medicinal herbs, ornamental plants, tree products). UPA includes trees managed for producing fruit and fuelwood, as well as tree systems integrated and managed with crops (agroforestry) and small-scale aquaculture [1]. The lead feature of UPA which distinguishes it from rural agriculture is its integration into the urban economic and ecological system.

Urban agriculture is not a new concept; however, it seems that the problems arising from urbanisation and the safety of food provision to urban populations have given a new rise to its study and understanding of the driving forces beyond it. Recent publications regarding the benefits of urban agriculture have been presented by Germany [2], Greece [5] and Iran [3]. Urban agriculture requires a wider set of interconnecting parameters in order to be operational and sustainable compared to traditional rural agriculture. Urban agriculture coexists with the complex social, economic and environmental framework of the cities. More specifically, the parameters that should be taken into account for the successful implementation of urban agriculture are the following [7]:

- Location of land used (availability, legal status of land, solar exposure)
- Irrigation (availability and easy access to good quality water)
- Soil (safe, non-contaminated)
- Work force (aware of cultivation practices and techniques)
- Capital and operational costs
- Legal framework

Over the past couple of decades, urban agriculture has increasingly gained recognition as a viable intervention strategy for the urban poor to earn extra income. In addition, urban agriculture is one source of supply in urban food systems and only one of several food security options for households; similarly, it is one of several tools for making productive use of urban open spaces, treating and/or recovering urban solid and liquid wastes, saving or generating income and employment and managing freshwater resources more effectively. Urban agriculture therefore contributes to the sustainability of cities in various ways, socially, economically and environmentally [6].

16.3   The Advantages of Urban Agriculture

Urban agriculture is expanding rapidly lately, despite the disadvantages faced for the development of agricultural activities within the urban structure. Cities may gain a lot of financial, social and environmental benefits, while urban citizens have a lot to benefit in terms of their health and education improvement, culture and the enhancement of the aesthetics of the urban landscape [8].

Urban farming has a positive effect on the economic activity of a city by generating new jobs and new enterprises. It also improves the health of the participating citizens via the increased body activity required for the agricultural practices. It also increases the food security of cities. Finally, it enhances the aesthetic conditions within cities by introducing green spaces and improves the urban landscape [4].

Beyond the generation of food, agriculture generates or preserves a lot of other common goods, such as landscapes, biodiversity and cultural goods. Therefore, the added value of urban agriculture is much higher than the generated food items. The increased realisation of the positive financial, social and environmental benefits of urban farming poses a real opportunity for the enhancement of its legal framework and the integration of agricultural activities in urban planning and the urban food system [5].

16.4 Research Methodology

Municipal allotment gardens, initiated from 2011 onwards in several cities all over Greece, are now the most prevalent type of collective urban allotment gardens in the country [5]. Municipal authorities all over Greece have adopted this initiative as social policy projects to alleviate socioeconomic and psychological problems that vulnerable groups (low incomes, pensioners, singles parents, etc.) are facing due to the crisis [5].

In the present case study, the operation of the allotment garden of the municipalities of Ag. Anargiri-Kamatero and Fili will be presented. The operation of the allotment garden can be divided in two distinct time periods: the first one starts in the beginning of 2013 while the second one starts 2 years later, in the beginning of 2015. During the first period, the allotment garden was operated as a project for the relief of poverty operated by the two municipalities (Ag. Anargiri-Kamatero and Fili) with a total financial support of €74,000 by the European Union and national funds. The initial beneficiaries/participants in the allotment garden were selected by the municipalities based on their sociodemographic characteristics. During the first operational period, two NGOs were also involved. The two municipalities, during the first operational period, provided the key personnel for the operation of the allotment garden, namely, a social worker, an agronomist and a manager. During the second operational period, the allotment garden is operated by volunteering participants, without any financial or of any other kind of support from the municipalities.

The aim of the research was to identify the sociodemographic characteristics and the attitudes regarding the actual financial, environmental and social benefits of the active participants/cultivators of the aforementioned allotment garden. The research took place in the first semester of 2016 based on a structured questionnaire filled in by 39 participants. The replies were analysed in terms of descriptive statistics. Data regarding the actual agricultural production of the allotment garden were provided by the agronomist involved in the operation of the garden.

16.5   Results and Discussion

The allotment garden was created in an abandoned urban space along the border of the two municipalities. Soil was brought in the aforementioned abandoned space, in order to improve the cultivation conditions. Eighteen gardening sections were created. The area of each section was 6 m long by 3 m wide. Irrigation to the allotment garden was provided, free of charge, by the two municipalities. During the first operational period of the garden, due to the social criteria set by the municipalities, the annual minimum personal income for any applicant had to be less than €5500 in order to be eligible to participate in the activities of the allotment garden. This income limit was increased by €1200 if the applicant had a family.

First of all, data on the actual agricultural production of the allotment garden during 2014 will be presented (see Fig. 16.1). During the autumn and winter seasons, broccoli, cauliflowers, lettuces, cabbages, garlic, onions, chicory, spinach, peas, broad beans, radish, beetroots, celery, dill, carrots and potatoes were cultivated. During the spring and summer seasons, tomatoes, zucchini, cucumbers, green peppers, eggplants, watermelons, melons and green beans were grown, respectively.

It is evident that during the summer months (June to September), the production of the allotment garden was either very close or well above the targets set. Ninety-five percent of the respondents said that the agricultural production of each garden share (i.e. 20 m²) was enough to feed two to six persons (see Fig. 16.2). However, it

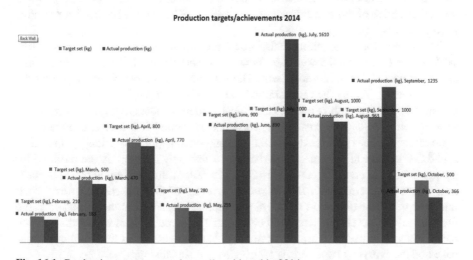

Fig. 16.1 Production targets set and actually achieved in 2014

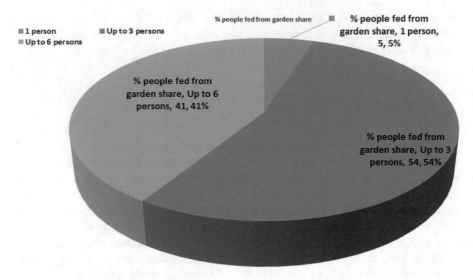

Fig. 16.2   % of people fed from each garden share (20 m²)

is clear that larger families had smaller per capita shares of food grown in the garden. Based on the initial agreement for the operation of the allotment garden, each participant gave 10% of its production for the activities of the grocery store operated by the social services of the two municipalities.

Moving on to the results of the questionnaire, the sociodemographic profile of the participating respondents is very variable in terms of age, gender, educational, family and professional status. More specifically, 82% of them are male while 72% of them are over 50 years of age. Eighty-seven percent of them are married and 23% of them have at least four children or more. Forty-nine percent of them are retired; 36% of them are unemployed, while 16% are actively working or have part-time jobs.

The reasons for participating in the programme are mainly financial, acquisition of skills and environmental awareness. Few of the participants jointed the programme aiming at socialising and amusement. The responses of the people indicate that the main reasons of joining the programme have been fully or partially fulfilled. The key outcome of our research is that the allotment garden operates to a great extent as an area of socialising, since 85% of the respondents replied that they exchange either agricultural services or knowledge. All the participants are now friends, since they spend 2–8 h in the allotment garden daily, of which only 2 h per day are required for the agricultural operations (see Fig. 16.3). As Partalidou and Anthopoulou [5] discuss, in a recent similar research from the northern part of Greece, the most important outcome of the involvement of people in allotment gardens is that all users value the garden and develop a sense of belonging. The same authors report that garden participants experience multiple emotions (joy, fulfilment) and relive rural memories based on the everyday experiences of growing and socialising.

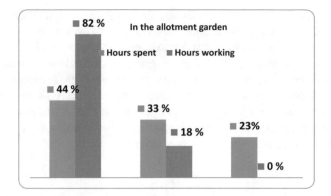

Fig. 16.3 Hours spent in the garden vs. working hours for the garden

In our case, most of the participants (52%), prior to their participation in the allotment garden programme, spent their time at home or at the cafeterias (40%). The respondents replied that their participation in the programme made them to spend their time more creatively, to learn about agriculture and to consume the products that they consume. The whole experience made them to feel better and have a greater interest towards agriculture.

Regarding the financial outcome resulting from their participation in the allotment garden, they replied that most of the benefits resulted as a side effect from the savings that they had by not spending their money at the cafeterias. Moreover, the production of food from the garden supports directly the family income, i.e. by not spending money for the procurement of food items that are grown in the garden. All of them replied that they participate in the exchange of their products, knowledge and services with other producers. The average annual cost for preserving the 20 m² for each participant is €120. The seeds for each cultivation period were either donated to them or taken from the collective seeding greenhouse that was constructed within the premises of the allotment garden.

Most of the respondents that participated in the programme replied that if they had the chance, they would get involved more actively with agriculture. Most of those that opposed this option were the elderly among the respondents. The participants replied that they exchange or donate to people in real need the products that they produce and cannot consume within their families.

Prior to the implementation of the programme, 79% of the participants didn't even know the existence of the allotment garden. After their involvement, they replied that they care about the garden, they want to improve it and that it's part of their everyday life.

Regarding the actual cultivation practices in the allotment garden, 92% of the respondents replied that they never used chemical fertilisers or pesticides and herbicides. The rest of them (8%) replied that they have used just small quantities.

All of the respondents replied that the local community has a very positive opinion about the allotment garden operation; 97% of the respondents replied that they would definitely suggest to other people to get involved with the activities of urban agriculture. Actually, the real problem towards the expansion of the activities is the lack of space, since a lot of people have expressed their interest in participating.

Overall, the allotment garden is an integral part of the local community of the two municipalities. For instance, schools during their environmental education activities, and social organisations for elderly people, are visiting the allotment garden premises in order to get informed about the cultivation practices and the benefits of urban agriculture. Most of the participants of the allotment garden project are actively involved in these networking activities.

16.6 Conclusions

Greece is now entering the seventh year of an economic crisis. Municipal allotment gardens, initiated from 2011 onwards in several cities all over Greece, are now the most prevalent type of collective urban allotment gardens in the country. The case study of an allotment garden, initially operated as a poverty relief action under the supervision of the central Greek government and the financial assistance of a European Union-funded programme for poverty alleviation, has been presented. Overall, our results indicate that we have the transformation of an empty urban space to an active productive work space combined with socialising of the participants while generating and sharing learning activities about the primary agricultural sector in an urban landscape.

References

1. FAO (2011) Food, agriculture and cities: challenges of food and nutrition security, agriculture and ecosystem management in an urbanizing world. http://www.fao.org/3/a-au725e.pdf
2. Hirsch D, Meyer CH, Klement J, Hamer M, Terlau W (2016) Urban agriculture and food systems dynamics in the German Bonn/Rhein-Sieg region. Int J Food Syst Dyn 7(4):341–359
3. Jamal M, Mortez SM (2014) The effect of urban agriculture in urban sustainable development and its techniques: a case study in Iran. Int J Agric For 4(4):275–285
4. Mogk KK (2010) Promoting urban agriculture as an alternative land use for vacant properties in the city of Detroit: benefits, problems and proposals for a regulatory framework for successful land use. https://digitalcommons.wayne.edu/lawfrp/109/
5. Partalidou M, Anthopoulou T (2017) Urban allotment gardens during precarious times: from motives to lived experiences. Sociol Rural 57(2):211–228
6. RUAF Foundation (2014) http://www.ruaf.org/
7. Tixier P, de Bon H (2006) Urban Horticulture. Chapter 11 of the RUAF publication. In: Cities farming for the future; urban agriculture for green and productive cities. RUAF Foundation, The Netherlands.
8. van Veenhuizen R (2006) Cities farming for the future: urban agriculture for green and productive cities. International Institute of Rural Reconstruction and ETC Urban Agriculture, Philippines

Chapter 17
Resilient Urban Design. Belgrade and Florence: Reconnect the Waters to the City

Chiara Odolini

Abstract

- Our mission
- Activities in Belgrade and in Florence
- Main "pillars" of future work
- Difficulties and future developments
- Expected products

Introduction

The cities, Belgrade and Florence, each important for different aspects, but both with an initial strong bond with a river, the Sava and the Arno. Rivers were once a source of wealth, connection, and exchange commonly perceived by everyone. Today designers and technicians plan infrastructures and settlements and either ignore or fear the rivers.

Climate change, floods, water bombs, and, at the same time, drought and scarcity of drinking water bring, these important waterways to the center of our attention.

Using an analogy, we can consider both the Sava and Arno as "veins" of a "body/city" that is our "cultural landscape" in Belgrade and Florence [1].

The study seeks to demonstrate the need for a structured method of approach to resilient urban design, equipped with specific tools and techniques.

Over the years, we passed from "designing into the nature" with the house on the waterfall of Frank L. Wright[1], to solar architecture and to high-tech architecture by

---

[1] Wright F.L., Organic architecture 1939. W. Intimately tied up to the concept of organic architecture for his attitude careful to the harmonic relationship among the parts and the everything, to the harmony among the man and the nature similar to what characterizes a living organism. The nature is assumed as fundamental external/inside reference of the organic architecture.

C. Odolini (✉)
UNIFI-DIDA, Florence, Italy

© Springer International Publishing AG, part of Springer Nature 2019   187
A. Sayigh (ed.), *Sustainable Building for a Cleaner Environment*,
Innovative Renewable Energy, https://doi.org/10.1007/978-3-319-94595-8_17

Renzo Piano and Norman Foster, from the 1997 Kyoto/COP3 Protocol to COP21 in Paris 2015, and from renewability to sustainability and resilience.

We entered Anthropocene, "In 2002, the Nobel Prize for Chemistry, Paul Crutzen, suggested that humanity has left the Holocene and has entered a new Era due to the global environmental effects of population growth and economic development. The term now entered into geological literature (though informally) to denote the current global environment now fully dominated by human activities, it is Anthropocene. What is still being discussed ... is the opportunity to amplify and extend the discussion of the effects described by Crutzen: It is wondering whether applying the criterion adopted to establish the coming into effect of a new era is really justified and if it is necessary to define a new term, and we are also wondering if we have clear the temporal boundaries of this Age and where can be located in time. However the magnitude of the changes taking place, even if only in their early stages, seems to many strong enough to be able to fix the boundary between the Holocene and the Anthropocene."[2]

Our vocabulary has enriched terminologies like sustainable architecture, management of buildings and of the environment, ventilation of the buildings, renewable energies in buildings and then in cities, ecological materials and new technologies, education and educative policies intertwined with financial ones, sustainable transport, urban agriculture and green urban infrastructures, and more.

"We have now entered in the semantic and operational field of NEXUS[3] [2] W.E.F. (Connection between Water, Energy and Food Resources), which represents a strong 'innovation' in the field of Sustainability. This approach has been structured mainly since 2011 with the Bonn Conference, although we can place it in the long path of maturing the assumptions of sustainability practices begun since the early twenty-first century and accelerating under the devastating effects of the climate change. Here the wretched question is not to ask whether we are now entering the turning point towards post-sustainability, as to ask if new conceptual horizons and new practices have emerged that overcome apories emerging in resource management after more than twenty years of experimentation of sustainability paradigms."[4]

Today, more than ever, designing means working in team and having multidisciplinary data and knowledge to manage and process. We are in the age of complexity, manipulation of images and data; we are in the era of virtual and digital. We talk about cognitive sciences, neurosciences, artificial intelligence, cybernetics, and the great "framework" given by the general theory of systems.

---

[2] To place well the terms of such approach, you see Trevisiol E.R., "New Paradigms, new Eras: the era of the Anthropocene," in *STEA News*, year 14, n. 40, Ediz.SOFT, Verona, 2014, p. 4. and Paul Crutzen, Steffen et al. also Hauls, Syvitski et al., Crossland, Andersson et al.

[3] Odolini C., "Integrated system morphogenesis for the Nexus," in U.I.D. (Unione Italiana del Disegno), *Research Line in the design area 2*, Aracne Edit., Roma, 2014.

[4] Trevisiol E.R., 2014, op.cit.

We are in an era where "the theory of complexity"[5] of E. Morin, De Angelis, and Battramm has now taken hold.

Complex systems require appropriate communication languages, and the design, graphics, and images are now masterful [3].

In the 1970s, McHarg, in *Planning with Nature*, applied examples of analysis and significant approaches to urban landscapes, such as those in the valleys and riverbanks, and introduced layer design at a time when GIS did not exist or was a tool for a few. Today, GIS becomes a tool for sustainable and resilient designing.

17.1   Does Sustainability-Complexity have specific Drawing Practices and Provides Specific Representations?[6]

GIS today becomes a necessary tool to face in a sustainable and resilient way the planning brought to modify the way to think, to analyze, and to represent: for layers, levels, and the whole.

Belgrade and Florence have been objects of laboratories of project from 2015 to 2017 of UNIFI-DIDA with the University of Belgrade, and also with the city of Florence that has undertaken a run of recovery of the suburbs.

We passed from Vitruvian Man to Le Corbusier's *The Modulor*, to Changeux's *Neuronal Man*, to *Tarzans in the Media Forest*, and to robotic cognitive design promoted by the IBM Watson program [4, 5]. We are in the digital age; social, BigData, and new terminologies, like coding, scratch, etc., and new needs arise. There are over 4.6 billion active smartphones, and about 2 billion people have access to the Internet. We see how the volume of data circulation has evolved: in 1986 data was 281 petabytes; in 1993 the data was 471 petabytes; in 2000 the data was 2.2 exabytes; in 2007 the data was 65 exabytes; and in 2014 an exchange of over 650 exabytes is expected.[7]

More and more archives are digital; BigData has been born, so extensive data in terms of volume and speed requires specific value-added analytical technologies and methods.

We are facing the need to interfere with multidisciplinary and interrelated data coming from potentially heterogeneous sources, so not just structured data such as databases but also unstructured such as images, e-mails, GPS data, and social networking information [6].

---

[5] The theory of complexity is a multidisciplinary science that takes into account elements of very different disciplines such as systems theory, cybernetics, meteorology, chaos theory, artificial intelligence (AI), artificial life, cognitive sciences, computer science, ecology, economics, and even circular economics, studies on evolution, genetics, game theory, immunology, linguistics, philosophy, social sciences, governance, etc.

[6] Odolini C., "Not only a new houses but a living place and a city: rendering sustainability and Anthropocene," in G. Damiani e D.R. Fiorino (by), AA.VV., *Military Landscapes. A future for military heritage*, ed. Skira, Milano, 2017.

[7] See Odolini C. op.cit. 2017.

The main features of BigData in 2012 have become the following [4]:

(a) *Volume*: represents the actual size of the dataset; the large volume of data that can be collected today may seem to be a problem. In fact the volume of BigData is a false problem, as cloud and virtualization help in the management of the large volume of data available, simplifying the processes of collection, storage, and access to data.
(b) *Speed*: refers to the speed of data generation; it tends to perform real-time or almost real-time data analysis.
(c) *Variety*: refers to the various types of data coming from different sources (structured and not).
(d) *Truthfulness*: refers to the reliability of the information.

In addition, over time the model is extended, going to add the characteristics of variability[8] and complexity.[9]

As part of the *analytics data*, new representation models[10] [7] can be created that can handle that amount of data with parallels of databases.

The design complexity, for us, has its own specific practices and requires new models: both for physical and logical data and for semantic-conceptual data[11] [8, 9]. A response to this is provided by the use of BD (databases), DBMS (database management systems), and BIM and GIS. Also we need to define the so-called logical database scheme, that is, to define the structure or "architecture" of work tools for sustainability [10].

The result was to set up a methodological and representative model for enhancing qualitative aspects in an urban regeneration agenda (UN 2030 Agenda). Metabolic, collaborative, and open-source model originated from functional, environmental, and social recovery, from the reclamation of natural and physical capital, and from data structuring (BigData) as common goods. The ongoing research is developing a guide for editing and displaying information and for clarifying the relationship between GIS and BIM analysis graphs using C-MAPs.

---

[8] Variability: this feature can be a problem and refers to the possibility of inconsistency of Data.

[9] Complexity: the greater the size of the Data set, the greater the complexity of the Data to handle. The most difficult task is to link information and get it interesting. It should be noted that the BigData volume and the widespread use of unstructured Data do not allow the use of traditional RDBMS.

[10] Distributed large Data collection architectures are provided by Google MapReduce and the open-source Apache Hadoop counterparty. With this system the applications are separated and distributed with parallel nodes and then executed in parallel (map function). The results are then collected and returned (reduced function).

[11] Alexander Christopher, structure and urban form. Notes on the synthesis of the form. (The city is not a tree), Milan, The Saggiatore, 1967, pp. 254; and Panofsky E. (op.cit.)

17.2 Evolution C-MAP About the Ecological Method for the District "Le Piagge" (Florence)

17.2.1 Methodology and Glossary

Glossary is a key project and work tool both for designers and the work team and for the users. Glossary is in four languages, Italian, English, Chinese, and Arabic/Persian, and next to the key keywords are their significant synthetic and reference sources. This tool is part of the tool kit called "writing and displaying information help guide."

Lynch's method[12] of study and project concerns perceptions of space and urban tissues in modern cities. The environment of the contemporary city is, according to the writer, less and less rich in sensations, especially perceptive, and increasingly rich in symbols and abstract meanings. The quality of an urban space is based on the clarity of the pictures that the inhabitants have of various places in the city. The idea of Lynch is that every citizen moves in the city following his mental maps of paths. These traces and reproductions of the exterior physical world that each person mentally possesses correspond to environmental images, which are the product of both an immediate idea and a memory of past experiences. The environment suggests various distinctions and relationships; the observer organizes and attributes meaning to what he sees. The image you create can be extremely personal and change for each viewer. Lynch is very pragmatic and realistic in his book with the ultimate result of finding the interaction between the urban scene and the observer, who elaborates his perception on the basis of his experience, his social disposition, and his culture.

The goal is to find the collective image of the city, "at Lynch," that is, the mental picture that most people carry daily with them.

A good urban design must have the high probability of evoking a vigorous image in every observer, that is, the figure[13] as defined by K.L.

The major problems are therefore metropolitan spaces whose design lacks clarity, simplicity of form, and hierarchy. For these reasons, Lynch in his book proposes a method of building an environmental image, aimed at urban design, created through the cognitive processes of the inhabitants, everyday uses, and their habits, all with the aim of making the cities readable.

To understand how environmental images develop in citizens of any city, Lynch conducted various researches on precise urban areas and then studied the behaviors of those who lived there.

---

[12] Lynch Kevin, *The image of the city*, 1960.

[13] The figurabilità is the ability to evoke in the observer a vigorous image of the reality in terms of extreme legibility, garishness, and visibility. It consists in the creation of environmental images vividly individualized, that is, the legibility, that is, the facility with which the parts of the urban landscape can be recognized.

Research was developed through studies on three American cities, Boston in Massachusetts, Jersey City in New Jersey, and Los Angeles in California, and in our study, we have tried "similarly" to find the same formal and environmental aspects in Florence.

To address the study of these cities, the unique method used was to systematically examine the area and subsequently to interview a sample of citizens. The descriptions of the three cities, the various relationships between the elements of urban tissues and their relation to the environmental images of individual citizens are analyzed. Being very different cities, with different characteristics, the collected data are very different. Lynch also studies a relationship between environmental images already owned by existing citizens, that is, in their memory and the relationship with physical reality. Another aspect that has been found by Lynch is the constant reference to various economic classes, for the subdivision and identification of neighborhoods in cities. The sample of the chosen inhabitants is as varied as possible so as not to interfere with the final results. Although cities are different at the end of the study, it is possible to find common aspects among the perceptions of the various citizens.

– The amplitude of views, in all cities for citizens, was important for the breadth of views.
– The empty space seems to attract attention even if not pleasantly.
– Natural aspects, cited by all citizens. Natural aspects such as rivers or vegetation are experienced as positive and qualitative characteristics of the urban environment.
– The paths perceived as the most influential elements of the structure and clarity of a city.

According to Lynch, there is a public image for each city, which is the overlapping of many individual images.

Each environmental image is individual and unique, though it is roughly similar to the public one, only taking into account the habits, and other influences on the ability of personal images (figurativity).

In urban images, content can be instrumentally classified into five types of elements: paths, margins, neighborhoods, nodes, and references.

Lynch highlights how an environment cannot be perceived as a single mental image but as a set of images arranged at different levels of organization and associated with McHarg leads us to think, reason, and draw for layers. The combined work of the various urban elements generates images with different degrees of organization and layers. Lynch's hypothesized and hypothetical elements (nodes, neighborhoods, margins, pathways) after some modification have been largely confirmed[14] by our search results.

The figurativity analysis was carried out using this procedure:

---

[14] Those have not been confirmed that these categories exist, but these are shown that I am able to coherently show the environmental images of the citizens.

- Generate a systematic lookout, with two or three observers covering the area on foot, by bicycle, or by car on public transport.
- Implement a collective interview and/or social research of a large sample, balanced so as to represent the characteristics of the inhabitants.

This sociological/visual approach to an urban center allows for an in-depth study of behaviors and urban displacements, starting with those who make the most of these spaces, namely, citizens. You can thus create unique maps of urban centers by understanding the internal operation.

17.2.2   Case Studies

The study aims to address the strategic role of open spaces within the widespread settlement system that characterizes the contemporary city. The areas under study are those still free or residual in the urban and peri-urban areas that maintain agricultural potential but are subject to real estate pressure for a possible construction. The themes of reflection are the protection of such areas, even with the aim of limiting the consumption of soil, the definition of any margins of the city, the improvement of the urban landscape, and the reunion between liquid landscape and built landscape.

17.2.2.1   Belgrade: Urban Water Sensitive Planning (Serbia)

In the case of Belgrade, the project invariants set out in this report attempt to configure a new identity in the urban confluence of the Sava River with the Danube (Belgrade, Serbia).

This project was born from the UNIFI-DIDA city laboratory starting from 2015. In particular, C. Odolini and E.R. Trevisiol[15] have been involved in the "urban water regeneration project," which overlooks the historic center of Belgrade. An exercise is strongly marked by a well-brained water design with the components of recovery of some of the destroyed artifacts, but well-present in the urban identity construct, and with a profitable dialogue with the melting riverine landscape of the city.

Belgrade was and is a lump of contradictions, clashes between civilizations, and confluence between waterways that have always linked Central Europe with its eastern margins (from the Black Forest to the Black Sea). The Danube is also the Connecting Line of Nations, now divided, which formed the former Yugoslav Federation. The Danube is the twenty-first European river by length. A large dragon axis powerfully structures the crossed territories, almost in perfect parallel with the

---

[15] Odolini C., Trevisiol E.R., *Firenze – Belgrade – Firenze, Kosancicev Venac 1941–2015*, Ediz. UNIFI-DIDA, Florence, 2015.

Sava, which tends between Ljubljana, Zagreb, and Belgrade, and is unfortunately characterized by its ruinous floods.

The river landscape of Belgrade saw 52 destructions and as many reconstructions of the city.

The intersection between two large bodies of water, both "primatial," is dominated by a rocky promontory that has always been fortified. Great "visual corridors" slip winding and steep from the city to the waters, from such an aulic and magniloquent urban fabric to be called the Balkan Acropolis to a deserted naturalness of the lonely island at the apex of the Sava/Danube confluence.

As the ancient Chinese science of topography in the water confluence states, a liquid and inverse y of such shape, it also carries great dangers.

Belgrade and its waters, its sometimes ridiculous and natural, and sometimes sadly postindustrial, riverine landscapes, are not easy to assume as design elements, their pins being the liquid urban landscape.

17.2.2.2   Florence-Le Piagge: Urban Riverine Design, the Urban Water Sensitive Planning (Italy)

The study case of the district of Piagge is located in the outskirts of Florence, close to the river Arno, which is bordered by the railway and the highway. It is regarded as a "shatter" of a city (a so-called neighborhood of the suburbs), which is a nucleus of dwellings with many nearby amenities (mall, swimming pool, schools, theaters, etc.). Common urban spaces play the role of "society," joining residential areas ("houses") and common spaces, which present numerous issues. The three large "ship-shaped" buildings identify the status and social rank at first sight, forming three architectural impact elements in the area, identifying it immediately as a popular area.

The coexistence of different ethnic groups that leads to groups and clans, who live in close proximity, "common residences," can create frictions, cultural clashes, micro-crime, and vandalism. The common spaces are not well defined; there are some public gardens, squares, and streets, but there is no link, a unifying theme among them, which allows use and makes these spaces of value.

By value it is not the economic one of goods, but it is related to the wealth that takes place once it is enjoyed by the inhabitants and external visitors, a place to recognize.

The first phase of the project approach consisted of a number of site inspections, under different conditions, such as the means used, daylight timing, and weather conditions (at Lynch and McHarg method) [3].

A formal and qualitative analysis of the physical, human/cultural, and natural matrices has been made. The internal dynamics of Q5, UOE, and the area related to the inhabitants and their behavior have been detected and analyzed, detecting and analyzing the present plants, natural and environmental resources, and regulatory aspects too. A large-scale framing table has been produced that allows the representation of territorial analyses at the municipal and supra-municipal level, identifying the main arteries, the boundaries, the zone/areas, the territorial park systems, the main historical areas and the neighboring fractions. Representation through the GIS of the physical and human matrix of Neighborhood 5, with the insertion of constraints, the areas of protection, and the transformation areas present in the RUC (municipal urban regulation) of Florence, allowed the display of regulatory plans.

By working on layers, the regulatory guidelines are noticed at a glance, to be followed before and during the design phase. Also given the particular ethnic wealth of the neighborhood, we analyzed the human matrix to understand the composition of Households, the percentage of foreigners, and their origin, whether European or extra-EU.

The second phase consisted in "seeing with the eyes of the inhabitants," thus giving importance to visual perceptions and mental maps (figurativity) that vary according to each individual, depending on the needs, gender, and age groups. Maps analysis, as theorized by K. Lynch,[16] will allow us to understand what are the nodes, the references, the margins, the paths, and the various dials. The study of these elements creates a set of networks, which constitutes a valid foundation on which to start designing. We have included a study of the array of smaller historical centers, villages, and suburbs, such as nodes, references, sub-neighborhoods: potential to be exalted during the design phase.

The survey and analysis of the natural matrix, its insertion into the existing ecological network, and how it penetrates the urban matrix and specifically the area of the Piagge highlight important points about connections with the river, with water, with green, and with its ecological continuity. GIS representation allows you to merge record, shapes, layers, vectors, and points. In our case, we have identified the ecological connections that will be fully exploited at the design stage [11]. On a more detailed scale and in harmony with the "public space map",[17] we have studied urban voids, i.e., the function of open spaces and unstructured, whether urbanized, classifying them into subcategories.

The study of visual perceptions,[18] that is, the visuals of passersby, of the one who travels the area, taking into account buildings, trees, paths, and impacting elements such as the railway, the highway, the river, landscape elements, and anthropic elements, has been subject to a subsequent survey and analysis/representation. A further step was the relief and census with tree species charting, identifying their health status, used plant strategies, and maintenance (otherwise poor). All of these elements that returned to DBMS systems and linked to GIS were used in the summary table: criticality/values. Through SWOT analysis we have identified strengths, weaknesses, opportunities, and threats. Finally, the following actions, guidelines, and scenarios have been identified too.

The synthesis represents in the design process the time of re-elaboration of the acquired knowledge, aimed at the interpretation of the problems, the criticalities in

---

[16] Lynch Kevin, *The image of the city,* 1960.

[17] Paper of the Public Space (in collaboration with A-habitat) 2013 – (A-habitat) adopted in Rome in the conclusive session of the Biennal exhibition of the 18 magio 2013.

[18] Arnheim Rudolf. Art and visual perception, Milan, Feltrinelli 1954.

the study area, and, at the same time, the potentialities and values present [12]. These elements are the ones that need to be corrected, improved, or enhanced through the final design.

Understanding the criticalities and values – which can be represented through C-MAPs, matrices, diagrams, and ideograms – is the identification of the strategic address of the project, by concept, meta-project, and scenarios [13].

17.3   Layering and SWOT Critical Synthesis Table/Values: Strategies/Actions

17.3.1   Final Conclusions

"To the question 'According to you who are we?' The Professor replied: 'We are the experience of others'... Experience includes the knowledge that 'exists' when an intelligence is able to use it. If this happens, in my opinion, the walls of the territorial, spatial, temporal boundaries of which are broken, and all acquires a form of particular knowledge endowed with a usefulness that can be used by all and forever."[19]

After a lot of important and analytical work, these are some critical aspects that emerged to improve the current situation. The city is not just an architectural space or an object of perception; it is a harmonious interweaving of physical spaces and inhabitants.

Recovering K. Lynch[20] and using the layering concepts introduced by McHarg, it is easy to locate and recognize readability, imageability, image of urban spaces, their "natural veins," and the quality of places, while with eidotip the evolution of the conceptual map of the places and the quality of the places can be traced through charting: readability, imageability, environmental image, and urban image.

- Readability is the ease with which parts of an urban landscape can be recognized.
- Figurability (building the image) is the quality of a physical object with high probability of evoking a vigorous image in every observer.
- Environmental image is the mental reconstructions of the citizens and the external physical traces.
- Urban image is creating through hypothesized elements (paths, nodes, margins, neighborhoods, referrals) of a conceptual map of the site.

Our cities have often developed around a river or coastline. By hiding our underground water systems, we lose the benefits of water management on the surface and to get the most water, we need to connect the water management to well-designed places, in areas near the rivers. It is important to design a new development that

---

[19] Corrado Balistreri T., *Ricordando Giuseppe Samonà ed. Egle Renata Trincanato, intervista in* www.Archimagazine.com *(visto 30.08.2017 h 10.12).*

[20] Lynch also coined the words "imageability" and "wayfinding." *Image of the City* has had important and durable influence in the fields of urban planning and environmental psychology.

redirects flood water using open space and strategic corridors to reduce damage to infrastructure and homes.

In summary urban water sensitive design:

- Reduces water pollution.
- Decreases the risk of flooding.
- Produces greater security of water reserves.
- Increases the health of the ecosystem.
- Helps communities to relate themselves with water.
- Eliminates the effect of the urban heat island.
- Constructs different disciplines to create an intelligent urban environment.

Metabolic design, using and connecting open spaces, could allow to create a new centrality in the Piagge as in Belgrade and has allowed to give order to the city by encouraging the union of today's urban tissue fragmented with legible environmental and urban images.

References

1. Odolini C (2011) Le geometrie della natura e dell'ambiente. Forme, rilievi e rappresentazioni degli ambienti d'acqua. Tesi di dottorato, Università degli Studi di Firenze, Dottorato di Ricerca in Rilievo e Rappresentazione dell'Architettura e dell'Ambiente, Ciclo XXIII
2. Odolini C (2014) Integrated system morphogenesis for the Nexus. In: U.I.D. (Unione Italiana del Disegno), Research Line in the design area 2, Aracne Edit., Roma
3. Mc Harg Ian L (1969) Progettare con la natura. Franco Muzio Editore, Padova
4. Dalla Mora T, Peron F, Cappelletti F, Romagnoni P, Ruggeri P (2014) Una panoramica sul building information modelling (BIM). AICARR, Milano
5. Longhi G (2017) Gli tsunami dell'innovazione e il rinnovo delle infrastrutture urbane. In AA.VV Gruppo di discussione Crescita, Investimenti, Territorio, innovazione e nuova strategia d'impresa, Mi, EGEA
6. Richard B, David J (2015) Graph analysis and visualization. VCH Verlag, Wiley, Weinheim (DE), Canada
7. Wong Dona M (2010) The wall street journal. Guide to information graphics. Norton, New York/London
8. Alexander C (1964) Note sulla sintesi della forma. il Saggiatore, Milano
9. Panofsky E (2010) Iconografia e iconologia. In: Il significato nelle arti visive. Einaudi, Torino
10. Bertocci S, Odolini C, Trevisiol RE (2016) Via Palazzuolo: survey and planning for the reshaping of the urban resilience Stronholds. In: AA.VV. Firenze e il suo fiume a 50 anni dall'alluvione DIspLUVIO, Angelo Pontecorboli Edit, Firenze
11. Gaffron P, Huisman G, Skala F (eds) (2005) Ecocity: book I. A better place to live. European Commission, Hamburg/Utrecht/Vienna
12. Clèment G (2014) L'Alternativa ambiente. Quodlibet, Macerata
13. Ferrara G, Campioni G (2012) Il paesaggio nella pianificazione territoriale. Ricerche, esperienze e linee guida per il controllo delle trasformazioni. Dario Flaccovio, Palermo, p 252

Chapter 18
Strategic Sustainable and Smart Development Based on User Behaviour

Shahryar Habibi ⓘ and Theo Zaffagnini ⓘ

Abstract It is clear that the field of artificial intelligence (AI) as a decision-oriented tool has recently proven to be a viable alternative approach to solve environmental challenges. For example, artificial neural networks (ANNs) and support vector machines (SVMs), which are a subset of artificial intelligence, are going to be widely used to predict energy consumption in the buildings. The work aims to explore the use of user behaviour and smart and passive systems to improve energy efficiency and indoor environmental quality (IEQ) in buildings. The presence of users within buildings can affect process improvement. For example, users can contribute to energy efficiency by switching off artificial lighting during daylight hours. Furthermore, they can reduce the use of energy by changing their behaviour to act according to principles of sustainable development. In order to evaluate the impact of user behaviour on energy consumption, development of an assessment model based on AI can be useful. On the other hand, the use of a new concept from artificial intelligence in assessment tools can not only explore the potential benefits of approach but also provide ways to achieve an optimum level of efficiency.

18.1 Introduction

In recent times, energy retrofit strategies and initiatives have significantly improved reductions in energy use and emissions. It is important to focus on the fundamental principles and practices of buildings retrofits to increase energy efficiency. In this context, the role and the importance of user behaviour on the performance of the buildings cannot be denied. User behaviour is a key factor for designing retrofit programmes. For example, Fabi et al. [1] studied the robustness of building design, with different operations of windows and movable shadings, and found that a description of user behaviour allows a better defining of robust building designs.

S. Habibi (✉) · T. Zaffagnini
Department of Architecture, University of Ferrara, Ferrara, Italy
e-mail: hbbshr@unife.it; theo.zaffagnini@unife.it

© Springer International Publishing AG, part of Springer Nature 2019   199
A. Sayigh (ed.), *Sustainable Building for a Cleaner Environment*,
Innovative Renewable Energy, https://doi.org/10.1007/978-3-319-94595-8_18

Users make significant contributions towards enhancing energy efficiency to such a degree that user behaviour can sometimes lead to significant energy savings. Furthermore, building users have a considerable impact on the performance of indoor environments. Therefore, to improve energy efficiency and indoor environmental quality (IEQ) in buildings, it is essential to understand and predict user behaviour. In fact, there are correlations between IEQ, user behaviour and energy consumption.

It is important to emphasize that user behaviour is one of the key factors which can be used in enhancing IEQ of workplace environments. Furthermore, user behaviour can have a significant impact on building adaptive performance. Therefore, user adaptive behaviours are needed in order to support adaptive building systems. Adaptive behaviours include different reactions and behaviours that can be categorized into three types, which are as follows:

1. Behavioural adaptation (adjustment)
2. Physiological adaptation (acclimatization)
3. Psychological adaptation (habituation) [2]

It is obvious that users adjust themselves to maintain and improve their well-being through physiological, psychological and behavioural reactions to the environmental stimuli.

Cena and Dear [3] carried out an analysis of the effects of indoor climates on thermal perceptions and adaptive behaviour of users in a hot-arid climate region. They showed that there is a significant correlation between thermal comfort responses of users and seasonal climate.

An investigation into the behaviour of user to maintain thermal comfort by adjusting their clothing was conducted by Parsons [4]. His study focused on a range of conditions (warm to cool) to investigate the effects of gender over 3-hour exposures in simulated office environments. The study found that it is important to consider individual requirements for users with physical disability.

The adaptive models are based on adaptive opportunities of users and are related to options of personal control of the indoor climate and psychology and performance [5]. Therefore, it is important that users are aware of which environmental conditions have the most impact on energy use and thermal comfort relative to the others.

The behavioural reactions can lead to actions such as switching, heating or cooling on/off, opening/closing windows or blinds, the putting on/taking off of clothes, switching lights on/off, etc., which are also influenced by user's expectations about their actions' effects [6].

For example, the study by Brittle et al. [7] sought to evaluate the impacts on mechanical ventilation and cooling energy when raising internal comfort temperatures. The study showed that higher mechanical cooling set point operations can be achieved when users have access to openable windows.

Although the results of these studies, in general, show support to the notion that users have a great role in operating systems associated and related to environmental and functional quality of buildings (indoor environmental quality, heating, ventilation and air conditioning (HVAC) systems, effectiveness of materials, layout and method of construction, etc.), there is still a need to explore the adaptation of building systems and dynamic components to new technologies through electrome-chanic devices which allow users to easily control environmental data. Furthermore, in order to achieve cost-effective energy efficiency by approaches described above, it is important to address the potential of building stocks with the use of these tools to monitor both environmental conditions and behavioural/physiological reactions of users.

In order to provide individual comfort controls on workplace level, environmental conditions, such as thermal environment, lighting, acoustics and air quality which mostly affect user satisfaction, should be taken into account. Although, air distribu-tion systems can provide individual comfort control, they may consume more energy than a conventional system. Therefore, individual comfort controls are needed to incorporate user behaviours in order to reduce energy consumption and environmental impacts.

However, due to variances in personal factors and behaviour patterns, individual comfort levels may be different. For example, in relation to individual thermal com-fort, Lian et al. [8] indicate that in general, different individuals have different scales of thermal comfort due to physiological difference and variety of subjective sensation.

In a comfortable thermal environment (based on the thermal comfort evaluation for most users), a user may still feel uncomfortable. In order to reduce these prob-lems, individual comfort controls allow users to control environmental conditions in space or zone, individually, and this can lead to improvement in thermal comfort. From an individual microclimate point of view, to achieve a comfort zone, users of the workplace need to monitor different environmental conditions. Therefore, it is of importance to understand the comfort needs of individuals and energy demands on workplace level. It might also be worth noting that function decomposition at workplace conditions can help develop individual microclimate control techniques in the process.

In this context, task/ambient conditioning (TAC) system is an alternative method to provide users with control of a local supply of air so that they can adjust their individual thermal environment. It can be defined as any space-conditioning system that allows thermal conditions in small, localized zones (e.g. regularly occupied work locations) to be individually controlled by building users while still automati-cally maintaining acceptable environmental conditions in the ambient space of the building (e.g. corridors, open-use space and other areas outside of regularly occu-pied work space) [9].

From an energy efficiency point of view, individual control of temperature may be a practical means to improve user comfort and reduce energy consumption. It is clear that the main aim of comfort control in the workplace is adaptation to ambient environmental conditions. In this regard, it is possible to claim that users should adjust ambient environmental conditions to maintain personal comfort levels.

Fig. 18.1 Some examples of Nest technologies [10]

In order to meet infrastructure requirements and individual user needs, a number of technologies and applications have been developed. One example of efficient applications is the Nest thermostat (Fig. 18.1). In this approach application, individuals are given the chance to control environmental conditions within their local environment.

18.2 Methodology

To reach high-performance buildings through climate-responsive and smart systems, it is important to pay attention to environmental parameters and their impact on the built environment. Sustainable smart behaviour is considered as a new method to investigate interaction between users and environmental parameters for improving comfort, efficiency and smart solutions in the built environment (Fig. 18.2). This concept can also be used to explain how users can make a place sustainable and smart. X-axis and Y-axis are considered as the built environment and efficiency goals in the figure. This also shows the importance of making a place sustainable, smart and then work on the behaviour of the user in the building to meet efficiency goals.

Sustainable smart behaviour can offer significant opportunities for developing smart and sustainable built environment. However, its main contribution is to highlight the importance of users in addressing sustainable development and smart growth.

The aim of sustainable smart behaviour methodology is to find optimal comfort conditions related to sustainable and smart systems. This can not only provide significant techniques for both new construction and retrofits but also improve environmental attributes of renewable resources and users' comfort towards fostering sustainable smart buildings. Sustainable smart behaviour methodology can be used for optimization of multi-energy systems and environmental parameters in buildings through sensorization or cost-effective smart systems.

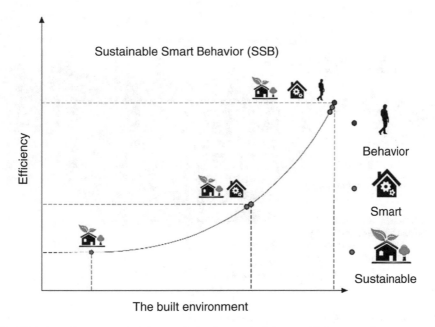

Fig. 18.2 Overview of sustainable smart behaviour methodology and influencing factors

18.3   Discussion and Result Analysis

In order to predict the behaviour of building systems, it can be useful to create a clearer connection between building information modelling (BIM) and smart sensor systems.

The current work has been carried out with the aim in particular to develop innovative devices for improving IEQ in the workplaces in office buildings.

In order to investigate user satisfaction and its relationship with parameters of IEQ, it is essential to understand environmental factors that influence user comfort in the built environment considering the fact that user satisfaction has been defined as an independent criterion for IEQ. However, user satisfaction in office buildings is associated with indoor environmental quality (thermal, visual, acoustic environment and air quality) and workspace and building features including size, aesthetic appearance, furniture and cleanliness [11].

In order to measure indoor environmental conditions, a prototype sensor system was developed based on Arduino microcontroller which can obtain and monitor environmental data in real time (Fig. 18.3). Arduino is an open hardware and software platform based on a microcontroller board. In the current work, it is used to produce an indoor quality apparatus to collect stand-alone environmental parameters such as humidity, temperature, lighting and ambient noise.

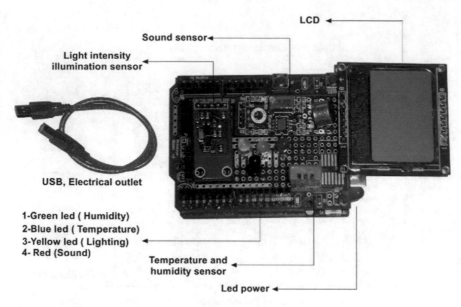

Fig. 18.3 Indoor quality apparatus developed

The goal of the developed apparatus is its use in collecting and then processing data of indoor environmental quality parameters which can lead to understanding environment conditions. The developed apparatus is equipped with different sensors and led lights to carry out monitoring practices in the indoor environments. It also can be used to sense and gather the data from certain places in the built environment. The data are collected in real time from sensors and are visualized within both liquid-crystal display (LCD) and interfaces.

The developed apparatus is programmed to interface with Rhino/Grasshopper and a MATLAB's graphical user interface to visualize real-time environmental data (Fig. 18.4). It can be powered by both electricity and universal serial bus (USB).

The most important function of the developed apparatus is to monitor, visualize and define optimum comfort conditions. The developed apparatus has a light-based alarm that uses lights to alarm user when the temperature, humidity, lighting and sound fall outside the comfort zone. Since Arduino systems are open-source software programmable, it is possible to generate the actual programme that is required to attain project objectives. In this respect, a programme is written in standard C/C++ code and uploaded on the developed apparatus through the Arduino IDE (integrated development environment). It is desirable to optimize data collected from the temperature/humidity, light and sound sensors to achieve optimal indoor environmental quality. In order to perform a multi-criteria optimization between the input data of sensors, a genetic algorithm (GA) can be considered. The optimization of the data is made with a genetic algorithm (GA) within Grasshopper (graphical algorithm editor) that is connected with the developed apparatus (Fig. 18.5). This process is performed when visualizing and collecting the data in real time.

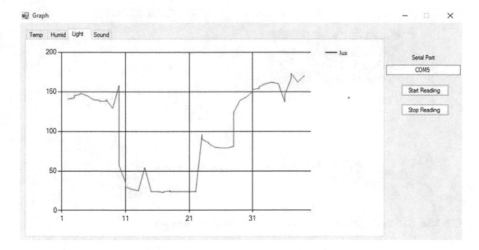

Fig. 18.4 Real-time interface monitoring tool based on temperature, humidity, lighting and sound level

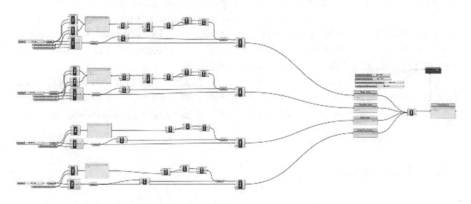

Fig. 18.5 The genetic algorithm used for the developed apparatus

The proposed genetic algorithm is based on the scoring algorithm. For example, scores for each of the data collected from the developed apparatus (temperature/ humidity, light and sound sensors) are scaled to values between 0 and 100 according to recommended comfort ranges.

In order to find the most efficient data from sensors, it is useful to initially optimize four parameters under certain indoor environment and send them to a log data which can be used for improving a building control system in order to provide the optimal environmental conditions for users. Since all parameters use a separate algorithm to optimize their objectives, there are no conflicts in determining the optimal data.

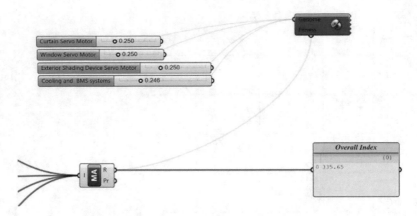

Fig. 18.6 The final optimization algorithm with Galapagos component

After data processing and optimization, Galapagos component is used to generate the final solutions (Fig. 18.6). In this context, overall index of the four optimized parameters transfers to fitness input in Galapagos component, and the inputs for the genome are used to control systems such as curtain, window, exterior shading device, cooling and building management systems (BMS). All portions of building control systems may not apply to genome inputs, but using them at the same time can be helpful to find the optimal indoor conditions.

Galapagos component can provide a generic platform for the application of optimization algorithms. It is a genetic algorithm component that uses and runs based on numeric fitness values. It can maximize the fitness value to achieve multiple objectives.

18.4 Conclusion

There is an increasing interest in real-time monitoring of environmental data to reduce energy consumption and to ensure comfortable conditions. ICT-related applications can provide a simplified communication between users within indoor environment.

The current work aimed at highlighting the importance of BIM and open-source software and hardware systems in the field of architectural technology. It has shown that open-source technologies can develop new design and technological possibilities within the architectural field. Furthermore, it introduced methods that can adapt and respond to changes in environmental conditions.

A sustainable smart behaviour method for sustainable development and smart growth has been developed. It encourages users to be involved and mindful of energy efficiency and IEQ. From a sustainable design point of view, user-centred analysis can indeed lead to find opportunities to foster sustainable behaviours.

References

1. Fabi V, Buso T, Andersen RK, Corgnati SP, Olesen BW (2013) Robustness of building design with respect to energy related occupant behavior. In: Proceedings of BS2013: 13th Conference of International Building Performance Simulation Association, Chambéry, France, 1999–2006
2. Brager G, de Dear R (1998) Thermal adaptation in the built environment: a literature review. Energ Buildings 27:83–96
3. Cena K, de Dear R (2001) Thermal comfort and behavioural strategies in office buildings located in a hot-arid climate. J Therm Biol 26:409–414
4. Parsons KC (2002) The effects of gender, acclimation state, the opportunity to adjust clothing and physical disability on requirements for thermal comfort. Energ Buildings 34:593–599
5. van Hoof J, Mazej M, Hensen JLM (2010) Thermal comfort: research and practice. Front Biosci 15:765–788
6. Liu J, Yao R, Wang J, Li B (2011) Occupants' behavioural adaptation in workplaces with non-central heating and cooling systems. Appl Therm Eng 35:40–54
7. Brittle JP, Eftekhari M, Firth SK (2016) Mechanical ventilation & cooling energy versus thermal comfort: a study of mixed mode office building performance in Abu Dhabi. In: Brotas L, Roaf S, Nicol F, Humphreys M (eds) Proceedings of the 9th Windsor conference. NCEUB: making comfort relevant, 7–10th April 2016, Windsor
8. Lian Z, Zhao B, Liu W (2007) A neural network evaluation model for individual thermal comfort. Energ Buildings 39:1115–1122
9. UC Berkeley Center for the Built Environment (2014) http://www.cbe.berkeley.edu/underfloorair/tacguidelines.htm. [Online; Accessed 23 May 2014]
10. Nest Company (2014) https://nest.com/thermostat/meet-nest-thermostat/. [Online; Accessed 10 May 2014]
11. Frontczak M, Schiavon S, Goins J, Arens E, Zhang H, Wargocki P (2012) Quantitative relationships between occupant satisfaction and aspects of indoor environmental quality and building design. Indoor Air Journal 22:119–113

Chapter 19
High|Bombastic: Adaptive Skin Conceptual Prototype for Mediterranean Climate

Omid Bakhshaei, Giuseppe Ridolfi, and Arman Saberi

Abstract *High|Bombastic* is part of a research on Living Building Envelopes still in progress at Mailab, a research center active for 7 years at the University of Florence on Multimedia, Architecture, and Interaction.

High|Bombastic identifies the development of a naturally ventilated double-skin adaptive envelope.

This paper will present the results of the first phase dedicated to the definition and the realization of its conceptual prototype involving dynamic components and materiality speculation.

19.1 Order Through Fluctuations

If we now consider instead of an isolated system, a system in contact with an energy reservoir ... we necessarily are confronted with open systems in which the exchanges with the external world play a capital role....

In all these phenomena, an ordering mechanism not reducible to the equilibrium principle appears. For reasons to be explained later, we shall refer to this principle as order through fluctuations. One has structures which are created by the continuous flow of energy and matter from the outside world. Their maintenance requires a critical distance from equilibrium, i.e. a minimum level of dissipation. For all these reasons we have called them 'dissipative structures'.

– Ilya Romanovich Prigogine (1974).

High|Bombastic is a research on additional external building envelope as a regulatory device/resource for adaptive architecture: a solution able to fit different climate conditions beyond the heating-dominated ones representing the best contextualization and the place of its first applications [10, 15]. The idea is to provide

O. Bakhshaei (✉)
Novin Tarh Studio, Tehran, Iran

G. Ridolfi · A. Saberi
Department of Architecture (DIDA), Università degli Studi di Firenze, Florence, Italy
e-mail: giuseppe.ridolfi@unifi.it; arman.saberi@unifi.it

© Springer International Publishing AG, part of Springer Nature 2019   209
A. Sayigh (ed.), *Sustainable Building for a Cleaner Environment*,
Innovative Renewable Energy, https://doi.org/10.1007/978-3-319-94595-8_19

ability to shape the envelope and actively participate energy flows between environment and users.

High|Bombastic looks to realize an efficient/active filter in opposition to the *Insulating Modernism* practice where sealing buildings, realizing the envelope as an insulated continuous barrier, and considering the building as a thermodynamic closed system persist to be used as the indisputable and universal solutions, due in particular to its robustness and simplicity [17].

Conversely, *High|Bombastic* rejects the simplistic idea that more insulation means greater well-being and economy [29]. It looks to get the best dynamic mediation between energy consumption and users' satisfaction using a reasonable amount of material and making the most of its features [27].

It looks to resiliency obtained by warping geometry and instancing physical properties according to climate zone, sun path, seasons' characteristics, hours, and environmental conditions of the day.

High|Bombastic looks for an object which can mutate, adapting itself in a dynamic way, which acts as a dissipative energy architecture incorporating some kind of intelligence [12] (under certain circumstances) able to select and to start actions in order to ensure reasonable metabolism of the living system [31].

19.2   Responsive Bio-Adaptation, Geometry, and Warping

The energy efficiency discourse of consuming less and minimizing dissipation ultimately discloses little about the role of people, buildings, and design in the thermodynamic evolution of urbanization but does finally amplify many neoliberal dynamics. To address the non-isolated, non-equilibrium, and non-linear thermodynamics that float the operations of buildings and of life itself, architects by now need a radically different epistemology — a different ethic of work in these systems — for energy and the energy designs that will engender maximal entropy production futures for civilization.

– Iñaki Abalos (2014)

The idea for a building skin able to assume variable characteristics dated back to 1981 as a radical new proposal for multifunctional glass building layers [5], and 15 years later, envelopes were recognized for their *permeability*: properties enabling adjustment functions and responsive adaptations in order to have a better approximation to the optimal energy balance and users' satisfaction [9, 22, 25, 28].

High|Bombastic adapts itself pursuing biologic (bioclimatic and passive design), incentivizing/blocking irradiation and air, and using mutant materiality able to be informed at different physical properties and geometry.

Based on a regular grid, the external layer is designed to assume accurate configurations through mechanical gears to get the maximum/minimum sun radiation and to resize the air cavity as it is an important regulatory device for natural ventilation [4, 6, 7, 19].

To allow movements, external elements are designed as on origami diagram made of rigid triangles able to warp the surface but maintaining the perimeter of the façade bound profile unchanged [30].

For the property of origami, the perimeter constrains required deformations in the façade surface that has been realized providing punctual elastic joints between its triangular components.

19.3   Joining and Modeling

The architect no longer designs the final form but rather creates an initial state, introduces a set of controlled constraints, and then allows the structure to be activated to find its form. – Farahi Bouzanjani [7]

The use of punctual joints is a temporary solution useful for conducting preliminary test on the prototype but not satisfactory from the standpoint of the façade sealing as the spaces between the elements, produced in the various configurations, at this point of the research are not negligible.

The research is currently investigating how these gaps can be used as a regulatory device on *permeability* [23]. On the other hand, materials and technologies with passive/active deformability and continuity on surface to ensure an airtight buffer are under study [2, 7, 8, 13].

According to typological classifications [11, 16, 21, 23], we are dealing with different "functional models" which go from *naturally ventilated buffer* up to the *airtight made box* trying to fit them on Mediterranean climate zone.

Parent Components   The triangular panels host three different parent components able to mutate thermal property, to activate/deactivate stack effects inside the air cavity, and to regulate energy flows. These components are:

Pneumatic-driven cushions
Memory alloy-driven flaps
Windows

For the dynamic cushions, two main alternatives are under investigation.

The first one is a three-layer Teflon-coated pillow where it can change its characteristics inflating/deflating the inner and outer air cavities. Taking advantage of this movement and the special treatment of the films, where opaque and transparent parts are alternated, it is possible to obtain a different alignment of the parts permeable to the light and consequently a precise regulation of the incoming natural light and connected internal irradiation.

The second one is based on standard ETFE pillows where the argon will replace the natural air used to inflate them in order to decrease its thermal conductivity of about 67%. In fact, and as well known, thermal conductivity of argon is 0,0168 W/mK where air is 0,025 W/mK. The different inflation of the cushion is then used to adjust the thermal energy of the system.

Another component of the façade is an opaque semirigid flap that can be opened up when the air cavity is overheated to increase permeability and its hot air dissipation capacity [19].

To operate the flap, *High|Bombastic* uses the property of nitinol (500 nm 55–75 °C wire), a metal alloy that changes its state and length when the temperature reaches a predetermined temperature.

Nitinol is the commercial name of the metal alloy of nickel and titanium able to assume, under higher temperature, a crystal cubic structure as well known as austenite and, when cooled, to reverse at its original configuration of martensite [26]. This structural transformation induced, on the wire, a difference in length of around 6–7% that can be used to pull and release the flap.

In short, when the wire is heated and the nitinol changes to its austenite state, a contraction occurs; conversely when the wire gets back to a normal temperature, the wire recovers its original length. A very interesting property of nitinol is also the fact that the changing state temperature can be predetermined by "cooking" it at a specific temperature in a way it is possible to operate the opening of the flaps when the air in the cavity reaches preassigned temperatures of excessive overheating.

As a result, flaps do not require the use of any detector and any actuator device requiring energy.

The last component is a traditional glazed surface that can be defined in its technical characteristic according to the required performance of the façade.

Windows positions represent the qualifying element of glazing where the sill is at the same level of the floor and the window height not exceeds two meters high. Considering that the double-skin façade is normally used in large buildings such as offices, commercials, hospitals, etc., [3] the research has set two fundamental goals: the first one is to provide a better visibility toward the outside for users sitting at working table or lying in bed and the second one is to provide a sensible glaring reduction in the room.

To meet these goals, and at the same time to ensure optimal energy behavior, the façade can take benefits from its geometric changes and the selective properties of the cushions on the sun's rays.

19.4   Configurations for Mediterranean Climate

The prototype assumed as reference the Mediterranean climate (3C|Mediterranean North) and is currently under evaluation for different configurations in relation to six different environmental conditions.

Sunny winter days
Cloudy winter days
Winter nights
Normal summer days
Hottest hours of the days
Summer nights.

In the first condition, the façade frame assumes a concave configuration in order to increase the capturing surface of the glazed components and to achieve an angle of incidence with the solar irradiation closest to the 90°. In this configuration cushions remain deflated in order to promote the thermal gain derived from the sun presence. Flaps are closed.

In the second condition, occurring during a cloudy winter day, the façade is fitted to completely seal the air cavity and limiting stack effects that normally take place inside this space. In this configuration cushions are inflated to increase thermal insulation, and flaps are closed.

The above configuration is the same that is assumed for the winter nights where the goal is to limit the flow of energy from the indoor space to the outdoor environment with lower temperatures. During the summer period, the façade predominantly assumes the convex configuration in order to obtain the maximum shading, and cushions can be inflated in order to increase insulation. In case of the three-layer solution, the cushions can be adjusted in such a way as to obtain the best compromise between thermal insulation and room illumination.

During the hottest hours and when the temperature inside the air cavity reaches high temperatures, flaps are activated in order to let the hot air to flow out. The last geometry occurs during the summer nights where the double-skin façade is well known as an effective device to extract excess temperature from the indoor environment. In this situation the façade is planar realizing a sealed chimney with lower and upper openings for cold air intake and hot air exhalation.

19.5   The Prototype and Energy Modeling

A prototype fabrication and energy modeling have been undertaken to verify the design hypotheses. Although energy simulation is primarily used in the design detail stage, a very useful benefit can come if applied in the early stage [1, 24]. Besides the fact that energy simulation is based on a computer program and the accuracy of results would largely depend on the inputs provided, many useful feed-backs can be gotten in the simplified mass modeling of the early stage [14, 18].

In addition, the physical prototype and the related digital modeling have also been entrusted with an educational role and used during Mailab's international workshop "Parametric Computation and Digital Prototyping for Environmental Responsive Envelopes" hold on 2016, Florence, Sep. 5th–16th.

In fact, we think that besides being a research tool, modeling is also a valid educational instrument giving intuitive perception about matter's behaviors. In this case, the model is used, not as a mere presentation of phenomena but as a cognitive artifact that allows students to interact and become familiar with the theoretical foundations that the prototype incorporates: expression of concrete thought and formalization of the traditional sketching in a way it can be now used as a shareable instrument of scientific research [20].

For this goal, numerical analysis was translated into more understandable and shareable infographics to give evidence of immaterial aspects of design such as temperature, wind, light, etc.

Therefore 3D visualizations and graphic animations were produced to show the changing state of the façade under different conditions of forms, building elements, weather, and time as the complementary entity of form and matter.

The model is composed of two main parts: a sample of the façade fabricated in a 1:10 scale robotic model and a screen for video projection showing the physical effects on the building.

Numerous researches and tests have been carried out in order to identify a material capable of supporting the geometric transformations of the façade.

At this stage the solution has been identified in the construction of rigid frames connected through casted joints made up of silicone with hardness characteristics between shore 00–20 and shore 00–40.

Silicone with harder characteristics (shore A-40 and shore A-80) was also tested for the flaps components but with unsatisfactory results that have decreed the use of common 3 mm plexiglass.

Because of the gaps between the elements produced by elastic joints, the extensive use of silicon membrane or other equivalent material is currently under test for the whole façade system.

The façade movements were made through the use of pistons driven by digital servo with a standing torque of 15 kg.cm at 6.0 volt.

In this preliminary phase, cushions were simulated using a thin silicone membrane, but a deeper research is currently under way in order to investigate potentiality and feasibility of polyester-coated films.

19.6   The Interactive Installation

To reinforce the didactic power of the model, an interactive installation was also realized.

A portable device is used as the users' interface to set a scenario between different environmental conditions. The choice is processed and sent to the computer via Wi-Fi using the OSC (Open Sound Control) protocol.

The core of the system is managed by Isadora, a graphic programming environment, that allows the real-time synchronization of video/audio contents and digital signals activating the mechanical parts of the model.

Another important component is represented by Arduino the most famous open-source digital prototyping platform used to realize interactive electronic objects. All the signals coming from Isadora are processed inside Arduino in a way servo can be activated in relationship to the selected environmental scenario.

At the same time, Isadora starts different video animations that are sent and projected on the side screen in order to show which kind of effects is produced inside the building.

Results coming out from analytical calculations have shown that the proposed solution, compared to simple sealed double-skin façade and to a normal solution on layer envelope, produces a better thermal performance and a better distribution of the natural daylight with a very low percentage of glaring during the day and the whole year.

In detail, results show that natural daylight reaches a better distribution with glaring slightly perceptible in the worst condition (≤ 15 in Unified Glare Rating scale); energy use intensity in different building program has always been better for a ratio of about 20–35%. Other benchmarkings, based on Performance Metric for Green and Smart Building [18], are currently under study.

References

1. AIA (2013) An architect's guide to integrating energy modeling in the design process. The American Institute of Architecture. https://www.aia.org/resources/8056-architects-guide-to-integrating-energy-modeli v: 2016
2. Beesley P, Khan O, Stacey M (2013) Interactive tensegrity structure. In: ACADIA 2013 adaptive architecture, proceedings of the 33rd annual conference of the association for computer aided design in architecture. Riverside Architectural Press, Cambridge
3. Musa BT (2016) Evaluating the Use of double-skin facade systems for sustainable development. Department of Architecture, Eastern Mediterranean University, Gazimagusa
4. Compagno A (2002) Intelligent glass facades., (5th revised and updated edition). Birkhäuser, Berlin
5. Michael D (1981) A wall for all season. Riba J 88(2), 135–139
6. Faist AP (1998) Double skin walls. Institut de technique du batiment. Department d'Architecture. École Polytechnique Fédéral de Lausanne (EPFL), Lausanne
7. Behnaz FB (2016) Alloplastic architecture. In: Fox M (ed) Interactive architecture. Princeton Architectural Press, New York
8. Russel F, Charles LD (2014) Kinetic architecture: design for active envelopes. The Images Publishing Group, Mulgrave
9. Thomas H, Norbert K, Michael V (eds) (1996) Solar energy in architecture and urban planning. In: Proceedings European conference on solar energy in architecture and urban planning, Berlin
10. Jaeggi A (1998) Fagus. Industriekultur zwisschen Werkbund und Bauhaus, Werlagsbüro, Berlin, tr. E. M. Schwaiger, Fagus. Industrial culture from Werkbund to Bauhaus, Princeton Architectural Press, New York, 2000
11. Kragh M (2000) Building envelopes and environmental systems. In: Modern façades of office buildings, proceedings. Delft Technical University, Delft
12. Kroner MW (1997) An intelligent and responsive architecture in automation in construction, vol 7. Elsevier, pp 5–6. http://www.sciencedirect.com/science/article/pii/S0926580597000174 v:jun 2017
13. Salford K (1993) Soft systems. In: Lab C (ed) Brian boigonm. Princeton Architecture Press, New York
14. Lawrence Berkeley National Laboratory Eleanor Lee (2006) High performance commercial building facades, California energy commission building technologies program. Environmental Energy Technologies Division, Ernest Orlando Lawrence Berkeley National Laboratory, University of California, Berkeley
15. Charles LD (2014) Ancestors of the kinetik facade. In: Russel F, Charles LD (eds) Kinetic architecture: design for active envelopes. The Images Publishing Group, Victoria, p 2014

16. Loncour X, Deneyer A, Blasco M, Flamant G, Wouters P (2004) Ventilated doubel facades. Classification & illustrations of facade concepts. Belgian Building Research Institute. Department of Building Physics, Indoor Climate & Building Services, Bruxelles
17. Moe K (2014) Insulating modernism. Isolated and non-isolated thermodynamics in architecture. Birkhäuser Verlag, Basel
18. Jadhav NY (2016) Green and smart buildings. Advanced technology options. Energy Research Institute Nanyang Technological University Singapore, Springer, Singapore
19. Oesterle E, Lieb R-D, Lutz M, Heusler W (2001) Double skin. Prestel, Munich/London/New York
20. Papert S (1996) A word for learning. In: Kafai Y, Resnick M (eds) Constructionism in practice: designing, thinking and learning in a digital world. Lawrence Erlbaum Association, Mahwah
21. Parkin S (2004) A description of a ventilated double-skin façade classification. In: International conference on building envelope systems &technology, Sydney
22. John P, Maurya MC (2000) The challenge of green building in Asia. Arup Facade Engineering, Sydney
23. Poirazis H (2004) Double skin façades for office buildings. Department of Construction and Architecture, Division of Energy and Building Design. Lund University, Lund Institute of Technology, Lund
24. Di Pongratz C, Rita PM (2000) Natural born caadesigners: young American architects. Birkhäuser, Basilea
25. READ (1996) European charter for solar energy on architecture and urban planning, proceedings, Berlin
26. Ritter A (2007) Smart materials in architecture, interior architecture and design. Birkhauser, Basel
27. Stec W, van Paassen AHC (2000) Integration of the double skin façade with the buildings, energy in built environment. Energy Technology, TU Delft
28. Streicher Wolfang (2005) Best practice for double skin façades, WP 1 Report "State of the Art" EIE/04/135/S07.38652
29. Sullivan CC, Barbara H (2011) Energy performance starts at the building envelope. In: Building design+construction. https://www.bdcnetwork.com/energy-performance-starts-building-envelope. v: 2016
30. Tachi T, Epps G (2011) Designing One_DOF mechanism for architecture by rationalizing curved folding. In: Algorithmic design for architecture and urban design, proceedings of the international symposium on algorithmic design for architecture and urban design, Tokyo, pp 14–16
31. Wigginton M, Jude H (2002) Intelligent skins. Butterworth-Heineann, Oxford, MA

Chapter 20
Quality of Healthcare: A Review of the Impact of the Hospital Physical Environment on Improving Quality of Care

Jazla Fadda

Abstract Quality of healthcareis not merely the technical care and professional performance: it is an integral concept measured by summing up different components. According to Donabedian's classification, quality can be assessed using three main components: structure, process, and outcomes. Structure is related to the physical environment, including the facility's design, technology, and equipment, besides the healthcare providers' credentials and qualifications. Process is the clinical care that considers patient safety, including effective infection control and mitigation of medical errors. Outcomes are assessed by the extent of improved patient health status, combined with patient satisfaction. This chapter generates a new conceptual model using previous evidence of hospital design as an input for quality of care.

Objectives (a) to review the main dimensions, metrics, and criteria to assess quality of care at hospitals; (b) to provide evidence of the impact of the physical environment ("hospital design") on the process, including staff and patient safety; (c) to provide evidence of the impact of the physical environment on the outcomes, including patient health status and patient satisfaction.

Methodology This paper uses a literature review to introduce a new conceptual model of correlating the physical environment as a structure, impacting the process component including safety aspects of patients and staff, along with assessing the outcome component represented by clinical outcomes and patient satisfaction. Evidence was synthesized, summed up, and interpreted to generate the proposed model. Hospital design includes privacy, noise reduction, lighting, increased square footage of patient rooms, safer design to mitigate falls, well-decorated rooms, well-chosen colors, and visitor-friendly amenities.

J. Fadda (✉)
American University in the Emirates- Block 7- Dubai International Academic City,
Dubai, United Arab Emirates
e-mail: Jazla.fadda@aue.ae

© Springer International Publishing AG, part of Springer Nature 2019   217
A. Sayigh (ed.), *Sustainable Building for a Cleaner Environment*,
Innovative Renewable Energy, https://doi.org/10.1007/978-3-319-94595-8_20

Key Findings Good hospital design, such as lighting and following a standardized design of patient rooms, adequate ventilation, better ergonomic design, good maintenance, and less noise, enhances staff and patient safety. This goal will eventually reduce medical errors and improve the efficiency and effectiveness of healthcare delivery. Considering the clinical outcome component of quality, the review supports that rooms with views have a noticeable influence on patient recovery, especially for postsurgical cases, with fewer doses of analgesics and shorter postoperative hospital stays. Colors at healthcare facilities have also an effect on patients and staff. Evidence was generated from different sources that selecting colors should go beyond the aesthetic design and should be more patient- and staff centered for better healing and performance. In regard to patient satisfaction as an outcome component of quality of care, patient satisfaction was correlated to hospital design reflected by privacy, noise reduction, WiFi, lighting, increased square footage of patient rooms, safer design to mitigate falls, well-decorated rooms, and visitor-friendly amenities. Nevertheless, further studies showed only a small effect of hospital re-design on patient satisfaction. Poor patient satisfaction is not related to aging buildings, but rather is more related to the healthcare providers, patient safety, patient-centered design, work flow, and efficiency of healthcare.

Conclusion Quality of care can be measured using different metrics: structure, process, and outcomes. Structure or physical environment is essential in impacting the healthcare delivery process and healthcare outcomes. However, it is not only the sophisticated or newly remodeled design that increases healthcare efficiency, effectiveness, and patient satisfaction. The design that meets patient and staff needs will strongly support staff performance and provide a better therapeutic and healing environment that eventually concludes in better quality of care.

20.1 Introduction

The healthcare industry has been rapidly evolving during the past few years. Many emerging and newly addressed trends could shape and outline different types of healthcare facilities that provide a variety of health services. Quality of care is now the common concept that cannot be compromised. Patients are better aware of the concepts of healthcare, their perceptions and expectations are better formed, and that entails continuous improving all aspects and components that affect the quality of care.

Health was defined by WHO as "a state of complete physical, mental, and social well-being, and not merely the absence of disease" (WHO Constitution, 1948). From that concept, we have witnessed a growing interest in using a wide range of environmental aspects as a part of the patient treatment process within hospitals.

The new trends emphasize the role of the built environment on improving patient health and better provision of healthcare services. This concept was addressed a long time ago by Hippocrates, who believed in the need of harmony between the individual and the social and natural environment, as a natural approach for treating diseases. The same concept was addressed in the nineteenth century by Florence Nightingale, who believed that a clean hospital environment is a major factor in improving the health of sick people [4].

20.1.1   Quality of Healthcare

"Quality is never an accident, it is always the results of intelligent efforts" (John Ruskin). According to Campbell et al. (2000) in an article published in the journal *Social Science and Medicine*, "Defining quality of care," quality of care is defined using different approaches, "generic" and "disaggregated." The generic approach means that the expectations and goals have been achieved; it also includes excellence. The disaggregated approach recognizes quality as a complex with multidimensional components: each component provides a partial picture. However, there was more focus on two main elements, access and effectiveness. The questions addressed are "Can an individual get the care they need when needed?" this is 'Access.' "Is the clinical care along with interpersonal care provided effective?" this is 'Effectiveness.' This approach was developed later to the 'System-Based Model of Care' including three components: "Structure, Process, and Outcomes" [8]. The Institute of Medicine defined quality as *"The degree to which health series for individuals and populations increase the likelihood of desired health outcomes and are consistent with current professional knowledge"* [6].

Quality of care was usually measured by patient satisfaction, which is mainly related to the outcomes triggered by getting medical care within the healthcare facilities. However, quality of healthcare is much more than a matter of patient satisfaction, and technical care; it should be a total sum of different components classified using a framework and special matrices or criteria.

According to Donabedian's classification, quality can be assessed using three main components: structure, process, and outcomes. Structure is related to the physical environment including the facility's building, design, technology, and equipment. Structure is also related to the credentials and qualifications of healthcare providers and staff. Structure refers to the organizational factors that define the health system under which it is provided [6], whereas process refers to the way of services are organized within the facility [9]. Process is also related to how efficient healthcare is provided by taking into consideration patient safety, embedding infection control, and reducing medical errors [4]. Finally, the outcomes are assessed by improving patient health status and patient satisfaction as well. On the other hand, poor quality of care is represented in three forms: overuse of unnecessary or ineffective treatment, or underuse that reflects failure to provide the care which works, and misuse, which means making mistakes that harm the patients ("medical errors"). Overuse occurs when care is provided when inappropriate, and underuse when care is not provided when necessary [2]. However, both forms are correlated with less effective care that results in unwanted outcomes, whereas misuse is correlated with poor process resulting in an inefficient quality of care. On the other hand, Maxwell addressed six dimensions of healthcare quality: effectiveness, efficiency, appropriateness, acceptability, access, and equity.

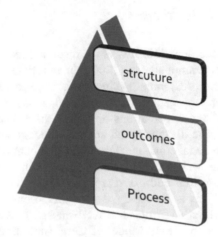

The classification by Donabedian generates another kind of quality measures, slotting healthcare into three parts: technical aspects ("process, interpersonal aspect" and "process and patient perceptions"), amenities, or the environmental aspects within which that healthcare is provided.

The structural features of a healthcare facility can have a direct impact on processes and outcomes, whereas the 'process' is also the interaction between users ("patients") and the healthcare structure: this approach is the main aim addressed in this chapter. Outcomes are the consequences affected by the structure and the process directly and indirectly [9].

Accessibility is a major part of defining quality of care. The most basic dimensioning of access in terms of structural access is geographical or physical access [3].

Structure includes the aspects of the environment of the hospital where patients are seen and treated [5]. Structure includes building design, equipment, beds, and number of clinics; process is clinical examinations, prescriptions, appointments, and nursing care. The interpersonal aspect of care is also considered. The outcomes are related to any change of patient status correlated with the intervention provided by the healthcare facility, such as restoration of function, relief of symptoms, improvement in current health, or future "life expectancy" [5].

Efficiency and effectiveness are the main aspects of quality. The former represents the process and the latter represents the outcomes. According to the Institute of Medicine, effectiveness is defined as providing services based on scientific knowledge to all who could benefit and refraining from providing services to those not likely to benefit (avoiding underuse and overuse). Efficiency is defined as "avoiding waste," in particular, waste of equipment, supplies, ideas, and energy [6]. Effectiveness is related to achieving the intended outcomes and results, or that all the targeted problems were solved without reference to cost. In the healthcare setting, effectiveness is about making sure that "the right things are done" [6], whereas efficiency is related to performance, process, to using the best methods without wasting. It is about "Doing things right" [6].

This report highlights the integration and interrelation between the three components of quality of care: structure, process, and outcomes. The structure is represented by the physical environment, including design and required healthcare inputs. The report further inspects how this structure will affect both the process component representing "efficiency" by highlighting staff safety and performance and patient safety in terms of infections, medical errors, and injuries. Furthermore, we discuss how the structure affects the outcomes component representing "effectiveness," by highlighting the impact of structure on clinical outcomes and patient satisfaction.

20.2   The Review Procedures

20.2.1   Objectives

The main objective of the review is providing an overview of variety of evidence related to the relationship of the physical environment and improved quality of care, through enhancing staff and patient safety, in addition to improving clinical outcomes and patient satisfaction. Specifically, the main sub-objectives of this paper follow:

(a) Reviewing the main dimensions, metrics, and criteria to assess quality of care at hospitals
(b) Providing evidence on the impact of the physical environment on the process including staff and patient safety
(c) Providing evidence on the impact of the physical environment on the outcomes including patient health status and patient satisfaction

20.2.2 Search Methods

A literature review was used to generate a new conceptual framework of the impact of physical design as an input on the other components of quality of care, process, and outcomes. The support is in generating evidence of the impact of the physical environment, the hospital design, on improving quality of care.

The review was intended to answer the following questions:

- Does the hospital design's "physical environment" enhance better workplace safety and reduce staff injuries?
- Does the hospital design's "physical environment" impact clinical care and reduce medical errors?
- Does the hospital design's "physical environment" impact the healing process or improvement of the patient's health status?
- Does the hospital design's "physical environment" have an effect in improving the patient's expectations and the patient's satisfaction?

Hospital design includes privacy, noise reduction, WiFi, better lighting, increasing the square footage of patient rooms, safer design to mitigate falls, well-decorated rooms, well-chosen colors, and visitor-friendly amenities.

A search has been conducted to generate a variety of sources, using computerized resources (Medline, Scopus, and pro-quest). Different keywords were used to synthesize the inputs, such as Quality of care, Quality of care metrics, Hospital design, Building constructions management, Physical environment and quality at hospital, Hospital architecture and quality, Colors impact at healthcare settings, Workplace safety, Patient safety, Patient satisfaction and hospital design, Healthcare outcomes, Efficiency, and Effectiveness.

Findings were synthesized, summed up, and interpreted to generate the evidence. Different keywords were used to synthesize the inputs, such as Quality of care, Quality of care metrics, Hospital design, Building constructions management, Physical environment and quality at hospital, Hospital architecture and quality, colors impact at healthcare settings, Workplace safety, Patient safety, Patient satisfaction and hospital design, Healthcare outcomes, Efficiency, and Effectiveness.

About 800 search results were generated from the previous search engine. The relevant studies were reviewed to match this chapter's questions to find the association between having a supportive facility structure and improving the process and outcomes components, which eventually lead to better quality of care.

20.2.3 Inclusion and Exclusion Criteria

This paper reviews the studies and journal articles that address the quality of care from structural perspectives or the physical environment, including architectural design, equipment and technology, lighting, colors, ventilation, privacy, noise, and ergonomic design, to investigate the influence of these factors on the process of providing healthcare, taking into consideration their impact on staff and patients, such as stress and fatigue, staff safety, and patient safety, with more emphasis on medical errors and efficiency of healthcare. Moreover, the chapter includes a search for the impact of the physical environment and architecture design on the process of patient healing and patient satisfaction. The structural inputs that focus on staff qualifications, as a driving force for medical errors and quality measurement, were excluded from this search. The social environment and types of relationships between staff and their patients were also excluded.

20.3 Literature Review and Key Findings

The search strategy generated 1009 papers, of which 715 papers were found to fit the inclusion criteria. The findings were then classified into subcategories, and the evidence was grouped in accordance with the related classified subtopics.

This section provides an overview of the literature that supported the author in addressing a new model of the impact of physical environment and design on the process and the outcomes of quality of care.

20.3.1 Physical Environment Impact on the Process of Quality of Care

20.3.1.1 Physical Environment and "Patient Safety"

Design and Patient Safety: "Infection Control"

Single-Bed Patient Room Versus Multi-Bed Room

There is a strong relationship between hospital design and the acquired infection rates at hospitals. More than 120 studies have shown the effect of the physical design on reducing infection rates [3]. The main routes of infections are direct contact or airborne. Moreover, the existence of handwashing sinks along with accessing alcohol hand rub dispensers have also show considerable reduction of contact contamination [69–71].

Different studies in this review have correlated multi-bed units at hospitals with having more infections. The cleaning aspects and the materials used was the main reason; many contaminated surfaces are found near infected patients, and contamination could possibly be transmitted by staff gloves while caring for different patients at the same ward. This point was addressed by Boyce et al. (1997) and Aygun et al. (2002) while investigating environmental contamination at hospitals by *Staphylococcus* and *Acinetobacter baumannii* bacteria [18, 26]. The concentration of microorganisms was high on some surfaces such as overbed tables, bed privacy curtains, rails, cuffs, chairs, and door handles [8]. Moreover, Palmer's (1999) study of bacterial contamination of curtains in clinical areas showed that ward bed curtains are more contaminated with methicillin-resistant *Staphylococcus aureus* (MRSA) compared with single-bed room curtains [52]. The existence of common infections by *Pseudomonas aeruginosa* was also compared in burn patients in single-bed rooms and in open wards, the latter having the highest incidence of such infections [48, 52].

More than 20 studies showed that the nosocomial infection rates are higher in multi-bed rooms than in a single-patient room. Three studies have shown that a single-bed room and better air quality can reduce the incidence and mortality among patients in the burn sections [13, 42, 58]. Moreover, putting patients in a single room decreases the possibility of infection by airborne diseases such as flu, measles, and TB [28, 41]. In a study investigating SARS (severe acute respiratory syndrome) contamination at one hospital in Toronto, Canada, 75% of the total infections was acquired from hospital settings. Multi-bed rooms recorded a higher rate of infections compared with single rooms [24]. Seven reviewed studies outlined the most contaminated places found at multi-bed rooms, which considered reservoirs of pathogens that facilitate the transmission of infections among patients and health workers such as "overbed tables, bed privacy curtains, computer keyboards, infusion pump buttons, door handles, bedside rails, blood pressure cuffs, chairs and other furniture, and countertops" [18, 21, 27, 44, 48, 54]. For instance, in a study on infected patients with methicillin-resistant *Staphylococcus aureus* (MRSA), displayed in multi-bed rooms at one hospital, 27% of the surfaces sampled in multi-bed rooms were contaminated, and 42% of nurses were infected with MRSA even though without direct contact with patients, just through touching the contaminated surfaces with their gloves. The reason behind was attributed to the difficulty in cleaning the multiple rooms thoroughly after discharging the patients [18]. As for the infections occurring in a single-patient room, having tight corners can affect the incidence and prevalence of diseases [23].

Hospital Design and Handwashing: "Use of Automated Sink"

Handwashing and infections are ever intercorrelated, especially at healthcare settings, and especially through health workers' hands. Larson et al. (1991) investigated the effect of an automated sink on handwashing practices in high-risk units at hospitals, finding that hands are washed thoroughly and better, but less often, at automated sinks [23]. Two studies conducted in acute units have shown that only one in seven of the staff is washing their hands after contacting patients. Compliance is considered acceptable between 15% and 35%, and exceptional above 40–50% [63, 68]. The reason for this low compliance rate was attributed to understaffing compared with high occupancy rate and high numbers of patients [88]. Three reviewed studies showed that educational programs in this regard have not produced any remarkable change in the compliance rate [65–67]. With no doubt, the high rates of morbidity and mortality taking place at hospitals are attributed to hospital-acquired infections, which entails more efforts for better compliance with handwashing among the staff [91]. Six studies showed that room design is correlated with better compliance of handwashing, for instance, increasing the ratio of accessible sinks or hand-cleaner dispensers within patient rooms [20, 63, 68, 71–73]. This impetus can be supported by displaying hand-rub dispensers beside guideline posters to remind the staff [72]. However, displaying the sink outside the patient's room does not show considerable change in compliance. Two studies showed that displaying the sink near beds increases the frequency of handwashing [71, 73] without any difference between automated and traditional sinks [69, 70]. Three studies suggested that having sinks in single rooms reduced the rates of nosocomial infections compared with displaying a few sinks in multi-bed units, especially for neonatal ICU and burn units [13, 42, 43].

Furniture and Spread of Infection

Several studies have shown a correlation between the materials used in a patient's room and the increased chance of infections. For example, greater concentrations of organisms were found on carpeted floors and the overlying air compared with bare flooring [53]. Moreover, different studies investigated the relationship between the fabric-covered furniture and vinyl-covered furniture of a patient room, finding that fabric furniture could be a reservoir of bacteria, especially of vancomycin-resistant enterococci (VRE), which was attributed to the difficulty of cleaning compared with vinyl-covered furniture [27]. Further investigation by Palmer (1999) compared patient bed curtain counts of bacteria with window curtains in clinical areas [48].

Ventilation and Airborne Infections

Two reviewed studies stated that the source of infections among patients is the air pathway, which is considered a threat to patient and staff safety within hospitals [78]. Two studies correlated hospital air quality with the existence of some pathogens such as fungal spores that increase the incidence of infections. Air quality was associated with different sources such as the ventilation system, air pressure and flow, filters, and humidity [38, 40]. Six studies showed further sources of airborne infections and the construction and maintenance activities taking place in the hospital [31, 32, 37, 45, 46]. Six studies have supported the effectiveness of using HEPA (high-efficiency particulate air) filters during construction in improving air quality and reducing infections [37, 39, 45, 46, 76, 77]. One study conducted by Oren et al. (2001) showed that a better ventilation system was the reason behind decreasing the rates of nosocomial diseases among leukemia patients with invasive pulmonary aspergillosis (IPA) [46]; the infection rate was 50% during hospital construction and renovation. In the following 3 years, no nosocomial infections were reported among the same categories of patients because of new ways, provided with HEPA. During construction, using portable HEPA along with sealing windows and installing barriers and enhance cleaning can be effective control measures [76].

Structure, Design, and Medical Errors

Patient Room Design and Medical Errors

Different studies have shown the relationship between standardization of the patient's room and reducing medical errors. According to Ulrich (1991), in his report "The Impact of Interior Design and Wellness" he highlighted the importance of standardization of patient room layout and equipment in reducing medical errors [91].

Lighting and Medical Errors

Another factor affecting medical errors is the lack of light. Buchanan et al. (1991) emphasized the importance of natural or electrical light in reducing errors [95]. Further, Booker and Roseman (1995) studied the frequency of medical errors at one hospital in Alaska; they found that medical errors are twice more likely to happen in December than in September, and 58% of total medical errors happen in the first quarter of the year [97]. Buchanan et al. (1991) compared three levels of illumination levels–480 lx, 1100 lx, and 1570 lx: they found that the highest illumination was associated with fewer errors, with a 2.6% rate, than the lowest illumination level, especially those related to prescription-dispensing errors [95]. Distraction was also a main factor of errors, especially in pharmacy settings, caused for instance by picking up a telephone call. Two studies suggested improving the environment will support the reduction of medical errors attributed to lighting and interruption issues [96, 98].

Further, changing room design can lower the transfer rates of the patients, which itself increases the possibility of medical error because of delays and discontinuities among staff [3]. In one study conducted in an Indian hospital, changing coronary intensive care units to single-bed rooms caused a 90% decline of patient transfer and a 67% decline in medical errors [3].

Structure/Design and Injuries: "Falls"

Many studies highlighted the risk factors of increasing falls among patients at hospitals. Patient falls impact two main factors, increasing both injury rates among patients and the cost of healthcare. The injuries caused are correlated with physical injuries, psychological impact, and prolonging length of stay (LOS) [89]. As for the cost, it was estimated that fall injuries among US elders cost more than 20.2 billion dollars per year, and are expected to reach 32.4 billion dollars in 2020 [88]. Injuries among patients are the result of the interaction between environmental factors and individual aspects. Different studies investigated the relationship of the falls happening among different type of diagnosed patients, such as mental disorder patients and musculoskeletal disorder patients; the former have a higher risk of falls compared with the latter [92].

Most falls were related to changing posture of the patients in the patient's room, and near the bed, especially on the way to the bathroom. Alcee (2000) found that 30% of total falls were related to the bathroom [95]; however, this was also investigated by Morgan et al. (1985), who found that half the total falls were related to moving to bathroom, as two thirds of falls happen near the toilet [92].

Brandis (1999), in his study of "A collaborative occupational therapy of nursing approach to falls prevention in hospital inpatients," suggested that taking preventive measures in designing the patient room could reduce the possibility of falls by as much as 17.3% [89]. Two studies showed that patient falls happen while they try to move without any assistance or observation [90, 94]. Three studies showed that having bedrails was not supporting the reduction of fall incidence among patients, and sometimes increased the severity [93, 207, 208], whereas having decentralized nurses' stations and single-bed rooms instead of multi-bed rooms that support family presence around patients will decrease fall incidence. As was measured by a study conducted at one hospital in Indiana, around a two thirds reduction of patient fall incidence was observed after changing unit design (from 6/1000 to 2/1000) [197].

20.3.1.2   Physical Environment and Staff Safety

Structure/Design and Infections

Ventilation

Four studies reviewed staff infections resulting from direct or indirect contacts with patients, along with those contracted by airborne routes [30, 35, 36, 80]. Ventilation is a key factor affecting the spread of infections among health workers. One study in China showed that rooms with good ventilation could decrease the transmission of SARS among health workers when associated with good pollution particles (PP) measures [80]. Environmental factors such as the ventilation supply and room temperature were the main cause of infecting 115 nurses with nasal symptoms while working in 36 nursing departments, as was shown in a study conducted in a Norway hospital [31] attributed to the presence of *Aspergillus fumigatus* in the ventilation supply, along with high room temperatures [31]. Another study conducted at the respiratory ACU at university hospitals in Canada showed a high rate of infection with TB among staff that was attributed to low ventilation and non-isolated patient rooms [81].

Structure/Design and Injuries

Medical Equipment: "Lighting"

One study reviewed showed a relationship between high-intensity surgical light during surgical operations and retinal damage among the surgical staff [104].

Ergonomic Design and Injuries

Alexandre et al. (2001) evaluated the programs that reduce nursing personnel back pain or the design solutions. Introducing adjustable bed height could enhance the quality of a nurse's spinal motion while conducting her or his tasks [107], leading to better musculoskeletal outcomes, as was supported by Petzall et al. (2003) in their study "Reducing back stress on nursing personnel: an ergonomic intervention in nursing home," who highlighted the importance of ergonomic intervention in reducing stress on nurses and lowering their future risks of musculoskeletal injuries and back pain. Ergonomic design is very important to decrease back injuries and stress among nurses [115] including the design of patients' beds and nurses' stations. For design of patient transferring devices, the study found that beds with standard small-diameter wheels could be easier to move in limited spaces [105], whereas beds having larger wheels would be more comfortable when moving patients for long distances. Modifying patient-transferring devices, toilets, and shower rooms has reduced back injuries among nurses by 50% [105].

Structure Design and Stress

Noise

Four reviewed studies showed that high noise levels can increase stress among nurses, affecting the process of care delivery, social support, and eventually quality of care, from decreased speech intelligibility [113, 172, 174, 191].

Understaffing

In a survey conducted in 1609 US hospitals, low levels of nursing staff caused 24% of injuries and mortality [7].

Ergonomic Factors

Better ergonomic interventions associated with air quality and less noise and adjusted light have significant impact on reducing stress among staff [105, 106].

Unit Layout: Fewer Trips

Six studies showed that nurses' trips within the hospital could be decreased for better effective care and decrease the stress caused by long walking [127, 118, 119, 189, times] with better layout of the radial and double corridor: 28.9% of nurses' time is spent walking, versus 59.9% spent in patient observation [110]. According to Shepley et al. (2003), radial units can be more effective in managing patient loads. Radial design showed fewer walking steps (4.7 steps) than rectangular units (7.9 steps) were required by nurses and was also preferred by staff [119]. Decentralized nurses' stations allowed less walking time and more time spent in patient care than centralized nurses' station [6, 210]. Thus, well-designed hospitals, by taking staff safety into consideration will improve the efficiency and effectiveness of healthcare and enhance better quality of care by reducing medical errors and increasing patient satisfaction.

20.3.2   *Physical Environment and Outcomes*

20.3.2.1   Physical Environment and Clinical Outcomes

Art in Healthcare Environments

A variety of studies investigated the relationship between the arts and psychological status. Ulrich and Giplin (2003), in their report "Healing Arts: Nutrition for Soul," discussed that artwork which includes images of nature, such as landscapes, flowers, water, or gardens, is psychologically appropriate and can reduce stress and improve outcomes such as relieving pain [141]. The same outcomes were generated from art that included positive gestures and facial expressions. However, the opposite results of negative reactions among patients were found with images that are abstract and ambiguous [157] and positive images were also reported as the patient's preference [138].

Nature Views and Clinical Outcomes

Much research has focused on the relationship between patient room views and clinical outcomes. Ulrich (1948) conducted a study of "View through a window may influence recovery from surgery," discussing the importance of window views of nature such as trees in having a shorter postoperative stay along with needing fewer drugs for pain relief, as compared with the reverse impact on those patients who have to look at brick building walls. This idea was supported by Verderber (1986) in "Dimensions of person–window transaction in hospital environments," in which he investigated the preference of having nature views by patients and staff [154]. Another study by Keep and Inman (1980) compared patients in ITUs with and without windows, noting that patients with a windowless ITU had a less accurate memory of length of stay and twice as often reported hallucinations and delusions [130].

Further studies investigated pain control among patients with and without nature-distracted interventions. Diette et al. (2003) and Schneide et al. (2004) found that better pain control and reduced pain were found when exposing patients to nature-distracted interventions [145, 148]; this was also supported by Shofield and Davis (2000) in their report of stimulations versus relaxation as a potential strategy for the management of chronic pain. Ulrich et al. (2003) found that viewing nature was also correlated with more positive clinical outcomes, such as low blood pressure and low pulse rates of blood donors, when exposed to a nature tape compared with those exposed to a TV tape of urban settings [152].

These authors also showed the impact of exposing patients to nature and faster recovery and less stress in another report "Stress recovery during exposure to natural and urban environments" [138]. Another study by Miller et al. (1992) exposed burn patients to videotapes of nature, showing significant reduction of anxiety and pain [129]. In another study conducted in Swedish hospital ICUs, heart surgery patients exposed to landscape and water pictures reported less anxiety, less stress, and fewer doses for pain relief than those who were not [142]. The same results were found with patients having abdominal surgeries, who reported better recovery rates when exposed to window views than those looking at brick walls [91]. *"Nature serves as a positive distraction that reduces stress and diverts patients from focusing on their pain or distress"* [142]. Diette et al. (2003) conducted a randomized control trial on patients undergoing bronchoscopy procedures exposed to nature scenes on the ceiling and a control group that was exposed to a blank ceiling; the former reported less pain [145]. Moreover, walking through nature scenes within the hospital will reduce the anxiety and distress of female cancer patients as well [148]. The same results were found with patients suffering from dementia or Alzheimer's [155].

Access to Daylight and Clinical Outcomes

Access to daylight has proved its effectiveness for better clinical outcomes. This paper viewed different studies investigating this point. Eastman et al. (1998) investigated the effectiveness of bright light as a treatment for winter depression [157]. Bright sunlight has a positive impact as an antidepressant effect on winter depression. Moreover, comparing morning light and evening light treatment of patients with winter depression, Lewy et al. (1998) found that morning light is two times more effective [154]. This point was supported by Beauchemin and Hays (1998) in their report "Dying in the dark" in which they investigated the impact of sunshine on myocardial infraction patients; sunny rooms led to a shorter stay compared with dim rooms, with about 3.67 days reduction of LOS; the former reported on average 16.6 days of stay, compared with 19.5 days in a dim room [162. The Choi et al. (2012) study of impacts of indoor daylight environment on patient length of stay in a healthcare facility supports that high illuminance in the morning has better impact on patients compared with afternoon light [161]. Seven studies showed that bright light, especially in the morning, can improve depression, sleep quality, circadian rhythm, and agitation and decrease length of stay for dementia patients or SAD (seasonal affective disorders) [134, 120, 156, 158, 164, 166, 201, patients]. When tested in an experimental study on elders with dementia, their agitation was reduced when exposed to 2500 lx for 2 h/10 days [155]. Another study exposed cervical and lumbar spinal surgery patients to bright light after the surgical operations; those patients reported 22% less intake of pain relief medication compared with patients placed in dim rooms (Anjali 2006 on impact of light in healthcare settings) [3].

Less Noise and Better Clinical Outcomes

Acoustic comfort is strongly correlated with improving a patient's condition. Different studies were reviewed showing the impact of acoustic comfort and the process of healthcare as a partial measure of quality. Some studies found that less noise can be correlated with fewer medical errors in terms of the process and patient safety [3, 173]. As for the relationship between noise and less recovery-related outcomes [171, 175], noise is found to be correlated with increased levels of stress, leading to negative clinical outcomes and less chance for recovery. In another study [168], when investigating the noise level at post-anesthesia units, high noise levels occurred in these units that could be prevented.

Ron et al. (1996) showed that noise level at hospitals can strongly affect the quality and quantity of sleep, and the cause of disturbance is attributed to therapeutic procedures, staff talking, and environmental noise, which usually happen in multi-bed units [185]. Furthermore, noise is also caused by staff while causing erratic interruptions that affect patient quality of sleep [81, 187, 192]. A higher noise level can increase the need for oxygen support therapy by decreasing oxygen saturation and elevate blood pressure with increased heart and breath rates [85, 177, 181].

Hospital Gardens

The existence of gardens has a great positive impact on patients and staff, and for visitors as well. Barnhart and Perkins (1998) found that garden surroundings and natural open settings will impact both staff and patients [198]. Some studies correlated gardens with treating some disorders such as dementia [150, 228]. Ulrich in his report "Effects of interior design on wellness" highlighted the importance of reviewing gardens on mitigating pain and reducing stress along with increasing satisfaction of both patients and staff and visitors [126]. Moreover, hospital gardens and viewing nature can also enhance social support and better communication with reported reduction of staff stress [150, 225]. Cooper-Marcus and Barnes [196] ("Based on post occupancy evaluations of four hospital gardens in California") concluded that many nurses and other healthcare workers used the gardens for achieving pleasant escape and recuperation from stress.

Colors and Clinical Outcomes

In one study conducted in UK related to color design at hospitals, balanced colors had a measurable role in the well-being of staff and patients and affected both recovery rates and staff morale. Florence Nightingale observed that "a variety of form and brilliance of color in the objects presented to patients are an actual means of recovery" [5]. Furthermore, research on the effects of colors at healthcare facilities shows that selected colors should go beyond an aesthetically attractive design and be more patient focused for better comfort and as a supportive factor of the healing process. It was proved the healthcare environment can affect mood and behaviors and reduce stress; green was shown as having a stress-reducing effect compared with other colors such as orange in a stimulating experiment conducted in one healthcare setting. Research reveals that orange stimulates appetite, being a very good option to treat people with anorexia. However, orange also stimulates mental activity, and thus should be avoided in rooms of patients with intensive psychological conditions. Intensive care units should follow calming and restful colors, and using green and blue is recommended in the operation rooms to counteract the effect on the eye of prolonged views of the deep red open wound. Color contrast in the dementia patient room is more preferable, to help those patients who have cognitive and perception problems [143].

20.3.2.2   Physical Environment and Patient Satisfactions

Single-Bed Versus Multi-Bed Design: "Privacy"

Many studies showed that the most important factor associated with privacy is that healthcare staff frequently breach patient privacy by talking with patients in front of others [126]. Some studies of the relationship between a single-bed room and a multi-bed room and privacy found that patient wards have less security and less control of social encounters [217]. Hutton (2002) in his report on privacy needs of adolescents in hospitals found privacy correlated with a quiet space was important for adolescents [213]. In one survey of patient satisfaction data regarding hospitalization in a single room or with a roommate, patients in single rooms showed more satisfaction in terms of comfort, especially when having more visitors and their families within their rooms [205]. This idea was supported by three more studies emphasizing that roommates are linked with stress and noise and less privacy when receiving visitors [119, 152].

Single-bed rooms also support better communication between the staff and patients, which itself enhances better outcomes and increases patient satisfaction toward good communication with the staff. Satisfaction with privacy was 4.5% higher in single rooms than double ones [234] for reasons of avoiding any discussion or providing information about patient issues in the multi-bed rooms, to respect the patient's privacy in front of others [210]. Few studies have shown the importance on hospital design on saving patient confidentiality. Three studies conducted in the emergency department showed that curtain partitions afforded less visual and audio privacy than walls, which entails patient reluctance of providing full information of their medical history and refusal of full examination [202]. A study in one hospital found that 85% of nurses reported that single rooms are better for patient examination, and 82% confirmed that single rooms are better for collecting a patient history [19].

Layout Design of Pharmacy

Pharmacy layout in healthcare settings can strongly affect work flow and waiting time, which was supported by a study [211] that further equated pharmacy design with patient satisfaction. Two more studies compared decentralized versus centralized pharmacy systems and found that the centralized drug use system can increase time of drug delivery by more than 50% [109, 112].

Better Lighting

Lighting level was associated with less stress and more pain relief, as already mentioned, which eventually increases patient satisfaction.

Noise and Patient Satisfaction

The World Health Organization recommended a maximum noise level of 35 Db for inpatient rooms during daytime and 40 Db during nighttime [172]. However, one study showed that merely equipment and staff voices are exceeding 70–75 dB, as noisy as a restaurant [173]. Three studies showed that the source of noise in multi-bed room units is related to staff talking, visitors, equipment, and patient sounds

[61, 113, 194]. Single rooms reported a high satisfaction of patients across all patient categories and types of units, 11.2% higher. In judging these noise levels, it is worth noting that the decibel scale is logarithmic; each 10 dB increase represents approximately a doubling in the perceived sound level. A 60 dB sound, accordingly, is perceived as roughly four times as loud as a 40 dB sound [175].

Noise was categorized [276] at a hospital under two main categories: equipment and related sources. The research reviewed suggests that hospitals are excessively noisy for two general reasons and that environmental surfaces—floors, walls, ceilings—are not sound-absorbing or sound-reflecting surfaces, causing echoes. Sounds could be lowered by a sound-absorbing system, noiseless equipment, and single-bed room design [97, 276].

Social Supportive Design

Studies showed that design friendly to social support can improve patient clinical outcomes by enhancing interaction among others, for instance, in lounges, day rooms, and waiting rooms, as addressed in one study conducted in psychiatric wards. Proper displaying of movable seating for dining areas can support social interaction and improve eating among elders [131, 218]. Well-designed waiting rooms such as side-by-side chairs have shown effectiveness in enhancing social interaction as well [161, 219].

Wayfinding

Seventeen studies were reviewed highlighting the impacts of finding one's way within hospitals [117, 206, 220, 221, 224–232, 234–236, 240]. Clear and good cognitive mapping will be easier on visitors, the same as for a nurse who has worked within the hospital for 2 years [227]. Ulrich et al. (2004) in their report on the role of physical environment in the hospital of the twenty-first century found that using directional signs on major intersection, each 4.6–7.6 m or approximately every 150–250 feet apart, along with changing floor materials to recognize entering a new zone, will be a supportive measure for wayfinding [8].

The wayfinding problem can be stressful for outpatients and visitors. Hospitals should provide integrated supportive signs and directional methods. *"Signs and cues that lead to the hospital, especially the parking lot, need to be considered carefully, as they are the first point of contact of the patient with the hospital"* [235]. Informed wayfinding design can support patients and reduce asking questions of staff, which saves time and reduces stress on both patient and staff [117]. Using 'you are here' maps can decrease time spent to find ways and provide more accuracy [225]. Using signs and maps together can be faster than only using maps [221, 232], as confirmed by an experimental study that reported faster movements, less hesitancy, and lower levels of stress [235]. On the other hand, some studies such as the one conducted by John Hopkins Medicine and published in the *Journal of Hospital Medicine* in 2015 noted a small effect of hospital re-design on patient satisfaction. Poor patient satisfaction is not related to aging buildings, but is more related to healthcare providers, patient safety, patient-centered design, workflow, and the efficiency of healthcare. Another study conducted in one hospital in Sweden showed that patient satisfaction is a combined impact of process change such as a changing hospital system and protocols, followed with space design. Another study published on PubMed, related to changes in patient satisfaction by a hospital's new clinical building, showed that moving to a new building impacts only 25% of the overall patient satisfaction measures, compared with 75% of patient satisfaction that was correlated with clinical care.

20.4 Main Results and Discussion

20.4.1 Impacts of Patient Single-Bed Room Design on Healthcare Quality

- Lower nosocomial infection rates
- Fewer patient transfers and associated medical errors
- Far less noise
- Much better patient privacy and confidentiality
- Better communication from staff to patients and from patients to staff
- Superior accommodation of family
- Consistently higher satisfaction with overall quality of care.

More than 700 studies have linked one or more aspect of hospital physical environment-related designs to healthcare process components or resulting outcomes. This paper can generate the evidence of the impact, outline and classify the underlying factors, starting from the main Donabedian's classification of quality of care–structure, process, and outcomes–then specify the main subcomponents, to relate each of them to the factors affecting design. Physical environment design has a significant impact on process and outcomes as major components of the quality of healthcare. Hospital design can support better delivery of care and enhance better outcomes. Good design can reduce hospital-acquired infection rates by controlling airborne and contact surface routes of infections, which will support both patient and staff safety, and reduce the LOS, along with reducing nosocomial infections.

Single-bed room-based design has shown a significant impact on reducing nosocomial infections, "post hospitalization acquired infections," and patient transfer, which itself protects patients from exposure to infections. Moreover, a single-bed room can be less noisy, providing better privacy and supporting better communication with the staff and families, which entails better patient satisfaction.

Moreover, design can increase patient satisfaction through providing comfort, pleasant art, and colors that relieve the pain and stress of patients and improve clinical outcomes. Quiet hospitals through providing some support of interventional methods to control noise have reported a high rate of improvement of patient clinical outcome, with least stress on patients and staff, which decreases medical errors, improves patient satisfaction, and eventually improves quality of care.

20.4.2 Impact of the Hospital Physical Environment on Patient Safety

Design and Acquired Infections Hospital design has effects on the process of healthcare delivery by affecting staff and patient care. Infection rates were strongly correlated with multi-bed room, poor ventilation, poor design, room furniture, existence of automated sinks, difficult-to-clean surfaces, and tight corners.

> **Design-Related and Acquired Infections-Related Factors: Quality of Care**
> Multi-bed rooms /open wards
>
> - Indoor quality: "Air quality," Less ventilation
> - Carpeted floor
> - Fabric furniture
> - Bed curtains
> - No automated sink
> - Tight corners
> - Difficult-to-clean surfaces
> - Overbed table

Design and medical errors were associated with multi-bed rooms because of understaffing or increased numbers of patients. Changing room design can lower the transfer rates of the patients, which in itself increases the possibility of medical error because of delays and discontinuities among staff. Decreased privacy also affects the effective communication between staff and patients and may be a reason behind medical errors. Noise found in multi-bed rooms causes stress among staff and leads to medical errors. Low lighting and high noise increase the possibility of errors among staff. Standardized patient rooms showed a low rate of medical errors. Moreover, poor ventilation will affect the health of healthcare workers, which in turn affects the efficiency of healthcare. Distractions and interruptions, especially in the pharmacy setting, increase the possibility of errors in medical dispensing, and poor design will lead to long waiting times and low patient satisfaction.

Design- and Medical Error-Related Factors
- Single-room versus multi-room design
- Low lighting
- Noise
- Ventilation
- Decreased privacy
- Distraction and interruption

Design, Injuries, and Falls Patient falls have two main impacts: increasing injury rates among patients and affecting the cost of healthcare. Injuries among patients are the result of changing position. Most falls happen near beds and in bathrooms. Well-designed patient rooms can decrease the rate of falls, such as having decentralized nurses' stations and single-bed rooms instead of multi-bed rooms that support family presence around patients, which decrease fall incidence.

Design and Injuries: Falls
- Design layout
- Lighting
- Glare flooring
- Slippery flooring
- Incorrect placement of rails
- Toilet and furniture height
- Control system

20.4.3 Impact of the Physical Environment on Staff Safety

Design and Staff Infections Infections were attributed to direct or indirect contacts with patients, or through airborne routes. Ventilation supply and room temperature, and poor isolation of infected patients, are the main causes of staff infections.

Design and Staff Infections
- Low ventilation
- High temperature
- Poor isolation of patients

Design and Staff Injuries High-intensity surgical light can cause damage to the retina among the staff during surgical operations. Poor ergonomic design causes back pain and increases stress among nurses. Applying some measures such as adjustable bed height and supportive ergonomic design in nurses' stations, with supportive design of patient-transferring devices with hospital beds, and modifying toilet and shower room design, can enhance the quality of nurse spinal motion and lead to better musculoskeletal outcomes, lowering the future risks of having musculoskeletal injuries such as back pain. Using good ergonomic design can increase efficiency and effectiveness in care delivery by increasing staff safety and satisfaction, which in turn will reduce errors and support patient safety.

Design and Injuries
- High-intensity surgical light
- Bad ergonomic design of patients' beds
- Bad ergonomic design of patient-transferring devices
- Bad design of toilet and shower rooms

Design and Staff Stress Well-designed hospitals that take staff safety into consideration will improve the efficiency and effectiveness of healthcare and enhance better quality of care by reducing medical errors and increasing patient satisfaction. High noise levels are associated with stress among staff, affecting the process of care delivery. Understaffing was also a cause of stress affecting the quality of care and leads to more injuries and mortality. Better ergonomic interventions associated with air quality and less noise and adjustable light levels have a significant impact on reducing stress among staff. Radial units allowed fewer walking steps compared with rectangular units, which can be more effective in managing patient load. Decentralized nurse stations showed less walking time and more time spent in patient care than a centralized nursing station.

Design and Staff Stress
- Noise
- Understaffing
- Poor ergonomic design
- Unit layout (fewer radial units and more rectangular units)
- Long trips
- Centralized nursing stations

20.4.3.1   Impact of the Physical Environment on Clinical Outcomes

A well-designed hospital can have positive clinical outcomes and patient satisfaction, which outline better quality of care. Different factors are related to enhancing design such as noise absorption, single-bed room design, natural views, well-chosen colors and art, and access to daylight, along with better wayfinding design. These factors can reduce stress and pain, and enhance quality of sleep and better recovery, which in turn reduces the length of stay and increases patient satisfaction. Artwork with nature images has a positive psychological impact on patients compared with abstract ones. Window views enhance decreased stress, less pain, better recovery, and shorter postoperative stay, also allowing less need for pain relief drug intake. Viewing nature was also correlated with more positive clinical outcomes such as low blood pressure and low pulse rates and reduction of anxiety and pain. Moreover, access to daylight has proved its effectiveness for realizing better clinical outcomes. Bright sunlight has a positive impact as an antidepressant effect on winter depression. Sunny rooms lead to a shorter stay compared with a dimly lit room. High illuminance in the morning has a better impact on patients compared with afternoon light; morning light is two times more effective.

Design and Outcomes: Related Factors
- Less noise in single-bed rooms
- Window views of nature
- Gardens
- Self-control system in patient rooms
- Well-chosen colors
- Art with natural images: landscapes
- Daylight access
- Wayfinding

Acoustic comfort is strongly correlated with improving patient condition; less noise can be correlated with fewer medical errors in terms of the process and patient safety. Moreover, noise can affect patient recovery as it increases the levels of stress and leads to negative clinical outcomes and then less chance for recovery. Higher noise levels can increase the need for oxygen support therapy by decreasing oxygen saturation, and elevate blood pressure, besides increasing heartrate and breathing rates. Noise also affects quality and quantity of sleep. Most sources of noise are therapeutic procedures, staff talking, and staff's erratic interruptions, in addition to environmental noise, which usually happen in multi-bed room units.

Hospital gardens have great positive impacts on patients, staff, and visitors. As well, gardens can mitigate pain and reduce stress as increasing the satisfaction of patients, staff, and visitors and can contribute positively in treating some disorders such as dementia. Hospital gardens and views of nature can also enhance social support, better communication, and reduction of staff stress.

As for the effect of colors, balanced colors have a measurable role in promoting the well-being of staff and patients and affecting recovery rates and staff morale. Selection of colors should go beyond an atheistic attractive design; it should be more patient focused for better comfort and a supportive factor of the healing process. Using green has proved its effectiveness in stress relief compared with other colors such as orange. Orange stimulates appetite and it is a good option to treat people with anorexia. However, it should be avoided in the room of patients with intensive psychological conditions as it stimulates mental activity. Intensive care units should follow calming and restful colors, and using green and blue is recommended in the operating rooms to counteract the effect on the eye of the prolonged view of the deep red of the open wound. However, color contrasts are preferable in the dementia patient's room.

20.4.4   The Impact of Physical Environment on Patient Satisfaction

Many contributing factors affect patient satisfaction in terms of patient single-bed or multi-bed rooms. Patient wards have less security and less control of privacy. Patients are more satisfied with single-bed rooms, in terms of comfort and privacy. Single-bed rooms allow them to receive their visitors and families, whereas the presence of roommates is linked with stress, noise, and less privacy. Moreover, single-bed rooms can also support patient–staff communication for better outcomes and increase the patient's satisfaction toward communication with the staff. Staff can discuss patient issues openly with the patient without any privacy constraints such as occur when having a roommate. Patients showed 4.5% higher satisfaction with single rooms than double ones.

Design and Patient Satisfactions: Related Factors
- Single- versus multi-bed design
- Privacy-based design
- Noise reduction
- Better lighting
- Increased square footage of patient room
- Well-decorated rooms
- Safe design
- Process change
- New building
- Garden views
- Social supportive design
- Pharmacy layout design
- Private emergency design
- Room self-control system

At emergency rooms, use of curtain partitions reported less visual and audio privacy than use of walls, and that was a reason behind disclosing some information related to a patient's medical history, besides refusing full examination.

On the other hand, hospital pharmacy design and layout can strongly affect workflow within hospitals. Good design that supports better movement and accessibility of medical drugs will reduce errors and support easier flow. Moreover, a decentralized drug use system within pharmacies can decrease time of drug delivery by more than 50%. Access to daylight has proved its effectiveness for better clinical outcomes. Bright light has a positive impact on depression, sleep quality, circadian rhythm, agitation, and less intake of pain relief medication, and eventually decreasing the length of stay within hospitals. These associated health clinical outcomes have shown considerable impact on patient satisfaction as the second pillar of the outcomes of quality of care. Noise was seen as a major contributor to the stress and discomfort of patients, affecting their satisfaction toward healthcare services delivery within healthcare facilities. Using sound-absorbing systems, noiseless equipment, and a single-bed room design have proved better effectiveness to reduce noise within hospitals.

On the other hand, designs that consider social support can strongly contribute in improving patient clinical outcomes by enhancing interaction and communication with others, especially in lounges, day rooms, and waiting rooms. Social interactions within the hospital have proved to increase patient satisfaction toward healthcare facilities. Patient satisfaction was also correlated with wayfinding design within hospitals. Poor wayfinding design can be a stressful factor impacting both outpatients and visitors. Hospitals should consider providing an integrated wayfinding system by using directional signs on major intersection, each 4.6–7.6 m. Using signs and maps together can be faster than only using maps; changing floor materials can also support staff, inpatients, outpatients, and visitors to recognize entering a new zone, and work as a supportive measure of wayfinding. Moreover, informed wayfinding design can support patients and reduce asking staff to be more accurate and less stressful, and less time consuming for both patient and staff.

A new building impacts only 25% of the overall patient satisfaction measures, compared with 75% of patient satisfaction that is correlated with clinical care. Therefore, poor patient satisfaction is not merely related to aging buildings, but is more related to healthcare providers, patient safety, patient-centered design, workflow, and efficiency of healthcare. Patient satisfaction is a combined impact of process change such as changing hospital system and protocols, followed with space design.

20.5 Proposed Model: Conceptual Framework of the Impact of Hospital Design on Quality of Care

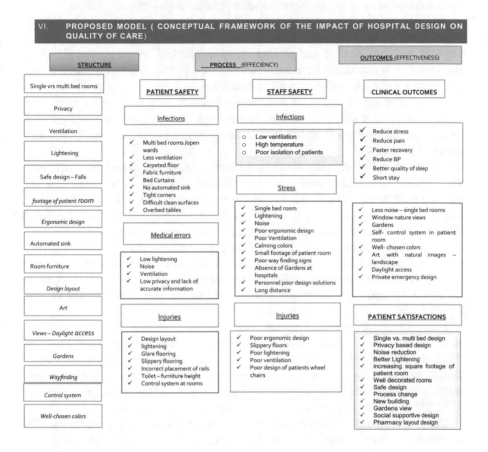

20.6 Conclusion

Hospitals are very complicated systems with many factors interacting and impacting each other. Therefore, isolating different factors and outlining them in a systematic way can support creating new guidance to support the quality of care within hospitals.

This chapter reviewed more than 700 studies that link one or more aspects of hospital physical environment-related design to healthcare process components or resulting outcomes.

This paper could generate evidence of the impact of the hospital physical environment on improving quality of care, to further suggest a new model that outlines and classifies all the underlying and impacting factors. The review adopted

Donabedian's classification of quality of care, structure, process, and outcomes, as a main starting point and the first step of this review. Then, the author specifies the main subcomponents, to relate each of them to the design-affecting factors.

As a conclusion, this paper highlights the importance of having a patient-/staff-friendly design to attain better efficiency and effectiveness within healthcare settings. In terms of process and efficiency, well-designed hospitals strongly affect patient safety by reducing infection rates and injury rates and promoting a stress-free environment. A well-designed hospital can in turn enhance staff safety by reducing infection courses and decreasing injuries, especially those happening because of poor ergonomic design. In turn, that provision directly affects staff morale and psychological status, entails better performance, and uplifts the efficiency of healthcare.

Patient-friendly design can also affect the main outcomes of healthcare delivery, forming the main pillar of healthcare quality, represented by clinical outcomes and patient satisfaction. Well-designed hospitals can strongly affect the healing process and enhance the fast recovery of the patient. Moreover, this will decrease the length of stay and be more cost-effective as another major targeted objective.

Healthcare outcomes can be also measured by the extent of patient satisfaction, which was correlated with different design-based factors, such as a privacy-based design that supports confidentiality and reduces noise and discomfort. Outcomes are also correlated with safe design and natural surroundings, which can be another supportive factor in the healing process.

Poor patient satisfaction is not related merely to aging buildings but is related more to healthcare providers, patient safety, patient-centered design, work flow, and efficiency of healthcare. Patient satisfaction is a combined impact of process change such as changing hospital system and protocols, followed by space design. A new building impacts only 25% of the overall patient satisfaction measures, compared with 75% of patient satisfaction being correlated with clinical care.

The proposed model of the impact of physical environment on improving quality of care needs further empirical investigation to try it in different contexts and to generate further rigorous evidence for it to be generalized and adopted in healthcare settings.

References

1. James BC (1989) Quality management for healthcare delivery. Research and Educational Trust, Chicago. ISBN 0-87258-537-9
2. Brook RH, McGlynn EA, Cleary PD (1996) Measuring quality of care. J Med 335:966–970. S 1996. https://doi.org/10.1056/NEJM199609263351311
3. Huiman ERCM, Morales E, van Hoof J, Kort HSM (2012) Healing environment: a review of the impact of physical environmental factors on users. Build Environ 58:70–80
4. Moss F (1995) Risk management and quality of care. Qual Health Care 4:102–107
5. Florence Nightingale's Rose Diagram (1858–January 1859)

6. Institute of Medicine (IOM) (2004) Keeping patients safe: transforming the work environment of nurses. National Academy Press, Washington, DC Institute of Medicine (IOM) (2001) Crossing the quality chasm: a new health system for the 21st century. National Academy Press, Washington, DC

7. Joint Commission on Accreditation of Healthcare Organizations (2002) Health care at the crossroad: strategies for addressing the evolving nursing crisis. Joint Commission on Accreditation of Healthcare Organizations, Oakbrook Terrace

8. Ulrich RS, Quan X, Zimring C, Joseph A, Choudhary R (2004) The role of the physical environment in the hospital of 21st century: a once-in-a-lifetime opportunity. Center for Health Design, Concord

9. Campbell SM, Roland MO, Buetow SA (2000) Defining quality of care. Soc Sci Med 51:1611–1625

10. WHO Preamble to the Constitution of the World Health Organization as adopted by the International Health Conference. New York, 19–22 June 1946/1948; signed on 22 July 1946 by the representatives of 61 states (Official Records of the World Health Organization, no. 2, p. 100) and entered into force on 7 April 1948

11. Alberti C, Bouakline A, Ribaud P, Lacroix C, Rousselot P, Leblanc T et al (2001) Relationship between environmental fungal contamination and the incidence of invasive aspergillosis in haematology patients. J Hosp Infect 48(3):198–206

12. Chang VT, Nelson K (2000) The role of physical proximity in nosocomial diarrhea. Clin Infect Dis 31(3):717–722

13. McManus AT, Mason AD Jr, McManus WF, Pruitt BA Jr (1992) Control of Pseudomonas aeruginosa infections in burned patients. Surg Res Commun 12(1):61–67

14. Skoutelis AT, Westenfelder GO, Beckerdite M, Phair JP (1994) Hospital carpeting and epidemiology of Clostridium difficile. Am J Infect Control 22(4):212–217

15. Archibald LK, Manning ML, Bell LM, Banerjee S, Jarvis WR (1997) Patient density, nurseto-patient ratio and nosocomial infection risk in a pediatric cardiac intensive care unit. Pediatr Infect Dis J 16(11):1045–1048

16. Ben-Abraham R, Keller N, Szold O, Vardi A, Weinberg M, Barzilay Z et al (2002) Do isolation rooms reduce the rate of nosocomial infections in the pediatric intensive care unit? J Crit Care 17(3):176–180

17. Birdsong C, Leibrock C (1990) Patient-centered design. Health Forum J 33(3):40–45. Joseph A (2006) The impact of the environment on infections in healthcare facilities. The Centre for Health Design, Concord

18. Boyce JM, Potter-Bynoe G, Chenevert C, King T (1997) Environmental contamination due to methicillin-resistant Staphylococcus aureus: possible infection control implications. Infect Control Hosp Epidemiol 18(9):622–627

19. Chaudhury H, Mahmood A, Valente M (2003) Pilot study on comparative assessment of patient care issues in single and multiple occupancy rooms (Unpublished report): the Coalition for Health Environments Research

20. Cohen L (2003) Computer keyboards and faucet handles as reservoirs of nosocomial pathogens in the intensive care unit. Am J Infect Control 28(6):465–471

21. Devine J, Cooke RP, Wright EP (2001) Is methicillin-resistant Staphylococcus aureus (MRSA) contamination of ward-based computer terminals a surrogate marker for nosocomial MRSA transmission and handwashing compliance? J Hosp Infect 48(1):72–75

22. Dharan S, Pittet D (2002) Environmental controls in operating theatres. J Hosp Infect 51(2):79–84. 30 30

23. Larson E, McGeer A, Quraishi ZA, Krenzischek D, Parsons BJ, Holford J et al (1991) Effect of an automated sink on handwashing practices and attitudes in high-risk units. Infect Control Hosp Epidemiol 12(7):422–428

24. Farquharson C, Baguley K (2003) Responding to the severe acute respiratory syndrome (SARS) outbreak: lessons learned in a Toronto emergency department. J Emerg Nurs 29(3):222–228

25. Arlet G, Gluckman E, Gerber F, Perol Y, Hirsch A (1989) Measurement of bacterial and fungal air counts in two bone marrow transplant units. J Hosp Infect 13(1):63–69
26. Aygun G, Demirkiran O, Utku T, Mete B, Urkmez S, Yilmaz M et al (2002) Environmental contamination during a carbapenemresistant *Acinetobacter baumannii* outbreak in an intensive care unit. J Hosp Infect 52(4):259–262
27. Noskin GA, Bednarz P, Suriano T, Reiner S, Peterson L (2000) Persistent contamination of fabric covered furniture by vancomycin-resistant enterocci: implication for upholstery selection in hospitals. Am J Infect Control 28(4):311–313
28. Gardner PS, Court SD, Brocklebank JT, Downham MA, Weightman D (1973) Virus crossinfection in paediatric wards. Br Med J 2(5866):571–575
29. Goldmann DA, Durbin WA Jr, and Freeman J (1981) Nosocomial infections in a neonatal intensive care unit. J Infect Dis 144(5):449–459 Gotlieb JB (2000) Understanding the effects of nurses, patients' hospital rooms, and patients' perception of control in the perceived quality of a hospital. Health Mark Q 18(1/2):1
30. Smedbold H, Ahlen C, Nilsen A, Norbäck D, Hilt B (2002) Relationships between indoor environments and nasal inflammation in nursing personnel. Arch Environ Health 57(2):155–161
31. Humphreys H, Johnson EM, Warnock DW, Willatts SM, Winter RJ, Speller DC (1991) An outbreak of aspergillosis in a general ITU. J Hosp Infect 18(3):167–177
32. Iwen PC, Davis JC, Reed EC, Winfield BA, Hinrichs SH (1994) Airborne fungal spore monitoring in a protective environment during hospital construction, and correlation with an outbreak of invasive aspergillosis. Infect Control Hosp Epidemiol 15(5):303–306
33. Jernigan JA, Titus MG, Groschel DH, Getchell-White S, Farr BM (1996) Effectiveness of contact isolation during a hospital outbreak of methicillin-resistant Staphylococcus aureus. Am J Epidemiol 143(5):496–504
34. Kromhout H, Hoek F, Uiterhoeve R, Huijbers R, Overmars RF, Anzion R et al (2000) Postulating a dermal pathway for exposure to anti-neoplastic drugs among hospital workers: applying a conceptual model to the results of three workplace surveys. Ann Occup Hyg 44(7):551–560
35. Kumari DN, Haji TC, Keer V, Hawkey PM, Duncanson V, Flower E (1998) Ventilation grilles as a potential source of methicillin-resistant Staphylococcus aureus causing an outbreak in an orthopaedic ward at a district general hospital. J Hosp Infect 39(2):127–133
36. Sehulster L, Chinn RY (2003) Guidelines for environmental infection control in healthcare facilities. Recommendations of CDC and the Healthcare Infection Control Practices Advisory Committee (HICPAC). MMWR Recommend Rep 52(RR-10):1–42
37. Loo VG, Bertrand C, Dixon C, Vitye D, DeSalis B, McLean AP et al (1996) Control of construction-associated nosocomial aspergillosis in an antiquated hematology unit. Infect Control Hosp Epidemiol 17(6):360–364
38. Lutz BDJ, Rinaldi J, Wickes MG, Huycke BL, Mark M (2003) Outbreak of invasive Aspergillus infection in surgical patients, associated with a contaminated air-handling system. Clin Infect Dis 37(6):786–793
39. Mahieu LM, De Dooy JJ, Van Laer FA, Jansens H, Ieven MM (2000) A prospective study on factors influencing aspergillus spore load in the air during renovation works in a neonatal intensive care unit. J Hosp Infect 45(3):191–197
40. McDonald LC, Walker M, Carson L, Arduino M, Aguero SM, Gomez P et al (1998) Outbreak of Acinetobacter spp. bloodstream infections in a nursery associated with contaminated aerosols and air conditioners. Pediatr Infect Dis J 17(8):716–722
41. McKendrick GD, Emond RT (1976) Investigation of cross-infection in isolation wards of different design. J Hyg (Lond) 76(1):23–31
42. McManus AT, Mason AD Jr, McManus WF, Pruitt BA Jr (1994) A decade of reduced gram-negative infections and mortality associated with improved isolation of burned patients. Arch Surg 129(12):1306–1309
43. McManus AT, McManus WF, Mason AD Jr, Aitcheson AR, Pruitt BA Jr (1985) Microbial colonization in a new intensive care burn unit. A prospective cohort study. Arch Surg 120(2):217–223. 33 33

44. Neely AN, Maley MP (2001) Dealing with contaminated computer keyboards and microbial survival. Am J Infect Control 29(2):131–132
45. Opal SM, Asp AA, Cannady PB Jr, Morse PL, Burton LJ, Hammer PG (1986) Efficacy of infection control measures during a nosocomial outbreak of disseminated aspergillosis associated with hospital construction. J Infect Dis 153(3):634–637
46. Oren I, Haddad N, Finkelstein R, Rowe JM (2001) Invasive pulmonary aspergillosis in neutropenic patients during hospital construction: before and after chemoprophylaxis and institution of HEPA filters. Am J Hematol 66(4):257–262
47. Panagopoulou P, Filioti J, Petrikkos G, Giakouppi P, Anatoliotaki M, Farmaki E et al (2002) Environmental surveillance of filamentous fungi in three tertiary care hospitals in Greece. J Hosp Infect 52(3):185–191
48. Palmer R (1999) Bacterial contamination of curtains in clinical areas. Nurs Stand 14(2):33–35. Parsons R, Hartig T (2000) Environmental psychophysiology. Cambridge University Press, New York
49. Passweg JR, Rowlings PA, Atkinson KA, Barrett AJ, Gale RP, Gratwohl A et al (1998) Influence of protective isolation on outcome of allogeneic bone marrow transplantation for leukemia. Bone Marrow Transplant 21(12):1231–1238
50. Pegues DA, Woernle CH (1993) An outbreak of acute nonbacterial gastroenteritis in a nursing home. Infect Control Hosp Epidemiol 14(2):87
51. Preston GA, Larson EL, Stamm WE (1981) The effect of private isolation rooms on patient care practices, colonization and infection in an intensive care unit. Am J Med 70(3):641–645
52. Palmer R (1999) Bacterial contamination of curtains in clinical areas. Nurs Stand 14(2):33–35
53. Anderson RL, Mackel DC, Stoler BS, Mallison GF (1982) Carpeting in hospitals: an epidemiological evaluation. J Clin Microbiol 15(3):408–415
54. Roberts SA, Findlay R, Lang SD (2001) Investigation of an outbreak of multi-drug resistant Acinetobacter baumannii in an intensive care burns unit. J Hosp Infect 48(3):228–232
55. Rountree PM, Beard MA, Loewenthal J, May J, Renwick SB (1967) Staphylococcal sepsis in a new surgical ward. Br Med J 1(533):132–137
56. Shepley MM, Wilson P (1999) Designing for persons with AIDS: a post-occupancy study at the Bailey-Boushay House. J Architect Plann Res 16(1):17–32
57. Sherertz RJ, Sullivan ML (1985) An outbreak of infections with Acinetobacter calcoaceticus in burn patients: contamination of patients' mattresses. J Infect Dis 151(2):252–258
58. Shirani KZ, McManus AT, Vaughan GM, McManus WF, Pruitt BA Jr, Mason AD Jr (1986) Effects of environment on infection in burn patients. Arch Surg 121(1):31–36
59. Thompson JT, Meredith JW, Molnar JA (2002) The effect of burn nursing units on burn wound infections. J Burn Care Rehabil 23(4):281–286
60. Williams HN, Singh R, Romberg E (2003) Surface contamination in the dental operatory: a comparison over two decades. J Am Dent Assoc 134(3):325–330
61. Yinnon AM, Ilan Y, Tadmor B, Altarescu G, Hershko C (1992) Quality of sleep in the medical department. Br J Clin Pract 46(2):88–91
62. Hamilton K (2003) The four levels of evidence based practice. Healthc Des 3:18–26. Environmental quality and healing environments: a study of flooring materials in a healthcare telemetry unit. Doctoral dissertation, Texas A&M University, College Station
63. Albert RK, Condie F (1981) Hand-washing patterns in medical intensive-care units. N Engl J Med 304(24):1465–1466
64. Diette GB, Lechtzin N, Haponik E, Devrotes A, and Rubin HR (2003) Distraction therapy with nature sights and sounds reduces pain during flexible bronchoscopya: a complementary approach to routine analgesia. Chest 123(3):941–948. Dorsey ST, Cydulka RK, Emerman CL (1996) Is handwashing teachable? Failure to improve handwashing behavior in an urban emergency department. Acad Emerg Med 3(4):360–365
65. Cohen B, Saiman L, Cimiotti J, Larson E (2003) Factors associated with hand hygiene practices in two neonatal intensive care units. Pediatr Infect Dis J 22(6):494–499

66. Conly JM, Hill S, Ross J, Lertzman J, Louie TJ (1989) Handwashing practices in an intensive care unit: the effects of an educational program and its relationship to infection rates. Am J Infect Control 17(6):330–339
67. Dubbert PM, Dolce J, Richter W, Miller M, Chapman SW (1990) Increasing ICU staff handwashing: effects of education and group feedback. Infect Control Hosp Epidemiol 11(4):191–193
68. Graham M (1990) Frequency and duration of handwashing in an intensive care unit. Am J Infect Control 18(2):77 81
69. Larson E (1988) A causal link between handwashing and risk of infection? Examination of the evidence. Infect Control 9(1):28–36
70. Larson EL, Bryan JL, Adler LM, Blane C (1997) A multifaceted approach to changing hand-washing behavior. Am J Infect Control 25(1):3–10
71. Muto CA, Sistrom MG, Farr BM (2000) Hand hygiene rates unaffected by installation of dispensers of a rapidly acting hand antiseptic. Am J Infect Control 28(3):273–276
72. Pittet D, Hugonnet S, Harbarth S, Mourouga P, Sauvan V, Touveneau S et al (2000) Effectiveness of a hospital-wide programme to improve compliance with hand hygiene. Lancet 356(9238):1307–1312
73. Vernon MO, Trick WE, Welbel SF, Peterson BJ, Weinstein RA (2003) Adherence with hand hygiene: does number of sinks matter? Infect Control Hosp Epidemiol 24(3):224–225
74. Friberg S, Ardnor B, Lundholm R (2003) The addition of a mobile ultra-clean exponential laminar airflow screen to conventional operating room ventilation reduces bacterial contamination to operating box levels. J Hosp Infect 55(2):92–97
75. Gabor JY, Cooper AB, Crombach SA, Lee B, Kadikar N, Bettger HE et al (2003) Contribution of the intensive care unit environment to sleep disruption in mechanically ventilated patients and healthy subjects. Am J Respir Crit Care Med 167(5):708–715
76. Cornet M, Levy V, Fleury L, Lortholary J, Barquins S, Coureul MH et al (1999) Efficacy of prevention by high-efficiency particulate air filtration or laminar airflow against Aspergillus airborne contamination during hospital renovation. Infect Control Hosp Epidemiol 20(7):508–513
77. Hahn T, Cummings KM, Michalek AM, Lipman BJ, Segal BH, McCarthy PL Jr (2002) Efficacy of high-efficiency particulate air filtration in preventing aspergillosis in immunocompromised patients with hematologic malignancies. Infect Control Hosp Epidemiol 23:525–531
78. Bauer TM, Ofner E, Just HM, Just H, Daschner FD (1990) An epidemiological study assessing the relative importance of airborne and direct contact transmission of microorganisms in a medical intensive care unit. J Hosp Infect 15(4):301–309
79. Jiang S, Huang L, Chen X, Wang J, Wu W, Yin S et al (2003) Ventilation of wards and nosocomial outbreak of severe acute respiratory syndrome among healthcare workers. Chin Med J 116(9):1293–1297
80. Menzies D, Fanning A, Yuan L, FitzGerald JM (2000) Hospital ventilation and risk for tuberculosis infection in Canadian health care workers. Ann Intern Med 133(10):779–789
81. Meyer TJ, Eveloff SE, Bauer MS, Schwartz WA, Hill NS, Millman RP (1994) Adverse environmental conditions in the respiratory and medical ICU settings. Chest 105(4):1211–1216
82. Mulin B, Rouget C, Clement C, Bailly P, Julliot MC, Viel JF et al (1997) Association of private isolation rooms with ventilator-associated Acinetobacter baumanii pneumonia in a surgical intensive-care unit. Infect Control Hosp Epidemiol 18(7):499–503
83. Moore MMN, Nguyen SP, Robinson D, Ryals SP, Imbrie B, Spotnitz JZ, William (1998) Interventions to reduce decibel levels on patient care units. Am Surg 64(9):894
84. Sherertz RJ, Belani A, Kramer BS, Elfenbein GJ, Weiner RS, Sullivan ML et al (1987) Impact of air filtration on nosocomial Aspergillus infections: unique risk of bone marrow transplant recipients. Am J Med 83(4):709–718
85. Slevin M, Farrington N, Duffy G, Daly L, Murphy JF (2000) Altering the NICU and measuring infants' responses. Acta Paediatr 89(5):577–581

86. Alcee DA (2000) The experience of a community hospital in quantifying and reducing patient falls. J Nurs Care Qual 14(3):43–45

87. Maies S, Spielberger CD, Defares PB, Sarason IG (eds) Topics in health psychology. John Wiley, New York, pp 193–203. van Leeuwen M, Bennett L, West S, Wiles V, and Grasso J (2001) Patient falls from bed and the role of bedrails in the acute care setting. Aust J Adv Nurs 19(2):8–13

88. Wong S, Glennie K, Muise M, Lambie E, Meagher D (1981) An exploration of environmental variables and patient falls. Dimens Health Serv 58(6):9–11

89. Brandis S (1999) A collaborative occupational therapy and nursing approach to falls prevention in hospital inpatients. J Qual Clin Pract 19(4):215–221

90. Uden G (1985) Inpatient accidents in hospitals. J Am Geriatr Soc 33(12):833–841

91. Ulrich RS (1984) View through a window may influence recovery from surgery. Science 224(4647):420–421

92. Morgan VR, Mathison JH, Rice JC, Clemmer DI (1985) Hospital falls: a persistent problem. Am J Public Health 75(7):775–777

93. Van Leeuwen M, Bennett L, West S, Wiles V, Grasso J (2001) Patient falls from bed and the role of bedrails in the acute care setting. Aust J Adv Nurs 19(2):8–13

94. Vassallo M, Azeem T, Pirwani MF, Sharma JC, Allen SC (2000) An epidemiological study of falls on integrated general medical wards. Int J Clin Pract 54(10):654–657

95. Buchanan TL, Barker KN, Gibson JT, Jiang BC, Pearson RE (1991) Illumination and errors in dispensing. Am J Hosp Pharm 48(10):2137–2145. Bures S, Fishbain JT, Uyehara CF, Parker JM, Berg BW (2000)

96. Flynn EA, Barker KN, Gibson JT, Pearson RE, Berger BA, Smith LA (1999) Impact of interruptions and distractions on dispensing errors in an ambulatory care pharmacy. Am J Health Sys Pharm 56(13):1319–1325

97. Booker JM, Roseman C (1995) A seasonal pattern of hospital medication errors in Alaska. Psychiatry Res 57(3):251–257

98. Kistner UA, Keith MR, Sergeant KA, Hokanson JA (1994) Accuracy of dispensing in a high volume, hospital-based outpatient pharmacy. Am J Hosp Pharm 51(22):2793–2797

99. Buchanan TL, Barker KN, Gibson JT, Jiang BC, Pearson RE (1991) Illumination and errors in dispensing. Am J Hosp Pharm 48(10):2137–2145

100. Garg A, Owen B (1992) Reducing back stress to nursing personnel: an ergonomic intervention in a nursing home. Ergonomics 35(11):1353–1375

101. Caboor DE, Verlinden MO, Zinzen E, Van Roy P, Van Riel MP, Clarys JP (2000) Implication of an adjustable bed height during standard nursing tasks on spinal motion, perceived exertation and muscular. Ergonomics 43(10):1771–1780

102. Fox RA, Henson PW (1996) Potential ocular hazard from a surgical light source. Australas Phys Eng Sci Med 19(1):12–16

103. Garg A, Owen B (1992) Reducing back stress to nursing personnel: an ergonomic intervention in a nursing home. Ergonomics 35(11):1353–1375. Gast PL, Baker CF (1989) The CCU patient: anxiety and annoyance to noise. Crit Care Nurs Q 12(3):39–54

104. Petzall K, Petall J (2003) Transportation with hospital beds. Appl Ergon 34(4):383–392

105. Daraiseh N, Genaidy AM, Karwowski W, Davis LS, Stambough J, Huston RL (2003) Musculoskeletal outcomes in multiple body regions and work effects among nurses: the effects of stressful and stimulating working conditions. Ergonomics 46(12):1178–1199

106. Alexandre NM, de Moraes MA, Correa Filho HR, Jorge SA (2001) Evaluation of a program to reduce back pain in nursing personnel. Rev Saude Publica 35(4):356–361

107. [85]

108. Burgio L, Engel B, Hawkins A, McCorick K, Scheve A (1990) A descriptive analysis of nursing staff behaviors in a teaching nursing home: differences among NAs, LPNs and RNs. The Gerontologist 30:107–112

109. Hibbard FJ, Bosso JA, Sward LW, Baum S (1981) Delivery time in a decentralized pharmacy system without satellites. Am J Hosp Pharm 38(5):690–692

110. Nelson-Shulman Y (1983) Information and environmental stress: report of a hospital intervention. J Environ Syst 13(4):303–316
111. Norbeck JS (1985) Perceived job stress, job satisfaction, and psychological symptoms in critical care nursing. Res Nurs Health 8(3):253–259
112. Reynolds DM, Johnson MH, Longe RL (1978) Medication delivery time requirements in centralized and decentralized unit dose drug distribution systems. Am J Hosp Pharm 35(8):941–943
113. Baker CF (1984) Sensory overload and noise in the ICU: sources of environmental stress. Crit Care Q 6(4):66–80
114. Shepley MM (1995) The location of behavioral incidents in a children's psychiatric facility. Child Environ 12(3):352–361. Shepley MM (2002) Predesign and postoccupancy analysis of staff behavior in a neonatal intensive care unit. Child Health Care 31(3):237–253
115. Shepley MM (2002) Predesign and postoccupancy analysis of staff behavior in a neonatal intensive care unit. Child Health Care 31(3):237–253
116. Sturdavant M (1960) Intensive nursing service in circular and rectangular units. Hosp JAHA 34(14):46–48. 71–78
117. Nelson–Shulman Y (1983–1984) Information and environmental stress: Report of a hospital intervention. J Environ Syst 13(4):303–316. Trites DK, Galbraith FD, Sturdavant M, and Leckwart JF (1970) Influence of nursing-unit design on the activities and subjective feelings of nursing personnel. Environ Behav 2(3):303–334
118. Trites DK, Galbraith FD, Sturdavant M, Leckwart JF (1970) Influence of nursing-unit design on the activities and subjective feelings of nursing personnel. Environ Behav 2(3):303–334
119. Volicer BJ, Isenberg MA, Burns MW (1977) Medical-surgical differences in hospital stress factors. J Hum Stress 3(2):3–13
120. Walch JM, Rabin BS, Day R, Williams JN, Choi K, Kang JD (2004) The effect of sunlight on post-operative analgesic medication usage: a prospective study of spinal surgery patients. Psychosom Med 67:156–163
121. Devlin AS, Arneil AB (2003) Health care environments and patient outcomes. A review of literature. Environ Behav 35(5):665–694
122. Beauchemin KM, Hays P (1996) Sunny hospital rooms expedite recovery from severe and refractory depressions. J Affect Disord 40(1–2):49–51
123. Burton E, Torrington J (2007) Designing environments suitable for older people. CME Geriatr Med 9(2):39–45
124. Weisman GD, Cohen U, Ray K, Day K (1991) Architectural planning and design for dementia care. In: Coons DH (ed) Specialized dementia care units. John Hopkins University Press, Baltimore, pp 83–106
125. Cacioppo JT, Tassinary LG (eds) (2000) Handbook of psychophysiology, 2nd edn. Cambridge University Press, New York, pp 815–846
126. Ulrich RS (1991) Effects of interior design on wellness: theory and recent scientific research. J Health Care Inter Des 3:97–109
127. Tanja-Dijkstra K, Pieterse ME (2011) The psychological effects of the physical healthcare environment on healthcare personnel. Cochrane Database Syst Rev 1:CD006210
128. Miller AC, Hickman LC, Lemasters GK (1992) A distraction technique for control of burn pain. J Burn Care Rehabil 13(5):576–580
129. Novaes MA, Aronovich A, Ferraz MB, Knobel E (1997) Stressors in ICU: patients' evaluation. Intensive Care Med 23(12):1282–1285
130. Peterson R, Knapp T, Rosen J, Pither BF (1977) The effects of furniture arrangement on the behavior of geriatric patients. Behav Ther 8:464–467
131. Codinhoto R, Tzortzopoulos P, Kagioglou M, Aouad G (2009) The impacts of the built environment on health outcomes. Facilities 27(3–4):138–151
132. Ulrich RS, Zimring C, Barch XZ, Dubose J, Seo HB, Choi YS et al (2008) A review of the research literature on evidence-based healthcare design. HERD 1(3):61–125
133. Ulrich RS (1992) How design impacts wellness. Health Forum J 35(5):20–25

134. Rountree PM, Beard MA, Loewenthal J, May J, and Renwick SB (1967) Staphylococcal sepsis in a new surgical ward. Br Med J 1(533):132–137. Rubin H, Owens AJ, and Golden G (1998) Status report (1998): An investigation to determine whether the built environment affects patients' medical outcomes. Center for Health Design, Martinez

135. Rubin H, Owens AJ, Golden G (1998) Status report (1998): an investigation to determine whether the built environment affects patients' medical outcomes. Center for Health Design, Martinez

136. Swan JE, Richardson LD, Hutton JD (2003) Do appealing hospital rooms increase patient evaluations of physicians, nurses, and hospital services? Health Care Manag Rev 28(3):254

137. Ubel PA, Zell MM, Miller DJ (1995) Elevator talk: observational study of inappropriate comments in a public space. Am J Med 99(2):190–194

138. Jonas WB, Chez RA (2004) Toward optimal healing environments in health care. J Altern Complement Med 10(Suppl. 1):S1–S6

139. Kecskes I, Rihmer Z, Kiss K, Vargha A, Szili I, Rihmer A (2003) Possible effect of gender and season on the length of hospitalisation in unipolar major depressives. J Affect Disord 73(3):279–282

140. Kinnunen T, Saynajakangs O, Tuuponen T, Keistinen T (2002) Regional and seasonal variation in the length of hospital stay for chronic obstructive pulmonary disease in Finland. Int J Circumpolar Health 61(2):131–135. 32 32

141. Ulrich RS, Gilpin L (2003) Healing arts: nutrition for the soul. In: Frampton SB, Gilpin L, Charmel P (eds) Putting patients first: designing and practicing patient-centered care. Jossey-Bass, San Francisco, pp 117–146. 36 36

142. Babin SE (2013) Color theory: the effects of color in meidcal environemt. Honors college the Aquila Digital Community

143. Wallenius MA (2004) The interaction of noise stress and personal project stress on subjective health. J Environ Psychol (2):24, 167–177. Health Design, Concord CA (2006) Issue paper 1

144. Dulux Trade (2011) Transforming the healing environment Choosing colours and products that make a difference for patients. Trade Contract Partnership

145. Diette GB, Lechtzin N, Haponik E, Devrotes A, Rubin HR (2003) Distraction therapy with nature sights and sounds reduces pain during flexible bronchoscopy: a complementary approach to routine analgesia. Chest 123(3):941–948

146. Maies S, Spielberger CD, Defares PB, Sarason IG (eds) (1988) Topics in Health Psychology. Wiley, New York, pp 193–203

147. John Wiley & Sons Inc, Schneider SM, Prince-Paul M, Allen MJ, Silverman P, Talaba D (2004) Virtual reality as a distraction intervention for women receiving chemotherapy. Oncol Nurs Forum 31(1):81–88

148. Tse MMY, Ng JKF, Chung JWY, Wong TKS (2002) The effect of visual stimuli on pain threshold and tolerance. J Clin Nurs 11(4):462–469

149. Ulrich RS (1999) Effects of gardens on health outcomes: theory and research. Heal Gardens Ther Benefits Des Recommend 27:86

150. Ulrich RS, Simons RF, Miles MA (2003) Effects of environmental simulations and television on blood donor stress. J Architect Plann Res 20(1):38–47

151. Van der Ploeg HM (1988) Stressful medical events: a survey of patients' perceptions. In: Maes S et al (eds) Topics in Health Psychology. John Wiley, New York

152. Verderber S (1986) Dimensions of person-window transactions in the hospital environment. Environ Behav 18(4):450–466

153. Whall AL, Black ME, Groh CJ, Yankou DJ, Kupferschmid BJ, Foster NL (1997) The effect of natural environments upon agitation and aggression in late stage dementia patients. Am J Alzheimers Dis Other Demen 12:216–220

154. Lewy AJ, Bauer VK, Cutler NL, Sack RL, Ahmed S, Thomas KH et al (1998) Morning vs. evening light treatment of patients with winter depression. Arch Gen Psychiatry 55(10):890–896

155. Terman JS, Terman M, Lo ES, Cooper TB (2001) Circadian time of morning light administration and therapeutic response in winter depression. Arch Gen Psychiatry 58(1):69–75

156. Benedetti F, Colombo C, Barbini B, Campori E, Smeraldi E (2001) Morning sunlight reduces length of hospitalization in bipolar depression. J Affect Disord 62(3):221–223
157. Eastman CI, Young MA, Fogg LF, Liu L, Meaden PM (1998) Bright light treatment of winter despression. Arch Gen Psychiatry 55(10):883–889
158. Federman EJ, Drebing CE, Boisvert C, Penk W, Binus G, Rosenheck R (2000) Relationship between climate and psychiatric inpatient length of stay in Veterans Health Administration hospitals. Am J Psychiatr 157(10):1669
159. Hebert M, Dumont M, Paquet J (1998) Seasonal and diurnal patterns of human illumination under natural conditions. Chronobiol Int 15(1):59–70
160. Holahan CJ (1972) Seating patterns and patient behavior in an experimental dayroom. J Abnorm Psychol 80(2):115–124
161. Choi J-H, Beltran LO, Kim H-S (2012) Impacts of indoor daylight environments on patient average length of stay (ALOS) in a healthcare facility. Build Environ 50:65–75
162. Beauchemin KM, Hays P (1998) Dying in the dark: sunshine, gender and outcomes in myocardial infarction. J R Soc Med 91(7):352–354
163. Lovell BB, Ancoli-Israel S, Gevirtz R (1995) Effect of bright light treatment on agitated behavior in institutionalized elderly subjects. Psychiatry Res 57(1):7–12
164. Van Someren EJW, Kessler A, Mirmiran M, Swaab DF (1997) Indirect bright light improves circadian rest-activity rhythm disturbances in demented patients. Biol Psychiatry 41(9):955–963
165. Wallace-Guy G, Kripke D, Jean-Louis G, Langer R, Elliott J, Tuunainen A (2002) Evening light exposure: implications for sleep and depression. J Am Geriatr Soc 50(4):738–739
166. Love H (2003) Noise exposure in the orthopaedic operating theatre: a significant health hazard. ANZ J Surg 73(10):836–838
167. Allaouchiche B, Duflo F, Debon R, Bergeret A, Chassard D (2002) Noise in the postanaesthesia care unit. Br J Anaesth 88(3):369–373
168. Babwin D (2002) Building boom. Hosp Health Netw 76(3):48–54 Baker CF (1992) Discomfort to environmental noise: heart rate responses of SICU patients. Crit Care Nurs Q 15(2):75–90
169. Bayo MV, Garcia AM, Garcia A (1995) Noise levels in an urban hospital and workers' subjective responses. Arch Environ Health 50(3):247–251
170. Berg S (2001) Impact of reduced reverberation time on sound-induced arousals during sleep. Sleep 24(3):289–292
171. Berglund B, Lindvall T, Schwela DH (1999) Guidelines for community noise: protection of the human environment. World Health Organization, Geneva
172. Blomkvist V, Eriksen CA, Theorell T, Ulrich RS, Rasmanis G (2004) Acoustics and psychosocial environment in coronary intensive care. Occup Environ Med 62(3):e1
173. Cmiel CA, Karr DM, Gasser DM, Oliphant LM, Neveau AJ (2004) Noise control: a nursing team's approach to sleep promotion. Am J Nurs 104(2):40–48
174. Couper RT, Hendy K, Lloyd N, Gray N, Williams S, Bates DJ (1994) Traffic and noise in children's wards. Med J Aust 160(6):338–341
175. Zahr LK, de Traversay J (1995) Premature infant responses to noise reduction by earmuffs: effects on behavioral and physiologic measures. J Perinatol 15(6):448–455
176. Falk SA, Woods NF (1973) Hospital noise: levels and potential health hazards. N Engl J Med 289(15):774–781
177. Hagerman I, Theorell T, Ulrich RS, Blomkvist V, Eriksen CA, Rasmanis G (2004) Influence of coronary intensive care acoustics on the physiological states and quality of care of patients. Int J Cardiol 98(2):267–270
178. Hilton BA (1985) Noise in acute patient care areas. Res Nurs Health 8(3):283–291. Hodge B and Thompson JF (1990) Noise pollution in the operating theatre. Lancet 335(8694): 891–894
179. Johnson AN (2001) Neonatal response to control of noise inside the incubator. Pediatr Nurs 27(6):600–605

180. Wallenius MA (2004) The interaction of noise stress and personal project stress on subjective health. J Environ Psychol 24(2):167–177
181. McLaughlin A, McLaughlin B, Elliott J, Campalani G (1996) Noise levels in a cardiac surgical intensive care unit: a preliminary study conducted in secret. Intens Crit Care Nurs 12(4):226–230
182. Morrison WE, Haas EC, Shaffner DH, Garrett ES, Fackler JC (2003) Noise, stress, and annoyance in a pediatric intensive care unit. Crit Care Med 31(1):113–119
183. Nott MR, West PD (2003) Orthopaedic theatre noise: a potential hazard to patients. Anaesthesia 58(8):784–787
184. Robertson A, Cooper-Peel C, Vos P (1998) Peak noise distribution in the neonatal intensive care nursery. J Perinatol 18(5):361–364
185. Ron JN, Carlisle CC, Carskadon MA, Meyer TJ, Hill NS, Millman RP (1996) Environmental noise as a cause of sleep disruption in an intermediate respiratory care unit. Sleep 19(9):707–710
186. Shepley MM, Davies K (2003) Nursing unit configuration and its relationship to noise and nurse walking behavior: an AIDS/HIV unit case study. Retrieved 26 Mar 2004, from http://www.aia.org/aah/journal/0401/article4.asp 35 35
187. Topf M (1985) Noise-induced stress in hospital patients: coping and nonauditory health outcomes. J Hum Stress 11(3):125–134
188. Topf M and Davis JE (1993) Critical care unit noise and rapid eye movement (REM) sleep. Heart Lung 22(3):252–258. Topf M and Dillon E (1988) Noise-induced stress as a predictor of burnout in critical care nurses. Heart Lung 17(5):567–574
189. Topf M, Dillon E (1988) Noise-induced stress as a predictor of burnout in critical care nurses. Heart Lung 17(5):567–574
190. Topf M, Thompson S (2001) Interactive relationships between hospital patients' noise-induced stress and other stress with sleep. Heart Lung 30(4):237–243
191. Walder B, Francioli D, Meyer JJ, Lancon M, Romand JA (2000) Effects of guidelines implementation in a surgical intensive care unit to control nighttime light and noise levels. Crit Care Med 28(7):2242–2247
192. Parthasarathy S, Tobin MJ (2004) Sleep in the intensive care unit. Intensive Care Med 30(2):197–206
193. Schnelle JF, Ouslander JG, Simmons SF, Alessi CA, Gravel MD (1993) The nighttime environment, incontinence care, and sleep disruption in nursing homes. J Am Geriatr Soc 41(9):910–914
194. Southwell MT, Wistow G (1995) Sleep in hospitals at night: are patients' needs being met? J Adv Nurs 21(6):1101–1109
195. Hendrich A, Fay J, Sorrells A (2002) Courage to heal: comprehensive cardiac critical care. Healthc Des:11–13
196. Cooper Marcus C, Barnes M (1995) Gardens in healthcare facilities: uses, therapeutic benefits, and design recommendations. Center for Health Design, Martinez
197. Heath Y, Gifford R (2001) Post-occupancy evaluation of therapeutic gardens in a multi-level care facility for the aged. Act Adapt Aging 25(2):21–43
198. Hendrich A, Fay J, Sorrells A (2004) Effects of acuity-adaptable rooms on flow of patients and delivery of care. Am J Crit Care 13(1):35–45
199. Whitehouse S, Varni JW, Seid M, Cooper-Marcus C, Ensberg MJ, Jacobs JR et al (2001) Evaluating a children's hospital garden environment: utilization and consumer satisfaction. J Environ Psychol 21(3):301–314
200. Mlinek J, Pierce J (1997) Confidentiality and privacy breaches in a university hospital emergency department. Acad Emerg Med 4(12):1142–1146
201. Capezuti E, Maislin G, Strumpf N, Evans LK (2002) Side rail use and bed-related fall outcomes among nursing home residents. J Am Geriatr Soc 50(1):90–96
202. Chang JT, Morton SC, Rubenstein LZ, Mojica WA, Maglione M, Suttorp MJ et al (2004) Interventions for the prevention of falls in older adults: systematic review and meta-analysis of randomised clinical trials. Br Med J 328(7441):680–680

203. Ganey, Inc (2003) National patient satisfaction data for 2003. (Provided by Press Ganey, Inc. at the request of the authors for this research review)
204. Hanger HC, Ball MC, Wood LA (1999) An analysis of falls in the hospital: can we do without bedrails? J Am Geriatr Soc 47(5):529–531. 31 31
205. Haq S, Zimring C (2003) Just down the road a piece: the development of topological knowledge of building layouts. Environ Behav 35(1):132–160. Harris, D. (2000)
206. Harris PB, McBride G, Ross C, Curtis L (2002) A place to heal: environmental sources of satisfaction among hospital patients. J Appl Soc Psychol 32(6):1276–1299
207. Hendrich A (2003) Optimizing physical space for improved outcomes: satisfaction and the bottom line. Paper presented at "Optimizing the Physical Space for Improved Outcomes, Satisfaction, and the Bottom Line," minicourse sponsored by the Institute for Healthcare Improvement and the Center for Health Design, Atlanta
208. Johns Hopkins Medicin (2015) Hospital design has little effect on patient satisfaction, study finds
209. Kaldenburg DO (1999) The influence of having a roommate on patient satisfaction. Satisfaction Monitor, January–February (www.pressganey.org). Kaplan LM, McGuckin M (1986) Increasing handwashing compliance with more accessible sinks. Infect Control 7(8):408–410
210. Pierce RA 2nd, Rogers EM, Sharp MH, Musulin M (1990) Outpatient pharmacy redesign to improve work flow, waiting time, and patient satisfaction. Am J Hosp Pharm 47(2):351–356
211. Press Ganey, Inc (2003) National patient satisfaction data for 2003. (Provided by Press Ganey, Inc. at the request of the authors for this research review)
212. Zishan K, Siddiqui MD et al (2015) Changes in patient satisfaction related to hospital renovation: experience with a new clinical building. J Hosp Med 10(3):165–171
213. Hutton A (2002) Privacy needs of adolescent in hospitals. J Pediatr Nurs 17(1):67–72
214. Barlas D, Sama AE, Ward MF, Lesser ML (2001) Comparison of the auditory and visual privacy of emergency department treatment areas with curtains versus those with solid walls. Ann Emerg Med 38(2):135–139
215. Prochansky HM, Ittelson WH, Rivlin LG (1970) Freedom of choice and behavior in a physical setting. In: Prochansky HM, Ittelson WH, Rivlin LG (eds) Environmental psychology: man and his psychical setting. Rinehart, Winston, Holt/New York, pp 173–183
216. Firestone IJ, Lichtman CM, Evans JR (1980) Privacy and solidarity: effects of nursing home accommodation on environmental perception and sociability preferences. Int J Aging Hum Dev 11(3):229–241
217. Melin L, Gotestam KG (1981) The effects of rearranging ward routines on communication and eating behaviors of psychogeriatric patients. J Appl Behav Anal 14(1):47–51
218. Sommer R, Ross H (1958) Social interaction on a geriatrics ward. Int J Soc Psychiatry 4(2):128–133
219. Carpman J, Grant M, Simmons D (1984) No more mazes: research about design for wayfinding in hospitals. The University of Michigan Hospitals, Ann Arbor
220. Brown B, Wright H, Brown C (1997) A post-occupancy evaluation of wayfinding in a pediatric hospital: research findings and implications for instruction. J Architect Plann Res 14(1):35–51
221. Butler D, Acquino AL, Hissong AA, Scott PA (1993) Wayfinding by newcomers in a complex building. Hum Factors 25(1):159–173
222. Carpman J, Grant MA, Simmons DA (1985) Hospital design and wayfinding: a video simulation study. Environ Behav 17(3):296–314
223. Carpman J, Grant M, Simmons D (1983-84) Wayfinding in the hospital environment: the impact of various floor numbering alternatives. J Environ Syst 13(4):353–364
224. Grover P (1971) Wayfinding in hospital environments: UCLA hospital disorientation pilot case study. Graduate School of Architecture and Urban Planning, University of California, Los Angeles, Los Angeles
225. Levine M, Marchon I, Hanley G (1984) The placement and misplacement of you-are-here maps. Environ Behav 16(2):139–157

226. Moeser SD (1988) Cognitive mapping in a complex building. Environ Behav 20(1):21–49
227. Ortega-Andeane P, Urbina-Soria J (1988) A case study of wayfinding and security in a Mexico City hospital. Paper Pres EDRA Environ Des Res Assoc US 19:231–236
228. Passini R, Rainville C, Marchand N, Joanette Y (1995) Wayfinding in dementia of the Alzheimer type: planning abilities. J Clin Exp Neuropsychol 17(6):820–832. 34 34
229. Peponis J, Zimring C, Choi YK (1990) Finding the building in wayfinding. Environ Behav 22(5): 555–590. Peterson R, Knapp T, Rosen J, Pither BF (1977) The effects of furniture arrangement on the behavior of geriatric patients. Behav Ther 8:464–467
230. Weisman J (1981) Evaluating architectural legibility: Wayfinding in the built environment. Environ Behav 13(2):189–204
231. Wright P, Hull AJ, Lickorish A (1993) Navigating in a hospital outpatients' department: the merits of maps and wallsigns. J Architect Plann Res 10(1):76–89
232. Zimring C (1990) The costs of confusion: non-monetary and monetary costs of the Emory University hospital wayfinding system. Georgia Institute of Technology, Atlanta
233. Zimring C, Templer J (1983-84) Wayfinding and orientation by the visually impaired. J Environ Syst 13(4):333–352
234. Carpman J, Grant M (1993) Design that cares: planning health facilities for patients and visitors, 2nd edn. American Hospital Publishing, Chicago
235. Christensen KE (1979) An impact analysis framework for calculating the costs of staff disorientation in hospitals. School of Architecture and Urban Planning, University of California, Los Angeles
236. Leather P, Beale D, Santos A, Watts J, Lee L (2003) Outcomes of environmental appraisal of different hospital waiting areas. Environ Behav 35(6):842–869
237. Sallstrom C, Sandman PO, Norberg A (1987) Relatives' experience of the terminal care of longterm geriatric patients in open-plan rooms. Scand J Caring Sci 1(4):133–140
238. Sanderson PJ, Weissler S (1992) Recovery of coliforms from the hands of nurses and patients: activities leading to contamination. J Hosp Infect 21(2):85–93
239. Schneider LF, Taylor HA (1999) How do you get there from here? Mental representations of route descriptions. Appl Cogn Psychol 13:415–441
240. Shepley MM (1995) The location of behavioral incidents in a children's psychiatric facility. Child Environ 12(3):352–361
241. Marcus CC, Barncs M (eds) (1999) Healing gardens. New York, Wiley, pp 27–86
242. Ulrich RS, Lawson B, Martinez M (2003) Exploring the patient environment: an NHS estates workshop. Tht Stationery Office, London

Chapter 21
Enhancing Indoor Air Quality for Residential Building in Hot-Arid Regions

Ghanim Kadhem Abdul Sada and Tawfeeq Wasmi M. Salih

Abstract Indoor air quality (IAQ) in the building depends mainly upon the level of temperature and humidity, where it is an indication of the comfort and health of occupants. The enhancement of indoor air needs to take care of both internal and external conditions. The internal conditions required to control the variation of room temperature and humidity as well as the discomfort sources like kitchen, laundry, and bathroom. The external conditions required to reduce the climatic effects upon buildings. In this study, a test model has built in order to apply some important passive techniques including insulating materials, airtightness, and air ventilation system. The experimental work includes the improving of indoor air quality by source control, suitable filtration, and the use of ventilation system. This study gives good indication to the benefit of using the passive house criteria in hot-arid regions like Iraq where there are high level of solar radiation and large fluctuations of air temperature. The results in summer time show that the indoor temperature could be reduced to 31 °C instead of 42 °C for conventional house. In Iraq, due to low ambient humidity in summer, about 15–40%, there is no need to the dehumidification process. However, measurements show stability in the fluctuations of relative humidity in the passive model compared to that measured in the traditional one, where the indoor relative humidity was maintained less than 37%. Furthermore, a local simulation program is used to calculate the energy consumption and the greenhouse gas effect. The results show that energy consumption could be saved up to 80%. The fuel consumption that is used in the power plant to produce electricity for cooling could be saved from 32 L/m^2 for traditional house to 3 L/m^2 for passive house which means reducing by 29 L/m^2. The corresponding CO_2 emission is saved from 85 kg/m^2 for traditional house to 11 kg/m^2 for passive house which means reducing by 74 kg/m^2 or 87%.

G. K. A. Sada (✉)
Mechanical Engineering Department/Al Mustansiriyah University, Baghdad, Iraq

T. W. M. Salih
Materials Engineering Department/Al Mustansiriyah University, Baghdad, Iraq

© Springer International Publishing AG, part of Springer Nature 2019
A. Sayigh (ed.), *Sustainable Building for a Cleaner Environment*,
Innovative Renewable Energy, https://doi.org/10.1007/978-3-319-94595-8_21

21.1 Introduction

Indoor air quality (IAQ) refers to the air quality within the buildings and structures, especially as it relates to the health and comfort of building occupants. Understanding and controlling common pollutants indoors can help reduce the risk of indoor health concerns. The improving of indoor air quality could be done by controlling both air inside the building and air outside. Indoor air quality (IAQ) focuses on airborne contaminants, as well as other parameters like odor, health, and safety. The controlling of the air inside the building includes air-conditioning, infiltration, and ventilation system, while the controlling of outside air means to mitigate the climatic effect of summer overheating due to high solar gains. This can be avoided using simple components like blinds, overhangs, and roof shading.

Air-conditioning device provides cooling and heating, as well as removing moisture and airborne pollutants. Also, it supplies clean air and keeps human being healthy. The air-conditioning device produces comfort conditions in which the human beings tend to feel highly comfortable. If the room temperature is very high (hot), then person tends to get tired faster. If the room temperature is very low (cold), then person tends to be idle and lazy. Human is sensitive to humid air because the human body uses evaporative cooling as the primary mechanism to regulate its temperature. Relative humidity, expressed as a percent, measures the humidity relative to the maximum value at that temperature. Relative humidity levels below 25% increase discomfort and drying of the mucous membranes and skin, which can lead to chapping and irritation. High humidity levels can result in condensation within the building structure and the subsequent development of molds and fungi. It is observed that ideal indoor relative humidity levels are 35% in the winter and 50% in the summer [1].

Infiltration means the undesired air enters to a building typically through the cracks, windows, and doors for passage. Infiltration is caused by wind, negative pressurization of the building, and air buoyancy forces. Because infiltration is uncontrolled, it is generally considered undesirable except for ventilation air purposes. Typically, infiltration decreases the thermal comfort, increases energy consumption, and increases the dust. For all buildings, infiltration can be reduced via sealing cracks in a building's envelope and, for new construction or major renovations, by installing continuous air retarders. The resistance to inward or outward air leakage through unintentional leakage points or areas in the building envelope is defined as building airtightness. The controlled airtightness has several positive impacts on the building, where it decreases heat loss, with potentially smaller requirements for heating and cooling equipment capacities, and also reduces the chance of mold because moisture is less likely to enter and become trapped in cavities. According to the airtightness value, the building could be classified into tight, medium, and loose [2]. The ACH value is a scale for the airtightness of the building. The airtightness can be measured by blower-door test. Following the low-energy Passivhaus standards, it is required to satisfy air leakage lower than 0.6 ACH at 50 Pa (test pressure) [3]. However, Table 21.1 shows ACH for a typical house.

Table 21.1 ACH for residential house due to infiltration [2]

Outdoor design temperature, °C									
Winter									
Class	**10**	**4**	**−1**	**−7**	**−12**	**−18**	**−23**	**−29**	**−34**	**−40**
Tight	0.41	0.43	0.45	0.47	0.49	0.51	0.53	0.55	0.57	0.59
Medium	0.69	0.73	0.77	0.81	0.55	0.89	0.93	0.97	1.00	1.05
Loose	1.11	1.15	1.20	1.23	1.27	1.30	1.35	1.40	1.43	1.47

Note: Values are for 6.7 m/s (24 km/h) wind and indoor temperature of 20 °C

Summer						
Class	**29**	**32**	**35**	**38**	**41**	**43**
Tight	0.33	0.34	0.35	0.36	0.37	0.38
Medium	0.46	0.48	0.50	0.52	0.54	0.56
Loose	0.68	0.70	0.72	0.74	0.76	0.78

Note: Values are for 3.4 m/s (12 km/h) wind and indoor temperature of 24 °C

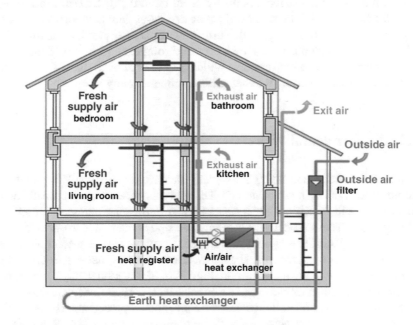

Fig. 21.1 Mechanical heat recovery ventilation system (MHRVS) [4]

Ventilation system in the building provides some forced ventilation, where air-conditioning designers typically choose to admit more external air to keep the air fresh and healthy. The Passivhaus standards satisfy IAQ with good air conditions (temperature, humidity) and supply ventilation rate more than 0.4 ACH [3]. The controlled fresh air can be drawn from outside and then passed through the ventilation system, usually called mechanical heat recovery ventilation system (MHRVS), as shown in Fig. 21.1. The benefits of MHRVS are many [4]: reheat the air by heat exchanging, fresh air, removing undesired air from bathrooms and kitchens, and reducing mold and fungus. In cold climate, a heat recovery ventilator (HRV) is used to supply fresh moderate air into the house by extracting the heat from the warm exhaust air to the incoming cold air without mixing [5]. According to the Passivhaus standards, the efficiency of heat recovery ventilator should exceed 75% [3]. Usually, in hot climate, heat recovery system is replaced by passive ventilation through wind catcher or underground tubes ventilation system to let fresh air into the home. This helps to provide cooling, as well as removing moisture and airborne pollutants. It will need to harness more air movement as humidity increases by orienting the home to catch the prevailing breeze and using pools and water features outside windows or in courtyards to cool air before it enters the house [6]. Earth cooling can be applied in Iraq by means of underground tubes. The outdoor air is drawn into the tubes where cooling is done due to low earth temperature. Then moderate air could be supplied naturally or forced into the home [7].

The controlling of outside air could be satisfied using suitable shading, where it helps in reducing the excessive gain by about a third [8]. In Darmstadt passive house, the frequency of summer overheating decreases with the increasing of over-hang length up to 1.6 m length [9]. Furthermore, planting provides flexible shade options, where deciduous trees let winter sun through and provide summer shade. A study carried out in a conventional building in Greece by Koliris M [10]. shows that tree shading reduces the internal surface temperature by an average of 7 °C.

21.2   Experimental Work

The aim of this study is to satisfy Passivhaus standards for Iraqi buildings including recommended IAQ by these standards. The study has been done practically across two models built in Kirkuk (35.5 °N, 44.4 °E) during the period 01 August–04 October, 2013. These models are located in an open area where no shaded region is near. The first model was built in traditional way, while the second model was built according to Passivhaus standards. Both models have the dimensions 1.8 m × 1.8 m × 2 m, as shown in Fig. 21.2. Each model has a window of 0.8 m × 0.5 m south orientation and a door of 1.9 m × 0.9 m north orientation.

The traditional model is built using the conventional procedure. The wall of the traditional room is constructed from 20 cm common hollow block, 2 cm external cement plaster, and 1 cm internal gypsum plaster, while the roof is constructed from 15 cm steel-reinforced concrete with 1 cm internal gypsum plaster.

The passive model has some innovative techniques. The foundation has built using standard 20 cm hollow block plastered by mortar. The outer face of the foundation should be insulated by cork of 4 cm thickness to prevent heat transfer from the surrounding area to the interior. Walls are built up to 2 m high using the standard 15 cm hollow block. Cementitious mortar is used for plastering the outer facade. On the inner facade, 1 cm gypsum plaster is used. The initial insulation is done by 2 cm cork enveloping the whole space. After that, the walls are covered by bubbly wrap of 4 mm thickness made up from polyethylene that is considered as

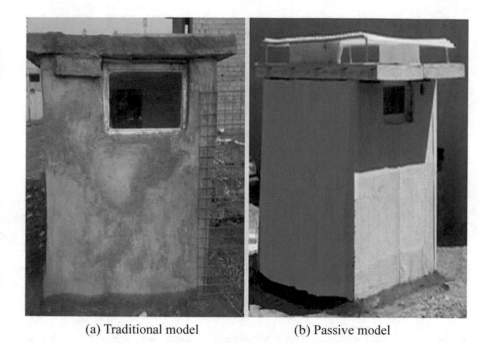

(a) Traditional model   (b) Passive model

Fig. 21.2 Constructed models used in the study. (**a**) Traditional model. (**b**) Passive model

thermal insulation and condensation barrier as well [11]. The walls are then insulated using polyurethane panels of 10 cm. Finally, the walls were wrapped with white sticker for the purpose of aesthetic. The overall U-value of the walls was 0.15 W/(m²·K). Roof is installed using standard polyurethane sandwich panel of 15 cm thickness, so the overall U-value for the roof was 0.19 W/(m²·K). Double-glazing closed-type window of plastic frame is used with south orientation. The overall U-value of the window is 1.1 W/(m²·K). The glazing is assembled locally with the following specification: two glass layers of low emittance, each layer of 4 mm thickness with 20 mm evacuated space. Silica gel desiccant is used as wet absorption material and it sealed by aluminum bar. Yellow glass is used to reduce absorption and increase reflection. The door of UPVC frame and double sheeting is used with overall U-value of 1.8 W/(m²·K).

The essential thing related to passive house standard is avoiding the thermal bridges around the wall, window, and door using EPS cork of 2 cm thickness. The airtightness is checked visibly using a lamp inside the room at night and determining leakage places which has processed by polyurethane spray. Since roof is exposed to high intensity of solar radiation throughout the day, hence a canopy of plastic (polyethylene) of 2 m × 2 m and 30 cm apart is used to shade the roof and to supply continuous air stream below. Controlled air enters to the passive model through a single slot connected to the underground cooling system and extracted from a diffuser at 2 m high, as shown in Fig. 21.3. Air flows through a 4 inch plastic tube down

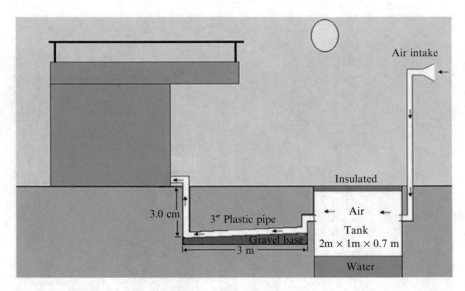

Fig. 21.3 Schematic diagram of underground air-cooling system

to the underground reservoir. The reservoir has dimensions of 2 m × 1 m × 0.7 m. Water level in the reservoir is about 25 cm and the rest is air. The reservoir is sealed by wood and suitable insulated materials. The mechanism of air cooling in the reservoir depends on the evaporating cooling which includes mass transfer of vapor at the water surface to the air which leads to increase the humidity content of it. This mechanism required involving in the details of mass transfer also the required time for cooling and the advantage of night flushing.

21.3   Results and Discussions

The study includes in-site measurements on selected buildings within 2 months (August and September) for both models (traditional and passive). It is advised to take more stable readings into consideration. Several instruments have been used in this project to measure the effective parameters in the building. Lutron data logger system has been used to measure outdoor temperature, indoor temperature, and indoor humidity each 15 min throughout the day and in different locations. The global solar radiation is measured as well.

Three ventilation cases (free ventilation, free ventilation daytime with night flushing, and forced ventilation) were examined within this study with different runs, as shown in Table 21.2. In free ventilation, the air is supplied to the passive room naturally due to temperature difference and in presence of underground cooling system. The results show that ambient air can be cooled down in the heat

Table 21.2 Different runs of work

Case	Duration	Specification of traditional model	Specification of passive model
1	01 Aug–11 Aug	No insulation, single-glazed window, traditional door, no overhang, no canopy, no ventilation system	Super insulation, double-glazed window, UPVC door, overhang, no canopy, free ventilation system
2	12 Aug–27 Aug	No change	Adding the canopy
3	28 Aug–29 Aug	No change	Adding night flushing
4	30 Aug–04 Sep	No change	Removing night flushing
5	05 Sep–07 Sep	No change	Adding forced ventilation
6	08 Sep–04 Oct	No change	Balanced forced ventilation

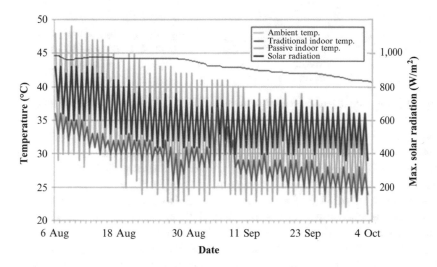

Fig. 21.4 Variation of indoor air quality during 06 Aug–04 Oct, 2013

exchanger to supply air of 34 °C temperature to the building. Night flushing ventilation could be used to deliver direct air when the ambient temperature becomes cold (lower than 30 °C). While in forced ventilation, the warm air from outside is withdrawn into the cooling system using a suction fan. In this case, it should take care to make a balance between the amount of cold air stored in the underground tank and the air flow rate supplied to the building because rapid velocity would be unuseful to avoid sucking the hot air during daytime which causes of rising the indoor temperature.

Figure 21.4 shows the variation of outdoor temperature (ambient), indoor temperature, and solar radiation within the period of experimental study for both traditional and passive models. In August, the outdoor temperatures are swinging daily between 30 °C (min.) and 48 °C (max.) with solar radiation up to 950 W/m^2 for 13 h, while for September, the outdoor temperatures are swinging daily between

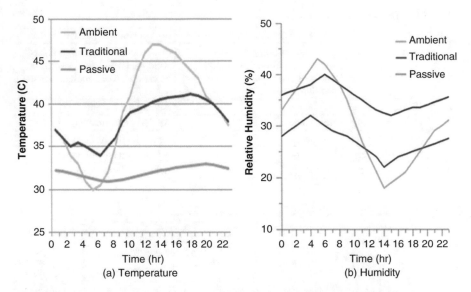

Fig. 21.5 Variation of indoor air quality on 15 Aug, 2013. (**a**) Temperature. (**b**) Humidity

24 °C (min.) and 40 °C (max.) with solar radiation up to 880 W/m² for 12.5 h. The results show decreasing in the indoor temperature for the case of passive model compared with that of traditional one. In August, the average indoor temperature of passive model was 32 °C with maximum value of 34 °C, while it was 40 °C with maximum value of 43 °C for traditional model. However, it is clear that there is a fluctuation in max-min temperatures about 3 °C in indoor temperature for passive model compared to that of traditional which is fluctuated about 7 °C, as shown in Fig. 21.5a. In Iraq, the ambient relative humidity in summer is moderate, about 15–40%; hence there is no need for dehumidification. However, in the present work, the indoor relative humidity is still comfortable, as shown in Fig. 21.5b.

Figure 21.6 shows the contribution of canopy in reducing the influence of direct solar radiation up on the roof hence to control the air temperature outside the building. This system provides shading and airstream which helps to remove the accumulated heat through the roof. It is found that the average reduction in the temperature of external surface of roof was 10 °C and the average reduction in the temperature of internal surface of roof was 3 °C with average decreasing in indoor temperature by 2 °C. The results show a decrement in the glass temperature by 6–8 °C when window overhang shading is used.

Furthermore, a local simulation program is used to calculate the energy consumption and the greenhouse gas effect. The results show that the annual energy consumption for the passive model is about 68 kWh/m²/yr. while for a traditional model is about 350 kWh/m²/yr. Then, the annual energy saving would be about 80%. The fuel consumption that is used in the power plant to produce electricity for cooling could be saved from 32 L/m² for traditional house to 3 L/m² for passive house which means reducing by 29 L/m². The corresponding CO_2 emission is saved from 85 kg/m² for traditional house to 11 kg/m² for passive house which means reducing by 74 kg/m² or 87%.

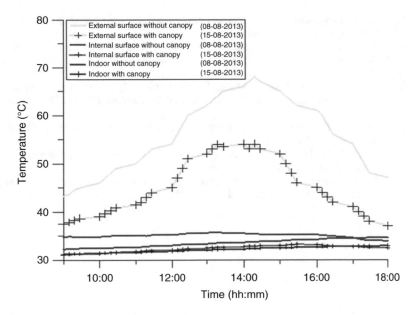

Fig. 21.6 Comparison for the effect of roof canopy

21.4   Conclusion

This method gives good indication for the ability of using the Passivhaus standards in hot-arid regions like Iraq, where high level of solar radiation and large fluctuation of air temperature, and to enhance the indoor air quality and increase the thermal performance of the building. In order to adopt passive criteria and decrease the climatic effects upon the building, some techniques were done like super insulation, avoiding thermal bridges, sufficient window and door, airtightness, and air ventilation system. Many conclusions can be obtained from this study which can be summarized as follows. The indoor temperature could be reduced to 31 °C instead of 42 °C for conventional house with relative humidity about 30–40%. The energy consumption could be saved up to 80%.

References

1. Healthy Heating (2014) Humidity: effects on the environment and occupants. Educational Resource Site. www.healthyheating.com
2. ASHRAE (1997) Chapter 27: Residential cooling and heating load calculations. In: ASHRAE Handbook SI (Fundamentals), ASHRAE, USA
3. Sustainable Energy Ireland (SEI) (2007) Passive homes, guidelines for the design and construction of passive house. Dwellings in Ireland, SEI, Ireland
4. ISOVER (2008) The ISOVER multi comfort house. www.isover.net
5. Ludeman C (2008) Passive house (Passivhaus) standard for energy efficient design. Available at www.100khouse.com

6. Smarter Homes Orginazation, New Zealand Government, Passive Cooling, Accessed 15 Jun 2016. Available at www.smarterhomes.org.nz
7. Amjed A, Asaad K, Ali A (2008) Towards a zero energy house strategy fitting for South Iraq climate. In: PLEA – 25th conference on passive and low energy architecture, Dublin, 22nd to 24th October 2008
8. Gandara M, Rubio E, Fernandez J, Munoz V (2002) Effect of passive techniques on interior temperature in small houses in the dry, hot climate of Northwestern Mexico. J Renew Energy 26:121–135
9. Wolfgang F (1998) Passive house summer climate study. Passivhaus Institute, Darmstadt. Available at www.passipedia.org
10. Koliris M (2003) The benefits of trees and their potential as shading devices in the office environment. MSc Thesis, University of London
11. Aerolam (2013) Bubble wrap insulation. Aerolam Insulation Private Limited Company. http://www.aerolaminsulations.com. Accessed 10 Dec 2013

Chapter 22
Performance of Solar Window Film with Reference to Energy Rationalizing in Buildings

Kamil M. Yousif and Alan Ibrahim Saeed

Abstract In the last decades the growing need for comfort conditions in residential buildings and the severe summer conditions typical of Mediterranean climate have led to an ever-increasing request for the installation of air-conditioning systems. This has led to rising consumption of electricity and has influenced the cost of electricity and the need for more power plants. Furthermore, the CO_2 equivalent emissions' factors from electricity will rise. In this study, two wooden cabinets (cabinets A and B) were built to be a model for two rooms in a building in city of Duhok, Iraq. These cabinets were of double-glazed window, insulated by air. Solar control window film (SCWF) has been attached on the window of one of the cabinets (cabinet B). Measurements of the following physical variables parameter were performed, outside and inside the cabinets: solar radiation (SR), temperature (T), and relative humidity (RH). The measurements were done in two different cases: one, when the surface color (SC) of both cabinets was white and, for the other, when the SC was black.

In this study, we investigate the performance of solar control window films based on a clear double glass window for one of the cabinets. The study is generalized for a large-scale practical case if a window film of similar properties is applied to the windows of rooms in a building in Duhok, Iraq. Thermal conductivity of each glass, wood, and SCWF has been measured. The absorbance and transmittance of the window glass as well as the solar control window film have been measured by using the UV, VIS, and NIR spectrophotometer instruments. It was found that the temperature of cabinets depends on surface color (SC) of the cabinet. In case of white surface (WS) cabinet, the temperature of cabinet was approximately 10 °C lower than black surface (BS) of cabinet, while the temperature difference between cabinet A [CA] and cabinet B [CB] was approximately equal to 5 °C. In case of black surface) BS) cabinet, the temperature difference between CA and CB was about 7 °C. It was observed that the blocking capacity of thin solar control film depends on the time of day and weather conditions. Utilizing window films in a building will lead to reduction in heat gain from solar radiation, i.e., it will reduce

K. M. Yousif (✉)
Department of Environmental Sciences, Zakho University, Zakho, Iraq

A. I. Saeed
Department of Physics, Zakho University, Zakho, Iraq

© Springer International Publishing AG, part of Springer Nature 2019
A. Sayigh (ed.), *Sustainable Building for a Cleaner Environment*,
Innovative Renewable Energy, https://doi.org/10.1007/978-3-319-94595-8_22

the building cooling load in summer, or using the solar control window films in buildings will lead to reduce the electrical energy consumption in summer. Also, this will contribute in reducing CO_2 and other greenhouse gases (GHG) emissions that cause global warming in the world. Further advantages in using such film are the reduction of glare, heat, and UV.

22.1 Introduction

Recently, considerable attention has justifiably been directed toward energy savings in buildings as they account for up to 20–40% of total energy consumption in developed countries [1]. Many developed and developing countries have established regulations aimed at the reduction of building energy consumption by significantly improving the energy efficiency of buildings [2]. Window panels are a major component of the building fabric with considerable influence on the façade energy performance and are accountable for up to 60% of a building's overall energy loss. Therefore, the thermal performance of glazing materials is an important issue within the built environment [3]. Windows are responsible for a large portion of the energy used in buildings. The heat energy lost or gained through windows directly affects the amount of energy required to maintain comfortable indoor conditions. Significant advances in fenestration technology have been made. Glazings and fenestration have always been a prominent element of architectural form and expression. Significant changes have already occurred in glazing and facade design over the last decades. New technology and nanotechnology will improve energy efficiency by filtering the transmittance of heat and light. Peak loads and excessive glare can be reduced with windows which have a low solar heat gain coefficient (SHGC). These windows prevent much of the solar radiation from entering the building. Window film is a thin laminate film that can be installed to the interior or exterior of glass surfaces in automobiles and boats and also to the interior or exterior of glass in homes and buildings. It is usually made from polyethylene terephthalate (PET), a thermoplastic polymer resin of the polyester family. Window films are generically categorized by their construction components (dyed, pigmented, metallized, ceramic, or nano), by their intended use (automotive, marine, or architectural), by substrate type (glass or polycarbonate), and/or by their technical performance (privacy, solar control, safety, and security). Window films are among the most cost-effective energy retrofits for existing residential and commercial buildings. Window films should be recognized as a viable solution to cost-effectively reduce greenhouse gases—GHGs [4]. Some of the benefits of window films include protection against the adverse effects of the sun (such as severe heat, fading, glare reduction, and ultraviolet radiation), enhanced security, increased privacy, and aesthetic purposes [5, 6]. In recent years, new materials have been developed that offer superior performance in glazing applications, compared to conventional sheets of glass. According to the International Window Film Association (IWFA), a pane of normal clear glass reflects about 6% of solar radiation, absorbs 5%, and transmits the remaining amount [4]. By using these new advanced glazing materials, the above values of reflect and absorb could be controlled. Some examples of these new advanced glazing materials include transparent insulation materials, such as honeycombs and aerogels, coated glass, electro-chromic materials, and angularly selective materials [4]. The global temperature is rising up because of the greenhouse effect. There remains little doubt that the building sector is one of the contributors in emitting $CO2$ to the global atmosphere. The focus of world attention on environmental issues in recent years has stimulated various responses in many countries. This has led to a closer examination of energy conservation strategies. Generally, one third of global $CO2$ emissions is attributed to the building, which emphasizes the important need for energy savings in consumption in buildings [7]. The environmental benefits and the reduction of ($CO2$) due to saving in electrical energy by window film were calculated and presented by Yousif [8] and Yousif et al. [9]. Dussault et al. [10] examined the energy-saving potential of incorporating smart window technologies on a double-glazed window pane of a typical low thermal mass office building in Quebec, Canada. The works of Li et al. [11] and Yin et al. [12] investigated the potential of solar window films to reduce the energy consumption of buildings in different climates. The literature review shows that studies have been carried out on systems with and without solar control films using software such as Window [13] and Energy-10 [14] which used a standard model for the calculation of thermal parameters (ISO-15099, [15]). Carriere et al. [16] studied the efficacy of window glazing in reducing a building's solar heat gain using the DOE-2 software. They found that cooling energy reduces with the number of glazing. Bahadori-Jahromi et al. [3] evaluated the impact of window films on the energy performance of an existing UK hotel. The simulation was conducted using a building energy simulation software (EDSL TAS). The building envelope in Iraq, including Kurdistan region, absorbs huge amount of solar radiation, and it induces the increase of exterior surface temperature, including roof and exterior side wall surface and interior surface temperatures. Then, indoor temperature rises accompanying a negative influence on thermally comfortable indoor environment. Consequently, excess electricity and energy are required for air-conditioning system to maintain the desired indoor temperature. In order to decrease huge building energy consumption as well as decrease greenhouse gases emissions, the building envelope ought to be well designed to minimize the unwanted heat gain from solar radiation so that the thermal performance of buildings can be improved and sustainability of built environment can be achieved.

Demand for electricity in the Kurdistan region of Iraq (KRGI) has increased from 482 MW in 2007 to around 3,000 MW in 2015 [17]. Electricity demand is subject to fluctuations on a seasonal basis, across the week, and during the day. Demand can also be influenced by irregular events, such as particularly extreme weather conditions. Electricity annual demand loads and supplied loads for Duhok, KRGI, during the period 2004 to 2014, are increased from 200 MW and 100 MW to 1300 MW and 550 MW, respectively, due to expansion and different development in Duhok. Also there is no sufficient supplied load in any year over all periods [18].

The aim of this study is to investigate the performance of thin window film in the wooden cabinet of double-glazed window pane for purposes of saving electrical energy, as well as the reduction in CO_2 emissions resulting from the drop in the building's energy utilization. The study includes an investigation of the energy balance inside the simulated model, calculations of solar heat gain reduction, and an estimation of CO_2 emission reduction.

22.2 Theoretical Calculations

Solar heat gain through fenestration is considered as the largest provider to building envelope cooling load and the most important parameter for the overall thermal transfer value determinations. The heat gain in building comes also from different other sources: walls, roof, floor, electrical domestic appliances, occupant, etc. An approximated equation was used by Alnaser [19] to model the solar heat gain in a car after exposure for time interval, t. That equation has been modified here in order to fit the case of solar heat gain $H1$ for a room in a building:

$$H1 \cong \tau I A_g + \left\{ p_{air} V_i C_{air} \left(\Delta T \right) \right\} / t + \left\{ \Sigma \left(m_i C_i \right) \right\} \Delta T / t + \varepsilon \sigma A_i \left(T_f^4 - T_i^4 \right)$$
$$+ A_d + w + r \left(T_f - T_i \right) / \Sigma \left(L_i / K_i \right) \qquad (22.1)$$

where τ is the glass transmissivity; I is the solar radiation (W/m^2); A_g is area of the glass windows (m^2); ρ_{air} is inside room air density (kg/m^3); V_i is the volume of the room interior (m^3); ε is the emissivity; ΔT is the difference between outside and inside room air temperature; C_{air} is the air-specific heat capacity (J/kg K); m_i is mass of the heat-generating appliances (kg); C_i is their compound-specific heat capacity (J/kg K); σ (= 5.6697 × 10^{-8} W/m^{2}K^4) is the Stefan-Boltzmann constant; A_i is the room interior area; T_f is furniture final temperature; T_i is furniture initial temperature (K); $A_d + w + r$ is the door, walls, and roof area (m^2); $T_f - T_i$ is the temperature difference between the door exterior and door interior; L_i is the total thickness of the door, wall, and roof (m); and K is the thermal conductivity (W/m.°C). When the selective absorbing window film is applied to the windows, it will act to reduce the solar radiation that will enter the room through the window. This will cause the solar heat gain of the building or room to be reduced to some lower value $H2$, and the difference in solar heat gain ΔH will be given by:

$$\Delta H = H1 - H2 \qquad (22.2)$$

According to Alnaser [16], when the window film is applied, the main change will occur in terms of the first, second, and fourth of Eq. 22.1. Then the difference in solar heat gain ΔH in the room before and after applying the film can be estimated as:

$$\Delta H = \left(\tau_b - \tau_a \right) I A + p_{air} V_i C_{air} / t \left\{ \left(\Delta T_b - \Delta T_f \right) \right\} + \varepsilon \sigma A_i \left(T_f^4 - T_i^4 \right) \qquad (22.3)$$

where τ_b is transmissivity of the window glass and τ_a is the transmissivity of the window glass after the application of the film.

The temperature difference ΔT_b will be given by [19]:

$$\Delta T_b = T_{fb} - T_{ib} \qquad (22.4a)$$

T_{fb} and T_{ib} are the final and initial temperatures in cabinet A, respectively.
Temperature ΔT_f will be given by [19]:

$$\Delta T_f = T_{ff} - T_{if}, \qquad (22.4b)$$

T_{ff} and T_{if} are the final and initial temperatures in cabinet B, respectively. Equation 22.3 was used to calculate the reduction in the solar heat gain in the two wooden cabinets under test.

22.3   Experiment and Research Method

Two identical wooden cabinets, A and B, are built, with double glass, and the distance between the two glasses is 16 mm, using the normal 4-mm-thick clear window glass.

Dimensions of both cabinets are as follows: 121 cm height, 61 cm length, and 36 cm width. In order to measure the effect of the window film on the solar heat gain in cabinets A and B, a solar control window film was applied to the windows of the cabinet (B) (Fig. 22.1). The cabinet consists of one chamber, and only a small wood shelf has been put on the middle of the chamber in order to put the sensor of the device on it (Fig. 22.1). The front side of the cabinet is directed to the south. Both wooden cabinets were put on the roof of a building. The tests were done by using wooden cabinets, of white color and black color.

Fig. 22.1 Two wooden cabinets during monitoring some physical variables

Cabinet B   Cabinet A

The following physical properties have been measured: the absorbance and transmittance of the window glass as well as the solar control window film; thermal conductivity of glass, wood, and solar control window film; temperature; the solar radiation; wind speed and direction; and humidity outside and inside cabinets (A) and (B) (except for wind). The solar radiation intensity and temperature were measured inside and outside of the two cabinets during 10 h from 8:00 a.m. to 18:00 p.m. local time. The solar radiation intensity for both cabinets, with film and without film, was measured during all the 10 h both outside and inside the cabinet using solar radiation intensity meter. The temperature was also measured for the same cases and period using ordinary thermometer. Readings of the solar radiation intensity, temperature, and humidity were recorded each half an hour. The absorbance and transmittance of the window glass as well as the solar control window film have been measured by using the UV, VIS, and NIR spectrophotometer instruments (Perkin Elmer, Lambda 25). The above instrument is the double-beam type. It was computerized for analyzing the recorded data. The selected wavelengths were between 192 nm and 1100 nm. Thermal conductivity of glass, wood, and solar control window film (SCWF) was measured by using an instrument heat conduction apparatus. The solar radiation for both inside and outside of cabinet (A) and cabinet (B) was measured by Voltcraft PL-110SM solar radiation measuring instrument. Its measurement ranged from 0 to 1999 W/m². The temperature of the outdoor was measured by DT-300 LCD-hand thermometer. For the inside of cabinet (A) and cabinet (B), thermocouple and ordinary thermometer were used. Hygrometer was used for measuring relative humidity. Measurements of wind speed were made by means of an anemometer.

22.4   Results and Discussion

Figure 22.2 illustrates the transmission of single and double glass, respectively. Around 300 nm, the transmittance starts to increase sharply up to around 84% (for single glass). It keeps almost constant at visible region, while it starts to decrease in the infrared region. A similar pattern happens for the double glass. The transmittance for single glass is higher due to smaller path (x), according to Beer-Lambert law ($I = I_o\,e^{-ax}$).

The transmission (T) for double glass, when the solar control window film (SCWF) is attached to double glass, is depicted in Fig. 22.3. T starts to increase sharply up to around 79% (for double glass). It keeps almost constant up to 600 nm region, while it starts to decrease in the infrared region. T for double glass with SCWF is much less. Around 310 nm, up to 600 nm, T is around 25% (at visible region). Then it starts to increase up to around 59% and then starts to decrease in the infrared region. The SCWF will lead to reduce the amount of solar radiation transmitted through double glass.

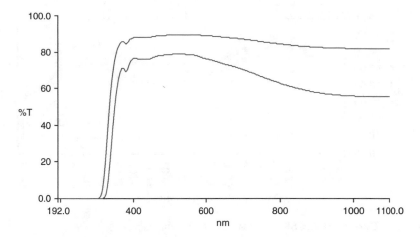

Fig. 22.2 Transmittance of single glass (up), and double glass (down)

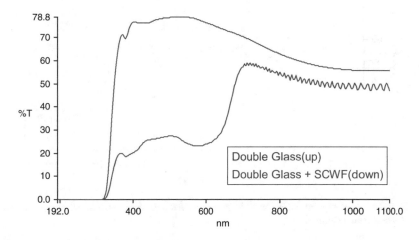

Fig. 22.3 Transmittance for double glass when SCWF is attached to double glass

Thermal conductivity of used glass was 0.92 w/m.°C. Thermal conductivity of used wood was 0.38w/m.°C. Thickness of SCWF sample was $X = 0.00003$ m. Thermal conductivity of used SCWF was 0.0095 w/m.°C .

22.5   Solar Radiation Blocking

Figure 22.4 shows the hourly average of Solar Radiation Blocking (SRB) for (CA) and (CB)during July 2015.

From Fig. 22.4, it can be seen that SRB for cabinet B is larger than cabinet A. The results show that SRB vary depending on the time of day. E.g., for cabinet B, they

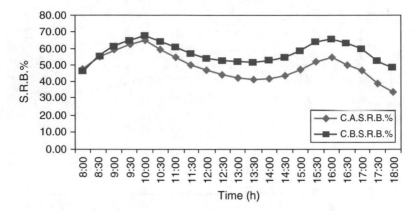

Fig. 22.4 Hourly average of SRB for (CA) and (CB) during July 2015

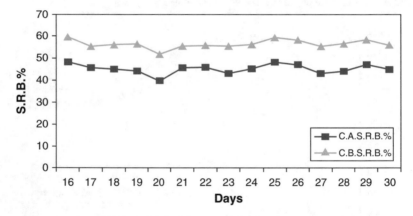

Fig. 22.5 Daily average of SRB for (CA) and (CB) from 16 Aug 2015 to 30 Aug 2015 (when using black surface cabinet)

are between 40 % and 50 % during midday (11 a.m. to 15 p.m.), because the sun is almost perpendicular to cabinets. Figure 22.5 illustrates the daily average of SRB for (CA) and (CB) from 16 Aug 2015 to 30 Aug 2015 (when using black surface cabinet). It is observe that SRB of cabinet A has lower value of SRB. The results show that the thin film blocks between 5 and 95% of solar radiation, depending on the time of day and weather conditions, e.g., clouds, haze, etc. A large reduction in solar radiation usually represents a sunny sky with a large portion of direct sun component.

In the case of white cabinets, the temperature difference between cabinet A and cabinet B is approximately equal to 5 °C due to SCWF at midday, while, in the case of black cabinets, the temperature difference between cabinet A and cabinet B is approximately equal to 7 °C due to SCWF at midday. Concerning the indoor and outdoor relative humidity, it was found that there was no noticeable change in the humidity both indoors and outdoors, with and without the use of the SCWF. Wind speed was in the range 0.7–3.5 m/s. Wall and roof constructions affect the thermal performance of buildings in which a high overall heat transfer coefficient (U-value) yields more heat flow by conduction inside the building through the walls and roof.

22.6   Calculations of Solar (Heat) Gain

The following are the calculations of reduction in solar heat gain.

The τ calculated by using a weighted coordinates method [20] and difference between the transmissivity of the window glazing of the two cabinet under test before applying the film (cabinet A) and after applying (cabinet B) is given by:

$$\tau_b - \tau_a = 0.726 - 0.327 = 0.399.$$

Total difference in solar (heat) gain will be:

$$\Delta H_{gain} = 220.248 + 2.564 + 94.91 = 317.72\,W.$$

The ratio of the glazing area to the total area is $A_g/A_i = 0.248$.

It is considerably high, and this led the first term of Eq. 22.3 to be dominant. The second term is almost negligible. For a large-scale practical case, if a window film of similar properties is applied to the windows of a lab at Duhok University (with dimensions: width = 10 m, height = 3 m, length = 12 m) containing five windows, of total net glazing area of 20 m^2. In this case the total difference in solar (heat) gain will be.

$$\Delta H_{gain} = 6384\,W + 80.68\,W + 12730.69\,W = 19195.37\,W.$$

22.7   Calculations of the Reduction in CO2 Emissions

According to the calculations and analysis regarding to solar (heat) gain, we can calculate the reduction in CO2 emissions. Modifying building envelope by using solar window films will significantly reduce energy required for heating and cooling. The total energy consumption of heating, ventilation, and air-conditioning (HVAC) systems will reduce by total difference in solar (heat) gain.

In the case of the test cabinet, ΔH_{gain} = 317.72 W which save about 1159.678 kWh/year. The estimated reduction in CO2 is 823.37 kg.CO2 or 0.8234 tons CO2/year, in the case study (i.e. in case of the WS cabinet).

This was calculated by using a factor of 0.71 kg.CO2 for every kWh [[21].

Likewise, for the lab case, total difference in solar (heat) ΔH_{gain} will be 19195.37 W, which saves 70063.1005 kWh/year. So, for the generalize case, the estimated reduction in CO2 is about 49.74 tons/year. Therefore, window films have a relatively good potential to impact GHG emissions when compared to other mains, especially when cost is taken into account. Hence, it is one of the most effective measures that can be considered.

22.8 Conclusions

It was found that the temperature (T) of cabinets depends on surface color (SC) of the cabinet. e.g. the T of white surface (WS) cabinet was ~ 10 °C lower than black surface (BS) cabinet. Using window films in a building, will lead to reduction in T inside a building. e.g. the temperature difference between CA and CB was ~5 °C and was ~ 7 °C due to SCWF.

For WS, the estimated reduction in CO2 is 823.37 Kg.CO_2 or 0.8234 tons CO_2/ year in the case study. For the generalize case, the estimated reduction in CO_2 is about 49.74 tons/year. Thus for ten labs in the university, the estimated reduction in CO2 emissions will be about 497. 4 tons/year.

For BS, the estimated reduction in CO_2 is 882.45 Kg.CO_2 or 0.882 tons CO2/year in the case study. For the generalize case, the estimated reduction in CO2 is about 57.479 tons/year. Thus for ten labs in the university, the estimated reduction in CO2 emissions will be about 574.79 tons/year.

Utilizing window films in a building will lead to reduction in heat gain from solar radiation, i.e., it will reduce the building cooling load in summer. Window films also demonstrate an effective means of reducing GHG emissions when used in retrofitting existing buildings.

References

1. Hee WJ, Alghoul MA, Bakhtyar B, Elayeb O, Shameri MA, Alrubaih MS, Sopian K (2015) The role of window glazing on daylighting and energy saving in buildings. Renew Sust Energ Rev 42:323–343
2. Sorgato MJ, Melo AP, Lamberts R (2016) The effect of window opening ventilation control on residential building energy consumption. Energ Buildings 133:1–13
3. Bahadori-Jahromi A, Rotimi A, Mylona A, Godfrey P, Cook D (2017) Impact of window films on the overall energy consumption of existing UK Hotel buildings. Sustainability 9:731. https://doi.org/10.3390/su9050731
4. IWFA- International Window Film Association (2012) Energy analysis for window films applications in new and existing homes and offices. Prepared by: CONSOL, February 7, 2012.7407 TAM O'SHANTER DR.
5. Glass and Glazing Federation (2012) Window film: application and solutions. Available online: www.ggf.uk/windowfilm. Accessed on 5 Nov 2016
6. Plummer JR (2015) Window film: a cost effective window retrofit. Available online, http://www.greenbuildermedia.com/buildingscience/window-film-a-cost-effective-window-retrofit. (Nov. 2016)
7. Moakher PE, Urmee T, Pryor T, Baverstock G (2015) A systematic process for investigating the influence of occupants energy use patterns on building energy performance GREEN FORUM – III (MGF) Mediterranean Green Buildings & Renewable Energy, Florence

8. Yousif KM (2015) Energy savings and Environmental benefits from solar window film for buildings in Kurdistan of Iraq. Published in Book (Renewable Energy in the Service of Mankind Vol-I), Sayigh A (ed), Springer International Publishing Switzerland, Chapter 57, pp 627–635. ISBN 978-3-319-17776-2. https://doi.org/10.1007/978-3-319-17777-9
9. Yousif KM, Yousif J, Mahammed MA (2013) Potential of using window films to support protection of the environment in Kurdistan of Iraq. Int J Eng Res Dev 7(3):26–32
10. Dussault JM, Gosselin L, Galstian T (2012) Integration of smart windows into building design for reduction of yearly overall energy consumption and peak loads. Sol Energy 86:3405–3416
11. Li C, Tan J, Chow T-T, Qiu Z (2015) Experimental and theoretical study on the effect of window films on building energy consumption. Energ Buildings 102:129–138
12. Yin R, Xu P, Shen P (2012) Case study: energy savings from solar window film in two commercial buildings in shanghai. Energ Buildings 45:132–140
13. Gueymard C, DuPont W (2009) Spectral effects on the transmittance, solar heat gain, and performance rating of glazing systems. Sol Energy 83:940–953
14. Chávez-Galan J, Almanza R (2009) Solar filters on iron oxides used as efficient windows for energy savings. Sol Energy 81:13–19
15. International Standard ISO 15099 (2003) Thermal performance of windows. In: Doors and shadings devices-detailed calculations
16. Carriere M, Schoenau GJ, Besant RW (1999) Investigation of some large building energy conservation opportunities using the DOE-2 model. Energy Conv Manag 40:861–872
17. Ekurd Daily, Ekurd.net. Kurd Net Group- Kurdistan Independent Daily. (OR Kurdishglobe. net). (Source:Kurdishglobe.net) Iraq Energy Institute (2015, February 23)
18. Saeed AI (2016) Performance of solar window film with reference to energy rationalizing in buildings and environment protection. M.Sc Thesis, University of Zakho, Iraq
19. Alnaser WE (2007) The solar gain in automobiles in Bahrain and its negative impacts. In: JAAUBAS Vol. 4 (Suppl.). Arab regional energy conference, pp 317–324
20. Wiebelt JA, Henderson JB (1979) Selected ordinates for total solar radiant property evaluation. J. Heat Transfer 101(1):101–107
21. Aboulnaga MM, Najeeb Mohammed AA (2008) Low carbon and sustainable buildings in dubai to combat global warming, counterbalance climate change, and for a better future. In: Sayigh A (ed) World Renewable Energy Congress (WREC-X), pp 566–575

Chapter 23
Visualizing the Infrared Response of an Urban Canyon Throughout a Sunny Day

Benoit Beckers, José Pedro Aguerre, Gonzalo Besuievsky,
Eduardo Fernández, Elena García Nevado, Christian Laborderie,
and Raphaël Nahon

Abstract The work presented here has consisted in placing a thermal camera in a street of the "Petit Bayonne," one of the densest districts of French cities, in order to obtain a double sequence of photographs (shortwave) and thermographies (longwave) on a sunny day. The next step will be to repeat this sequence by numerical simulation to see how the measurements are used to calibrate the simulation and how the simulation can help to interpret the measurements.

In order to achieve by simulation a spatial resolution similar to that of the digital camera, it is necessary to use a finite element-like method. This raises specific questions concerning the scene and sky meshing, the boundary conditions (shortwave and longwave sky model), the optical and thermal properties of materials, etc.

Throughout its history, architecture has always been very sensitive to innovations in representation: central perspective, projection of shadows, axonometries, descriptive geometry, solar diagram, realistic rendering, etc. Thermography is intended to integrate this series of tools in order to support the architectural and urban projects. For this reason, it is necessary to be able to simulate these "infrared renderings." We present here the possibilities offered by the recent advances in computing and measurement and the difficulties that remain, particularly in addressing the urban scale.

B. Beckers (✉) · C. Laborderie
Université de Pau et des Pays de l'Adour, Anglet, France
e-mail: benoit.beckers@univ-pau.fr

J. P. Aguerre · E. Fernández
Universidad de la Republica, Montevideo, Uruguay

G. Besuievsky
Universidad de Gerona, Girona, Spain

E. G. Nevado
Universidad Politécnica de Cataluña, Catalonia, Spain

R. Nahon
Université de Lille 1, Villeneuve-d'Ascq, France

© Springer International Publishing AG, part of Springer Nature 2019   277
A. Sayigh (ed.), *Sustainable Building for a Cleaner Environment*,
Innovative Renewable Energy, https://doi.org/10.1007/978-3-319-94595-8_23

23.1   Introduction

From the time of its discovery at the beginning of the Renaissance, the central perspective has been exerting a strong influence on architectural drawing, theater sets, buildings, streets, and finally on large urban and landscape scenes. It made it possible to structure the three-dimensional space by respecting its harmonic relations (regular pavement) and its cross ratios (coherence of the scene).

The perspective treatises of the seventeenth century emphasized the study of light as a projection in the projection. The architectural treatises of the following century set out to find the ideal acoustic form for large opera houses. The pioneering work of Athanasius Kircher on ray tracing for the study of specular reflection [1], however, is only very slow in practice, and it is only at the beginning of the twentieth century that it will be actually used in acoustics and in optics for the application of geometric methods to the different scales of the architecture. In the nineteenth century, the development of descriptive and projective geometries was accompanied by the improvement of a rendering theory that could accurately depict illuminated scenes, and in the last third of the twentieth century, these techniques were computerized [2]. Their influence has become considerable in the architectural projects of our time.

What about thermographies, these images representing the temperatures of the scene surfaces? Their use has become widespread in recent years. Will they in turn revolutionize the practice of architecture and urbanism?

In general, radiative exchanges in our environment involve a problem with two black bodies: the Sun, whose surface temperature is about 5700 K and which, according to Planck's law, emits in shortwaves (<4 microns), and the Earth, whose average surface temperature is 15 °C, i.e., 288 K, and which emits in longwaves (>4 microns). Since the two spectra are practically separated, the distribution of shortwaves on a scene at a given moment depends only on the source (the Sun and the sky) and the reflection coefficients of the scene surfaces.

The picture of Fig. 23.1 on the left shows the distribution of visible light (shortwave) on a street in Bayonne (France). Thermography, on the right, shows how the city responds in longwaves. Surface temperatures are directly dependent on the shortwave load (thus, on this image, the hottest surfaces are, as might be expected, those of the dark shutters exposed to the Sun) but also heat exchanges by conduction, convection, and radiation. The fact that the latter is strongly nonlinear (function of the fourth power of the temperature) further complicates the problem.

In both images, the optical device is the same, but the shortwave scene is a simple snapshot, whereas thermography depends on the previous moments (thermal inertia). Its simulation assumes to completely solve the heat equation, and, last but not least, the image shows surface temperatures that are deduced from the radiative fluxes, the only things to be actually measured by the camera. This translation presupposes that the whole scene is composed of black bodies or gray bodies all sharing the same emissivity, which is necessarily assumed to be nondirectional. Today, there are excellent references on thermography, but they all suggest using the

Fig. 23.1 Photograph and thermography of *Rue des Tonneliers* (Petit Bayonne, France)

camera in the simplest situations (a single façade examined frontally at dawn on a cloudy day), so as to minimize the difficulties of interpretation [3].

23.2 A Sequence of Urban Thermographies

On April 23, 2017, the sky was clear all day. We placed a FLIR B200 thermal imager at the western end of Rue des Tonneliers, a narrow street typical of Petit Bayonne, east-west oriented and lined by four level buildings, some of which having timber-fronted façades. A series of photographs and thermographies were carried out, every half hour, between 04:00 a.m. and 01:00 a.m. the following day. The air temperature varied between 10 °C and 20 °C and there was virtually no wind.

A piece of the sequence obtained is shown in Fig. 23.2. All the thermographies are on the same scale (color bar between 6 ° C and 36 ° C) and the emissivity of the surfaces is assumed to be unitary. In the first image, at 4:30 in the morning, the surface temperatures are substantially equal to that of the air. The common color bar does not favor details. Thus, for example, the heat emitted by the public lighting is scarcely visible on the image. At 8:30 a.m., the Sun comes from the east, from the camera, and heats the signs that face it. At 2:00 p.m., the south-facing façades are illuminated at the top, and their temperature reaches a maximum value of 36 degrees. Finally, at 11:00 p.m., the temperatures of the southern surfaces remain slightly higher than the temperature of the air, which is still 16 degrees. This is the consequence of thermal inertia.

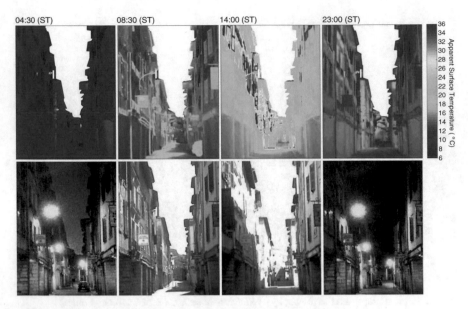

04:30 (ST)   08:30 (ST)   14:00 (ST)   23:00 (ST)

Fig. 23.2 Excerpts from the sequence made on April 23, 2017, in the *Rue des Tonneliers*

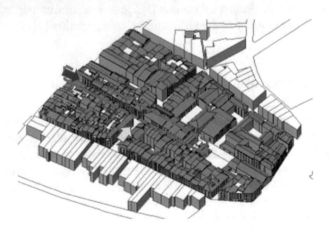

Fig. 23.3 3D model and position of the cameras

To refine the interpretation, a 3D model of the neighborhood was realized by the conventional means of CAD. It includes façades, arcades, roofs, and windows. In Fig. 23.3, the Rue des Tonneliers is surrounded by an orange dashed line. The camera was located on April 23 in the upper left and took the street in succession (red arrow). In a second series of thermographs carried out on May 8, a second camera (that of Fig. 23.5) was placed in a side street, so as to observe frontally the south façades (yellow arrow).

The following two figures are carried out using the software Heliodon2 [4]. To the left of Fig. 23.4, the direct sunlight was accumulated throughout the morning

Fig. 23.4 Cumulative sunlight (calculated with Heliodon2) and measured surface temperatures

Fig. 23.5 Distribution of sky view factors (Heliodon2) and night surface temperatures

until 2:00 p.m., when the right thermography was taken. In general, there is a strong similitude between the two distributions, from which it can be deduced that sunlight and inertia of sunny objects govern surface temperatures when the Sun is present. The slightly projecting roofs explain that the top of the façades remains colder. The lighter parts of the façades are also colder, because a large part of the incident solar energy is reflected instead of being absorbed. It can also be deduced that the lateral conduction on the façades is weak.

The thermography on the right of Fig. 23.5, taken on May 8 at 4:00 a.m., shows much more detail, thanks to a narrower color bar (between 4 °C and 16 °C). We can see the wooden structures, a material with lower thermal inertia and refreshing faster at night. The warmer windows (bottom left) seem to indicate that the corresponding studio is heated (the outside air temperature is about 10 °C). On the other hand, the south-facing façade of the right-hand part of the image most likely keeps its temperature higher because it overlooks a square and has been much more exposed to solar radiation during the previous day.

The distribution of temperatures on the other surfaces of the scene is, however, governed by another phenomenon: refreshment by radiative exchange with the sky. Whereas the temperature of the urban surfaces is very close to one another (with the exception of the ground and the roofs, the greatest temperature difference does not exceed 10°), the clear sky is much colder and even more when approaching the zenith (in this image, the minimum temperature of the sky is already −25 °C). So, the better a surface sees the sky (i.e., the higher its sky view factor is), the more it cools. This effect is particularly obvious on the oblique façade, for which the sky view factors were calculated in Heliodon2 (Fig. 23.5 on the left). It is clear that the most exposed areas are the coldest ones on thermography. Four points were chosen on the façade (top to bottom: a, b, c, d), and Table 23.1 shows the inverse ratio between the sky view factor and the surface temperature.

23.3 Toward Thermography Simulations

Beyond the general principles that have been illustrated in the previous paragraph – sunshine, inertia, and sky view factor – it is difficult to qualify the interpretation of the measured thermographies. Does the emissivity vary greatly in the scene (depending on the materials but also on the point of view), which would distort the images expressed in surface temperatures? What is the importance of shortwave reflections? What is the influence of heating or air conditioning, internal inputs, traffic, wind, clouds, etc.?

In order to answer such questions, it is necessary to use simulation. This one would also allow us to study scenarios for modifying the urban fabric: what would be the impact of a new building, the generalized use of insulation from the outside, etc.? Finally, the establishment of the complete energy balance of a street must be done over a whole year, which is only possible by simulation.

The existing tools for the dynamic thermal simulation of buildings or districts are all based on nodal models, which lead to a drastic simplification of the geometries, incompatible with obtaining simulated thermographies. It will therefore be necessary to use finite element methods (*FEM*). This implies six major problems to deal with:

Table 23.1 Calculated SVF and measured surface temperatures for the four points reported in Fig. 23.5

	a	b	c	d
SVF (%)	30.5	11.2	5.6	4.0
Surface temperature (°C)	9.1	10.4	11.8	12.2

1. *The construction of the geometric model.* To the traditional CAD tools, two more recent techniques are added: the detection of clouds of points (in particular, by drones) and the procedural methods. These different techniques should be combined to obtain very huge models based on the actual situation and parameterized to explore changes in the urban fabric.
2. *Meshing the geometric model.* There are currently no *FEM* mesh generators adapted to the specificities of urban geometries. Some simulations do not require a mesh (ray tracing), others are limited to a nonconforming surface mesh (radiosity), and the *FEM* assumes to mesh the 3D volumes, possibly with shells suitable for radiative exchanges. *CFD* simulations involve meshing the volume of air.
3. *The introduction of boundary conditions and loads.* The solids are characterized by their conduction coefficient and their thermal capacity and the surfaces by their coefficient of convection, their shortwave and longwave reflection coefficients (possibly distinguishing between specular and diffuse reflection), and their emissivity in longwaves. For meteorological data of the sky (shortwave and longwave radiations) and outdoor air (temperature, humidity, wind speed, and direction), measurements can be made, or, for complete annual balances, the constitution of standard meteorological years is also possible.
4. *The simulation.* Shortwave loading can be calculated separately (by ray tracing or radiosity techniques). The complete thermal calculation involves a *FEM*-like method. Thermal-aeraulic coupling can be seriously considered only at a later stage.
5. *Measurement for simulation.* In order to simulate the existing situation, it is necessary to carry out in situ measurements, in particular to determine parameters such as shortwave reflection coefficients strongly influenced by the time (aging, pollution). It is also important to have a nearby meteorological station.
6. *Experimental aspects.* Thanks to the progress of 3D printing, it has become much easier to produce an urban mock-up. However, the thermal and aeraulic measurements on models are difficult, and their transposition to the scale 1:1 is practically impossible. The development of an accurate simulation tool appears to be a necessary step in establishing similarities.

23.4   Conclusions

The first finite element tests that we have carried out show that it is possible to simulate urban thermography but that obtaining convincing results depends on a very careful definition of the geometric model, its mesh size, the boundary conditions, and the parameters of the simulation. In this sense, the measurement campaign described here is a first test case, which gives a goal to these simulations and which sets the steps for future measurement campaigns and for laboratory experiments.

These first ideas and first results open up a new field for *urban physics*, an emerging discipline that will soon be able to acquire the standard tools of calculation that had previously been lacking [5]. Through the precise knowledge of the thermal exchanges which take place in the urban environment, the architect and urban planner can then recover the spirit of urban design and composition, enriched by a sensitivity to the major problems of the contemporary city: the necessary densification, the control of the production and consumption of energies, as well as the comfort of urban life.

References

1. Kircher A (1646) Ars magna lucis et umbrae, Rome, Sumptibus Hermanni Scheus
2. Beckers B, Beckers P (2014) Reconciliation of geometry and perception in radiation physics. Wiley ISTE, Hoboken and London, p 192
3. Vollmer M, Möllmann K-P (2010) Infrared thermal imaging: fundamentals, research and applications, 2nd edn. Wiley
4. Beckers (2009) Heliodon2, software and user's guide, 2009. www.heliodon.net
5. FICUP (2016) In: Beckers B, Pico T, Jimenez S (eds) First international conference on urban physics, 26–30 September, 2016, Quito – Galápagos, Ecuador

Chapter 24
Meta-Design Approach to Environmental Building Programming for Passive Cooling of Buildings

Giacomo Chiesa and Mario Grosso

Abstract Sustainable design practices are being disseminated all around the world, thanks to a growing interest by users, builders, and politicians in facing the impact of climate changes and the need for a more sustainable future. Nevertheless, although design practices include currently green issues and technologies, these are applied mainly in the last design phases in order to comply with local and/or national regulations and requirements (e.g. minimum values for the energy demand to be covered by renewable sources and for the envelope transmittance). Instead, to integrate sustainable technologies in an energy- and cost-effective way, it is necessary to deal with them since the earliest design phases, i.e. building programming and site analysis. Furthermore, passive and hybrid technical building systems (TBS) are dependent on the specific project context, and this is even more apparent for cooling. In fact, while the performance of passive heating TBS is mainly related to solar access and reduction of energy losses, the one of space cooling TBS depends on other variables such as internal heat gains, heat capacity, and wind environment. The paper describes a methodology to assess the energy-saving potential of passive ventilative systems in the earliest design phases. Site and climate aspects, together with definitions of needs and requirements for building programming, will be described. Results from an application of a method based on Givoni-Milne bioclimatic chart to evaluate the climate-dependent potential of passive system are reported. Criteria for spatial and technological integration of passive cooling systems are also presented.

24.1 Introduction

The building sector is responsible for more than 40% in total energy consumptions [1, 2]. If we consider the cooling season, the global consumption for space cooling reached in 2010 1.25 PWh [3, 4], while the residential sector alone consumes for

G. Chiesa (✉) · M. Grosso
Politecnico di Torino, Dipartimento di Architettura e Design, Torino, Italy
e-mail: giacomo.chiesa@polito.it

© Springer International Publishing AG, part of Springer Nature 2019   285
A. Sayigh (ed.), *Sustainable Building for a Cleaner Environment*,
Innovative Renewable Energy, https://doi.org/10.1007/978-3-319-94595-8_24

space conditioning 11.5% of the total US energy consumptions [5]. Nevertheless, the amount of energy consumed for space cooling and ventilation has been constantly rising in the last decades in both industrialised and emerging countries with consequent increasing in the consumption of electrical energy in summer, in black-out risks, in the cost of energy peaks, and in the amount of GHG emissions. Between 2010 and 2015, the air-conditioning market registered an increase of 70% in the sole emerging country sector (source [6]). Furthermore, in Europe, the consumption for cooling is estimated to reach 44,430 GWh by 2020 [7], while the balance between cooling and heating consumptions in the building sector is changing [8].

Passive cooling techniques may constitute an alternative strategy to mechanical cooling acting on three main aspects: heat gain prevention (e.g. shading solutions), heat gain modulation (e.g. thermal mass effect on reducing peak intensity and delaying the thermal peak), and heat gain dissipation by thermal sinks (air, water, soil, and night sky). A detailed description of passive cooling solutions can be found in literature [9–12].

This paper describes a performance-driven approach to environmental and technological design of passive cooling strategies and systems in the earliest design phases, i.e. building programming and site design described within the whole design framework [13].

The scheme of Fig. 24.1 represents the design process framework, within which the above-mentioned performance-driven approach to environmental and technological design in the earliest design phases is included.

Section 24.2.1 deals with the proposed environmental and technological approach applied to passive cooling systems in building programming. Section 24.2.2 deals with the relation of local passive cooling potential to the phase site analysis. Specific indicators to assess climate-dependent passive cooling potential indicators are described in Sect. 24.3. In Sect. 24.4 spatial and technological requirements to integrate passive cooling dissipative systems in buildings are analysed considering the related environmental parameters.

24.2 Passive Cooling Dissipative Technologies in Building Programming and Site Analysis

24.2.1 Building Programming

24.2.1.1 Environmental and Technological Approach

Building design deals with complexity, facing different needs, constraints, and choices that need to be assess since the earliest design phase, i.e. building programming. During this phase, it is necessary to control responses to both the framework of needs (UNI 7867/4:1979 and further UNI 10838:1999) [15], even by considering feedback cycles, and to the user-activity programme [16].

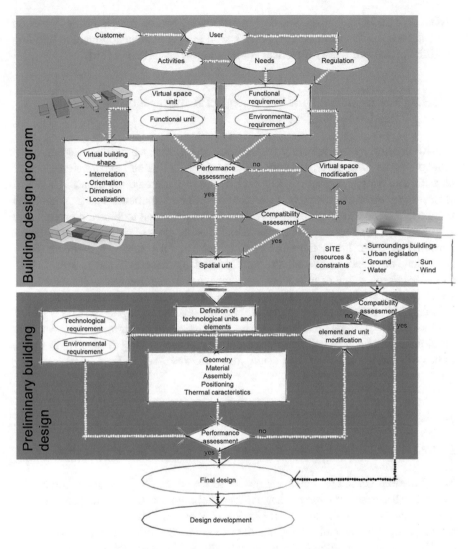

Fig. 24.1 Proposed approach in building programming considering environmental and techno-logical design issues. (Re-elaborated from [14])

A performance-driven approach to environmental and technological design includes the assessment of standard-related constraints (e.g. minimal space requirements) and the definition of user activities, needs, requirements, and expected performances to be further checked and improved. This adaptability/disadaptability analysis based on a selected set of indicators (see also [17] and [18]) may relate to the mandatory requirements of Directive CEE No. 106/1989 [19], the specific system definition of UNI 8289:1981 [20], and the eco-compatibility performance-driven framework of

UNI 11277:2008 [21]. A specific design methodology for school buildings based on a similar approach was described in [22], while an analogous approach to building facilities considering also didactical applications is described in [16].

24.2.1.2 Environmental and Technological Requirements for Assessing Passive Cooling Systems

Passive cooling systems mainly respond to the following building programming-related classes of needs: "perceptive and sensory wellbeing" and "rational use of climatic and energy resources". For each of the considered class of needs, a list of classes of requirements can be determined for both environmental and technological systems, based on the definition of technological units and technical elements (see UNI 7867/4). The former class of needs includes the class of requirements "thermal comfort" and, the latter, the class of requirements "use of thermal sinks" (for a definition of need, requirement, and related performance, see Sect. 24.4). Considering the scope of this paper, only the class of requirements "use of thermal sinks" will be analysed, while the other can be deducted from literature [12].

Table 24.1 reports a list of general requirements belonging to the class "use of thermal sinks", to be analysed in the building programming phase. Specific requirements related to building integration of passive cooling dissipative systems, to be analysed in the schematic design phase, are described in Sect. 24.4.

24.2.2 Site Analysis

24.2.2.1 A Method to Assess Passive Cooling Local Climate Potential

The climatic potential to reduce energy needs for space cooling and heating can be studied by using several approaches. Among these approaches, an analysis based on the Givoni-Milne bioclimatic chart [23, 24] is here considered, since it can roughly relate yearly typical meteorological data to expected indoor comfort/discomfort conditions. This technique is internationally diffused and consolidated. An updated version of the chart can be used based on the ASHRAE comfort model defined in the *2005 ASHRAE Handbook: Fundamentals*. An interesting tool to be used for climate data analysis is Climate Consultant 6.0 [25].

The Givoni-Milne bioclimatic chart is based on a psychrometric chart whose main processes are shown in Fig. 24.2. However, it is essential to remember that the use of bioclimatic charts does not consider specific building aspects (e.g. envelope data, shape, site-building layout, etc.) but allows for evaluating possible energy-saving strategies to be verified for a specific building using proper tools.

Table 24.1 General requirements for the class of requirements "use of thermal sinks" relative to the class of needs "rational use of climatic and energy resources" – building programming phase for both environmental and technological systems [12]

Class of needs:	Class of requirements	System	Requirements for technological units	Requirements for technical elements
"Rational use of climatic and energy resources"	Thermal sinks	Environmental	The spatial distribution of environmental units must follow microclimate and local context in order to maximise the potential use of local thermal sinks	Spatial elements have to be distributed and connected in order to maximise the potential use of local thermal sinks
		Technological	Massive internal surfaces (partitions including wall, slabs, ceiling) have to be considered in order to activate thermal masses for fulfilling night ventilation, increasing the attenuation factor and the time lag of thermal stored release	Devoted technical elements for passive cooling systems (e.g. openings, vents, massive walls with exposed surfaces) and relevant materials must allow for controlling in-out and internal airflows by correctly designing their characteristics including shape, dimensions, orientation, positioning, and thermal-physical aspects
			Openings devoted to ventilative cooling must be foreseen in order to allow for controlling both natural inlet and outlet air flows considering air temperature (coupled or not with other parameters, e.g. RH%)	
			Internal building partition has to assure the passage of natural air flows	
			When needed, integration between mechanical and natural systems has to be guaranteed	Dimensions and performance of mechanical systems have to guarantee the possible integration to passive techniques

24.2.2.2   An Application to the Givoni-Milne Method

An application of this approach was recently adopted in a provision of services for developing minimum building energy performance standards (MBEPS) and nearly zero-energy buildings (nZEB) approach for Turkey, funded by UNEP. This bioclimatic analysis was performed on the 81 Turkish provinces, while the main considered bioclimatic strategies based on Givoni-Milne approach are the following:

– High thermal mass (HTM)
– High thermal mass coupled with night ventilation (HTM & NV)

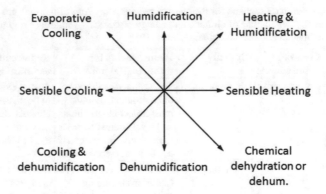

Fig. 24.2 Psychrometric chart processes

- Direct evaporative cooling (DEC)
- Controlled natural ventilation (CNV)
- Internal heat gains (IHG)
- Passive solar low mass – direct gains (SG LM)
- Passive solar high mass – direct gains (SG HM)
- Dehumidification (DEH)

In particular HTM & NV, DEC, and CNV are related to passive cooling dissipative systems, while the others relate to heat gain mitigation (e.g. the use of thermal mass) and other bioclimatic techniques, which may also be integrated to hybrid systems, as DEH with solar regeneration of absorbent beds. For each of these strategies, thresholds are needed to define a comfort boundary limit.

Table 24.2 reports a distribution of percentage of hours in which each specific bioclimatic strategy can be effective. Twelve reference locations were chosen in order to represent the major Turkish climate conditions [26].

This method can be easily applied to any design plot if local climate data (e.g. by the typical meteorological year) are known or assumed by devoted databases, e.g. derived from Meteonorm or EnergyPlus. Here, the comfort threshold is based on the simulation of the effects of ambient climate conditions to thermal comfort regardless of the building characteristics. Hence, this method is useful only in the building programming phase.

24.3 Environmental Indicators for Assessing Geo-Climatic Ventilative Cooling Potential

Two indicators of geo-climatic potential of ventilative cooling were recently developed by the authors to evaluate seasonal hourly comfort/discomfort conditions as well as the hourly intensity of discomfort reduction due to controlled natural ventilation (see, e.g [27, 28, 29].).

Table 24.2 Percentage of hours in which each considered bioclimatic strategy can be effective based on Givoni-Milne approach. Percentages are calculated by using [25]

	Comfort (%)	HTM (%)	HTM & NV (%)	DEC (%)	CNV (%)	FFV (%)	IHG (%)	SG LM (%)	SG HM (%)	DEH (%)	MC (%)	MH (%)
1. Istanbul Florya	16.8	1.5	1.9	0.3	5.8	6.6	22.7	3.2	3.0	3.6	18	30
2. Ankara	13.2	3.7	5.3	3.2	2.1	4.6	23.0	4.2	5.1	0.3	3	55
3. Izmir	19.1	2.3	2.9	5.1	4.6	9.0	23.8	4.2	3.5	0.3	2	57
4. Adana	16.0	2.7	3.8	1.8	7.5	9.9	23.9	5.3	1.8	1.9	7	42
5. Kayseri	10.4	4.6	4.9	2.5	1.9	3.8	23.2	4.5	5.0	0.2	0	72
6. Eskisehir	9.4	2.6	2.7	0.9	2.4	3.4	24.1	3.8	4.0	0.5	2	61
7. Sanliurfa	12.3	2.5	3.9	13.4	0.8	7.6	15.0	3.9	1.5	0.2	4	49
8. Denizli	18.3	4.4	6.0	7.1	1.4	7.3	21.5	2.0	2.0	0.3	10	36
9. Samsun	16.8	2.0	2.5	0.8	6.8	8.1	23.9	4.9	4.1	2.6	2	59
10. Van	12.8	2.7	4.0	2.8	2.0	4.2	17.8	8.5	7.4	0.0	3	47
11. Erzurum	7.8	3.1	3.1	1.8	1.4	2.7	14.9	5.1	3.1	0.1	13	39
12. Bingöl	12.5	5.7	8.3	5.0	0.9	4.7	17.3	4.0	3.8	0.1	1	59

where: *HTM* high thermal mass, *HTM & NV* high thermal mass coupled with night ventilation, *DEC* direct evaporative cooling, *CNV* controlled natural ventilation, *FFV* fan-forced ventilation, *IHG* internal heat gains, *SG LM* direct solar gains low mass, *SG HM* direct solar gains high mass, *DEH* dehumidification, *MC* mechanical cooling and air-conditioning, *MH* mechanical heating

These two indicators are based on cooling degree-hour (CDH) as an index to assess local hourly climate potential cooling needs. CDH is based on ambient dry-bulb temperature and a reference base air temperature, corresponding to the "virtual" effect of a building envelope and gains (solar and internal) on the expected indoor comfort – see Eq. 24.1.

$$CDH_{26} = \sum_{n} (DBT_{amb} - 26)[\text{only positive values}] \tag{24.1}$$

where:

DBT_{amb} = hourly ambient dry-bulb temperature.
n = number of hours in the considered period.

The two proposed indicators compare the ambient CDH with residual CDH values calculated by modifying the ambient temperature according to an expected cooling effect due to ventilative cooling. The first indicator considers the cooling effect of wind-driven cross ventilation for comfort ventilative cooling, while the second refers to a buoyancy-driven ventilation, considering a structural ventilative cooling effect due to daily temperature variations and internal heat capacity.

The reference base air temperature is assumed as a comfort threshold and may be based on fixed values or on variable values based on the adaptive comfort approach. However, in the latter case, the effect of passive systems is already included in calculation, and the application of the two above-mentioned indicators may overestimate ventilative cooling performance in some situations.

Equation 24.2 reports the calculation method used to define the residual CDH for wind-driven comfort cooling ventilation.

$$CDH_{res-wd} = \sum_n \left\{ (DBT_{amb} - \Delta\vartheta_{v,air}) - 26 \right\} [\text{only positive values}] \quad (24.2)$$

where:

$\Delta\vartheta_{v,\,air} = 2.319 \, (v_{wind} \times f_{rw}) + 0.4816$ [27] – considering a comfort upper limit for indoor air velocity generally assumed as $v_{air} = 1$ m/s; for $v_{wind} > 1.5$ m/s $\rightarrow v_{wind} = 1.5$ m/s f_{rw} = internal air velocity reduction factor based on an average discharge coefficient of external openings ($C_d = 0.6$).

Equation 24.3 shows the proposed methodology to calculate the residual CDH for structural ventilative cooling.

$$CDH_{res,svc} = \sum_{day} \left\{ \sum_{hs,cd} (DBT_{amb} - 26) + \sum_{hs,svc} (DBT_{amb} - 26) * 0.7 \right\} [\text{only positive values}] \quad (24.3)$$

where:

day = number of days in the considered cooling period (May to October).
hs, cd = hours in the specific day when $DBT_{amb} > 26$ °C (virtual cooling demand).
hs, svc = hours in the specific day when $DBT_{amb} < 26$ °C (virtual structural ventilative cooling potential).

These indicators were compared to the results of dynamic energy simulations showing a good correlation factor, especially for the indicator that refers to structural ventilative cooling [27, 28]. Differently, the wind-related indicator is less accurate being dependent on various parameters such as plan area density, aspect ratios, surrounding building heights, wind velocity profile exponent, and terrain roughness, all affecting the pressure coefficient (Cp) distribution on a building envelope, which is the main variable influencing wind-driven airflow through a building. Similar indicators were developed for other passive cooling dissipative techniques, such as evaporative cooling [29]. Furthermore, compatible indicators were developed considering indoor comfort, such as the CIDH (comfort indoor degree hours index [30]), or the cooling energy demand reduction, such as the CCR (cooling requirements reduction) and the SEERVC (seasonal energy efficiency ratio of the ventilative cooling system) [31].

24.4 Criteria for Technological and Spatial Integration of Passive Cooling Dissipative Systems in the Earliest Design Phases

The applicability potential of passive cooling systems is highly dependent on local microclimate as above described, but their actual performance is mostly related to their technological integration. This aspect, although defined in the schematic and development design phases, needs to be considered as well in earlier phases such as building programming and site analysis in order to avoid possible constraints for their application in the subsequent steps.

This integration aspect can be divided in two main categories:

- Building spatial integration, which refers to specific spatial aspects that influence passive cooling system applicability in a specific plot and context
- Technological integration, which considers specific technical choices influencing both building definition and technological applicability of the chosen passive cooling system

Table 24.3 reports the main aspects to be considered for integrating passive cooling dissipative systems in building programming and site analysis.

24.5 Conclusions

In this paper, an environmental and technological methodology assessing the potential of passive cooling systems in early design phases is described. The use of a performance-driven approach helps in defining passive cooling requirements for building programming and site analysis considering specific indicators. In particular, the described indicators are related to:

- A well-known climate-related potential approach based on the number of discomfort hours calculated by using a psychrometric chart and the Givoni comfort limits
- A new geo-climatic potential analysis including a hourly residual discomfort intensity calculation
- A potential ventilative cooling method that calculates the number of comfort/ discomfort hours according to a virtual building cooling demand

Furthermore, the paper introduces a list of aspects to be considered for integrating passive cooling dissipative systems in buildings.

Further studies are under development to identify a complete list of aspects for the applicability of passive cooling systems – for heat gains prevention, mitigation, and dissipation – considering environmental, technological, and operational requirements for both spatial and technical levels. Moreover, the integration framework for natural and hybrid systems will be also investigated considering the limitations of natural cooling systems in specific environmental conditions and the consequent need for hybrid solutions.

Table 24.3 Identification of the main aspects to be considered for passive cooling dissipative systems integration in building programming and site analysis

System	Site analysis		Building programming	
	Spatial integration	Technological integration	Spatial integration	Technological integration
CNV wind-driven	Plan area density Surrounding building heights Terrain roughness, wind direction	Aspect ratios, wind velocity profile exponent Wind velocity	Room exposure to prevailing winds Absence of indoor air flow barriers (opening-opening) Leeward positioning for rooms with air pollutant sources	Exposure of internal thermal mass for night cooling depending on activity Controllable window/vent openings depending on functions (position, orientation, and net area)
CNV stack-driven	Temperature difference between indoor and outdoor spaces (density of air)	Gravity force Orientation of top Openings to avoid conflict, between wind- and stack-driven flow	Absence of indoor air flow barriers (opening-chimney) Presence of chimney/tower adjacent to considered rooms depending on activity	Height difference between inlet and outlet openings Controllable window/openings position and net area depending on functions (chimney section)
Direct evaporative cooling (DEC)		Wet-bulb depression	Building type selection: Atrium (closed/open) Adjacent tower Detached tower Absence of indoor air flow barriers	Water recirculation based on water saving strategy Height of the tower
Earth-to-air heat exchanger (EAHX)	Presence of building surrounding empty lots for burring the pipes Buried pipe depth need of inlet air tower outside (1.5/2 m from the ground level)	Soil temperature at the given depth Ambient temperature Water vapour content of air Ground water level	Connection between EAHX fields and the building basement Direct connection to rooms to be cooled for simple domestic installations	Connection with building HVAC system depending on energy demand
Indirect radiative cooling	Daily shading of the radiation exchange surface system	Nocturnal sky exposure of roof surfaces	Direct/indirect connection between covering and cooled room	Thermal mass storage for night-day cycles depending on activities Use of glassless air collectors

References

1. Cuce PM, Riffat S (2016) A state of the art review of evaporative cooling systems for building applications. Renew Sust Energ Rev 54:1240–1249
2. Zouaoui A et al (2016) Open solid desiccant cooling air systems: a review and comparative study. Renew Sust Energ Rev 54:899–917
3. Santamouris M (2016) Cooling the buildings – past, present and future. Energ Buildings 128:617–638
4. Harvey LDD et al (2014) Construction of a global disaggregated dataset of building energy use and floor area in 2010. Energ Buildings 76:488–496
5. Logue JM, Sherman MH, Walker IS, Singer BC (2013) Energy impacts of envelope tightening and mechanical ventilation for the U.S. residential sector. Energ Buildings 65:281–291
6. Daikin Industries (2015) Air conditioning is on the rise. Retrieved from http://www.daikin.com/about/why_daikin/rise/
7. Adnot J (ed) (1999) Energy efficiency of room air-conditioners (EERAC), study for the directorate general for energy (DG XVII) of the Commission of the European Communities, Luxemburg
8. Isaac M, van Vuuren DP (2009) Modeling global residential sector energy demand for heating and air conditioning in the context of climate change. Energy Policy 37:507–521
9. Santamouris M, Asimakopolous D (eds) (1996) Passive cooling of buildings. James & James, London
10. Santamouris M (ed) (2007) Advances in passive cooling. Heartscan, London
11. Cook J (ed) (1989) Passive cooling. MIT Press, Cambridge, MA
12. Grosso M (2017) Il raffrescamento passivo degli edifici in zone a clima temperato, 4th edn. Maggioli, Sant'Arcangelo di Romagna
13. Grosso M, Scudo G, Piardi S, Peretti G (2005) Progettazione ecocompatibile dell'architettura. Esselibri, Napoli
14. Grosso G, Chiesa G, Nigra M (2015) Architectural and environmental compositional aspect for technological innovation in the built environment. In: Gambardella C (ed) Heritage and technology. Mind, knowledge and experience. Scuola di Pitagora, Napoli, pp 1572–1581
15. UNI 7867-4/1999. Edilizia. Terminologia per requisiti e prestazioni. Qualità ambientale e tecnologica nel processo edilizio. And the updated version UNI 10838:1999. Edilizia – Terminologia riferita all'utenza, alle prestazioni, al processo edilizio e alla qualità edilizia
16. Chiesa G, Grosso M (2017) Environmental and technological design: a didactical experience towards a sustainable design approach. In: Gambardella C (ed) Worlds heritage and disaster. Knowledge, culture and representation, La scuola di Pitagora, Napoli, pp 944–953
17. Cavaglià G, Ceragioli G, Foti M, Maggi PN, Matteoli L, Ossola F (1975) Industrializzazione per programmi. Strumenti e procedure per la definizione dei sistemi di edilizia abitativa. RDB, Piacenza
18. Chiesa G (2016) Model, digital technologies and datization. Toward and explicit design practice. In: Pagani R, Chiesa G (eds) Urban data. Tools and methods towards the Algorithmic City. FrancoAngeli, Milano, pp 48–81
19. Council Directive 89/106/EEC of 21 December 1988 on the approximation of laws, regulations and administrative provisions of the Member States relating to construction products (89/106/EEC) (OJ L 40, 11.2.1989, p.12)
20. UNI 8289:1981. Edilizia. Esigenze dell'utenza finale. Classificazione
21. UNI 11277:2008. Sostenibilità in edilizia – Esigenze e requisiti di ecocompatibilità dei progetti di edifici residenziali e assimilabili, uffici e assimilabili, di nuova edificazione e ristrutturazione
22. Chiesa G, Grosso M (2017) An environmental and technological approach to architectural programming for school facilities. In: Sayigh A (ed) Mediterranean green buildings & renewable energy. Springer, Amsterdam, pp 701–715

23. Milne M, Givoni B (1979) Architectural design based on climate. In: Watson D (ed) Energy conservation through building design. Mc Graw Hill, New York, pp 96–119
24. Givoni-Milne bioclimatic chart 1981. In Diamond S., Crow G. and Shafer B. (1993). Climate and site. In Watson (ed) The energy design handbook. the American Institute of Architects, Washington, DC, p 25
25. Liggett R, Milne M (2016) Climate Consultant 6.0, UCLA Energy Design Tools Group, Department of Architecture and Urban Design, University of California, Los Angeles – software, retrieved from http://energy-design-tools.aud.ucla.edu/climate-consultant/request-climate-consultant.php
26. Chiesa G (2017) Turkey MBEPS: Climate environmental strategies towards nZEB, IV research report, Politecnico di Torino Consultancy for the Provision of Services for developing minimum building energy performance standards (MBEPS) and nearly zero energy buildings (nZEB) approach for Turkey – UNEP
27. Chiesa G, Grosso M (2015) Geo-climatic applicability of natural ventilative cooling in the Mediterranean area. Energ Buildings 107:376–391
28. Chiesa G, Grosso M (2017) Cooling potential of natural ventilation in representative climates of central and southern Europe. Int J Vent 16(2):81–83
29. Chiesa G (2016) Geo-climatic applicability of evaporative and ventilative cooling in China. *Int J Vent* 15(3–4):205–219
30. Pellegrino M, Simonetti M, Chiesa G (2016) Reducing thermal discomfort and energy consumption of Indian residential buildings: model validation by in-field measurements and simulation of low-cost interventions. Energ Buildings 113:145–158
31. Flourentzou F, Bonvin J (2017) Energy performance indicators for ventilative cooling. In: Proceedings of 38th AIVC conference, 13–14 September 2017, Nottingham, (in press)

Chapter 25
Urban and Architectural Sustainability in the Restoration of Iranian Cities (Strategy and Challenges): Case Study of Soltaniyeh

Nazila Khaghani

Abstract Urban sustainability is a widely used term, and to help to refine and define the term in the context of this study, one can refer to the basic principles that are widely known from the 1987 Brundtland report: "Humanity has the ability to make development sustainable – to ensure that it meets the need of the present without compromising the ability of future generations to meet their own needs" (World Commission on Environment and Development 1987). The exclusive architecture style advent in Soltaniyeh Zanjan, in the wake of the construction of the glorious Soltaniyeh dome (one of the most colossal brick-made buildings in the world), the rampart of the city, and other historical buildings happened in the region of Oljiato (one of the Mongol chieftains). In the field of improving the quality of the historic urban landscape of Soltaniyeh to become a sustainable city, the role of the World Heritage sites as a significant factor in improving the quality of historical urban landscapes is emphasized. For this purpose, field study and analysis of documents, especially with consideration of the UNESCO 2011 recommendation in the analysis, of a part of the landscape surrounding the dome is the method of this inquiry. The strengths and city potentials of this town are identified with its expansion to the Soltaniyeh's historical urban landscape. Social human activities, visual effects, access routes, farms and partitioning the farmlands, irrigation systems, historical sites, buffer zones for historic areas, and also liability to create spaces, are covered here.

Soltaniyeh is a region unique to Iran and even the world, which unites historical, cultural, and natural values. Soltaniyeh Historical Sites, as one of the few remaining works of the Ilkhan civilization, include the work of the World Heritage Sultanate Dome and other valuable historical and natural elements adjoining it, especially the

N. Khaghani (✉)
University of Florence, Florence, Italy
e-mail: nazila.khaghani@unifi.it

© Springer International Publishing AG, part of Springer Nature 2019   297
A. Sayigh (ed.), *Sustainable Building for a Cleaner Environment*,
Innovative Renewable Energy, https://doi.org/10.1007/978-3-319-94595-8_25

Soltaniyeh Historical Grass. In recent decades, unsustainable urban development has caused some issues and visual and environmental anomalies in the urban landscape, which has been criticized by some historians, and cultural influences. In this chapter, while presenting the results of research on improving the quality of the historical urban landscape of Soltaniyeh, the importance of the global impact as a matter of quality improvement has been demonstrated in the context of urbanization. Therefore, the issue of information and communication, especially the inclusion of the 2011 UNESCO descriptive-analytical review, is part of the historical perspective of Soltaniyeh. In this case, the concept of the historic urban landscape and its constructive approach have been examined first. Then, by expanding it to the historic city of Soltaniyeh, the social-human activities, the visual effects of the environment, the quality of the network of existing roads, the fields and their network, its blueprint, the historical centers, the areas of protection of privacy, as well as the integration and activation of the site as the most important determinants of the world-class resilience of this concept, are identified. The final result of the research emphasizes the elements in relationship to historical urban monopoly, integrated as a form of identity and value.

25.1   Introduction

The land of Iran, with its wide area, has many examples on the historical record of global registration, which emphasizes the relevance of these effects on the livelihood of the land and the importance of this land. According to the definition of outstanding universal value, each of these works has unique characteristics from the time of human history, which reflects the evolution of human society and the residence of human beings over the course of time (Farredanesh 2010: 20). Valuable historical works, especially the effects of Universal Recording Architecture, have created a series of deep links to the surrounding area, which is impossible in one aspect, regardless of other facets. Thus, they are interconnected as one-body components, and this should be in perspective to preserve the life and lives of these works.

The surrounding area comprises these works, as well as the buildings themselves. In the case of the Soltaniyeh dome complex, it should also be noted that from the beginning the dome of the building was not created in isolation, but in a solid and beautiful bond with the collection of buildings and the natural surroundings. Soltaniyeh's dome is more important than paying attention to the Soltaniyeh Historical Grass very near the dome and the Soltaniyeh citadel. On the other hand, the existence of important historical, tourist, and cultural centers such as the ancient Hills of Light, the tomb of Kashan Kashi I, and the Khangah Chalipi Oghli, more than 700 years old, have placed the importance of Soltaniyeh's area in the domain of the dome, emphasizing the improvement of urban landscapes on this global site. In recent years, Soltaniyeh's general appearance has gradually changed as the result of contemporary developments and expansion. Inappropriate development, especially after the liberation of the citadel from the sub-district of contemporary Soltaniyeh in the early 1960s, has also led to the general appearance of the city in the wake of the Great Soltaniyeh World as a world heritage site and one of the largest brick domes worldwide, and some of the unique historical and natural capabilities of this city are threatened. The research questions are as follows:

1. What is the most important aspect of the 2011 UNESCO Recommendation on Historical Urban Landscapes and which elements of Soltaniyeh's historical metropolis?
2. What are the main needs in the process of development and landscape planning of the Soltaniyeh Historic Site? What about the historical axis between the Soltaniyeh dome and the grave of Molla Hasan Kashi?

The main purpose of the research and its results presented in this chapter is to identify the main building blocks of the historical urban landscape around the Soltaniyeh dome (especially the axis of the border between the Soltaniyeh dome and the grave of M. Hasan Kashi) based on UNESCO Recommendation 2011. This review and analytical-comparative study provide a means of paying attention to the diverse aspects of urban landscapes in the perimeter of the global influence of the Solomon's dome.

25.2   Research Methodology

The research was carried out on the basis of the descriptive-analytical method to determine and evaluate the most important elements of the historical urban landscape of Soltaniyeh. The area under consideration is the southern part of the city of Soltaniyeh, especially the axis of the border between the Soltaniyeh dome and the tomb of Hasan Kashi. In this chapter, first, the components of the historical urban landscape are explained; then, with their adaptation to the sampling, the architectural elements of the historical city of Soltaniyeh in the southern parts of the city are identified and studied, which provides part of the desired output in the form of tables and drawings. The need has been identified. Analysis of these natural, historical, and cultural elements has taken place with regard to reasonable documents such as library resources, aerial photographs, maps, Cultural Heritage Organization reports, the Soltaniyeh dome record book, and the laws, documents, and global recommendations on the past and present status of the site.

25.3   UNESCO Recommendation 2011 on Historical Cities

Following the discussions on the urban project for the development of the Wienmitte Railway Station in Austria at the 27th Paris Summit in 2003, the World Heritage Committee decided to convene a conference to discuss "the need for modernization" of historic towns with the cultural aspects of their heritage and their values. For this reason, in 2005, the World Heritage Committee hosted the "Vienna Memorandum" conference on the subject of Urban Historic Landscape Management in Vienna. The recommendations of this conference are the first guideline to address the issue of improving the coherent communication between preservation of historical cities and new developments for the preservation of the integrity of urban landscapes (Riahi Moghadam 2012). A series of specialized meetings of the World Heritage Committee and its colleagues, held from 2006 to 2010, eventually led to the establishment of a credible UNESCO Recommendation document in 2011, which provided an exact definition of the historical urban landscape and its constituent elements to provide guidance and useful policies and practices available to planners, policymakers of historical cities, and more officials of the World Heritage Site.

25.4 Urban Landscape and Historic Urban Landscape Layers in the 2011 UNESCO Recommendation

In most theories and definitions relating to the human relationship and the urban environment surrounding it, much emphasis has been placed on the natural and social components. For example, in the field of urban design, the model can be "components of quality: urban design," which includes three main aspects of functional qualities, experimental qualities of aesthetics, and environmental quality of the site (Golkar 1378: 51). Also, in the field of theoretical foundations, landscape architecture is a model of the values of social, environmental, and economic environment; their interaction is presented [5]. In the model presented for the elements of landscape design, a comprehensive ensemble, elements of human artifacts, and elements of the natural environment and human resources have been introduced as the main characteristics of the landscape constructor (Taghvaei 1391: 135). The 2011 UNESCO Recommendation focuses on the definition of a historic urban landscape, focusing on all aspects of city-specific artifacts, the natural and social dimensions. In this definition, the historical urban landscape is a city boundary that is understood in the wake of historical layering, values, and cultural and natural characteristics, which concept is beyond the understanding of historical centers and collections, and it also encompasses the landmark of its geographic landmarks (UNESCO [9]: article 8).

The scope of this platform includes the topography and gradient of the earth, the morphology of the earth, hydrology and natural phenomena, the historical and contemporary art environment, landscape infrastructure, open spaces, land use patterns and spatial organization, visual perception and communication, and other elements of urban structures. It also includes social and cultural behaviors and the values, economic processes, and other issues related to the heritage's diversity (UNESCO [9]: article 9).

It should be noted that the historic landscape includes a wide variety of residential buildings and areas that are deeply connected with their natural, artisanal, and social surroundings. The term "historical depiction" indicates the existence of the factor of time in this concept and the study of the components that make up the historic urban landscape in time, so that the slow and steady changes that make it evolve have been taken into consideration. Therefore, understanding the city or part of the city as the result of natural, cultural, economic, social, and spiritual processes is considered (UNESCO [8]: 14).

The historical landscape approach seeks to improve the quality of human environments. Therefore, the definition of a historical landscape highlights the wide range of influential components in understanding or shaping the historical urban landscape by emphasizing the importance of value relationships as much as physical relationships.

According to the eighth recommendation of the Recommendation and based on the definition from the historical point of view, two main layers of natural attributes and cultural features are identifiable. Part IX is a component of this main line, in which, by combining these factors, the historical point of view can be understood. Therefore, if any of the components has a problem or a pest, its negative impact on the entire set is visible. From the compatibility of these components with the model presented by the elements of landscape design, charting can be delineated that identifies the most important historical urban landscape layers with emphasis on the characteristics of the natural environment, artifacts, and human activities.

25.5 Soltaniyeh as an Historic Urban landscape

In the context of the global heritage of Gonbad-e-Sultaniyeh, the importance of the site is that "the planning of the main and ancient city of Soltaniyeh has been based on the satisfaction of the natural and social needs of the international community. Soltaniyeh is a rare and unique example of a complex architecture related to its perimeter landscape, which combines the needs of the nomadic people (Mongols and their successors) as well as the unicamous people (Iranians)." This collection meets the standards of the World Heritage Committee. With respect to the outstanding global values of Soltaniyeh, three criteria (second, third, and fourth) of the six criteria set in 2005 have been recorded. The city of Soltaniyeh is one of the examples where the numerous historical monuments of the city have been linked with natural life and social events, natural monuments, and natural values. In general, the landscape of this city consists of four main aspects: urban landscapes, ancient and historical monuments, farms, and the Historical Grass of Soltaniyeh (see Fig. 25.1).

Historical works and ancient sites are scattered around the inner city and its margins, and fields and natural landscapes are located around the city and the city of Baghdad. The status of the geographic area of the area has created a natural texture

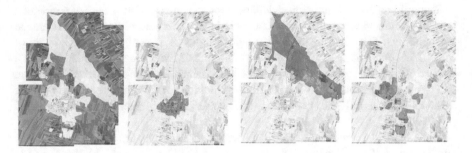

Fig. 25.1 Four main monuments of Soltaniyeh, from the right: historical monuments, Historical Grass of Soltaniyeh, the contemporary city of Soltaniyeh, and the surrounding countryside. (Source: Archives of the Documentation Center of the Dome World Base, Soltaniyeh, mid-1980s)

with an open and horizontal skyline such that, because of its flatness and the lack of physical effects, the Soltaniyeh dome is visible from all parts of the city and even miles away (Sobooti 1369; Kiani 1374). In the UNESCO Recommendation 2011 criteria and library and field reviews, the main aspects of Soltaniyeh's historical urban landscape are identified and classified according to the third charts. According to the studies and questions about the most basic needs in the development of the historic urban landscape of Soltaniyeh, identifying issues and damage in the urban landscape, especially issues related to the natural environment and the environment of the artifacts, and the human-social activities in the study area, are presented next.

25.6 Soltaniyeh Historical Urban Landscape Pathology in the Study Area

Despite the importance of dignity and despite all the efforts and approaches that have been developed in recent years in the development of tourism, the protection and identification of important historical and cultural influences, after the registration of the Soltaniyeh dome on the UNESCO list, the city is still built as is, and the need has not been addressed. The target axis of the research is located in the south of Soltaniyeh Dome and is about 2.5 km long at the grave of M. Hasan Kashi, the poet, and the world of the thirteenth century (Figs. 25.2 and 25.3). The main aspects of field studies and library studies are presented here. The landscape quality of the urbanization focuses on the study in a tabular format, which provides a summary and review of the overall problems of this city. Full discussion of the pathology of the city of Soltaniyeh on a large scale does not fit into this chapter; only the damage of Soltaniyeh's perspective in the Solarmatian Dome is addressed.

Fig. 25.2 The axis between the tomb of Hasan Taki and the Soltaniyeh Dome. (Source: Archives of the Documentation Center of the Dome World Base Soltaniyeh, mid-1980s)

Fig. 25.3 The top of the Soltaniyeh Dome and the tomb of Hasan Tile in the southern part of the city. (Source: Author 2017)

25.7   The Most Important Historical Sites of Soltaniyeh in the Study Area

25.7.1   Perception and Visual Aspects

Environmental perception is the process by which a person receives data and mental perceptions from the surrounding environment. Also, perception can be considered as a goal-oriented process in which one receives environmental transmissions based on the culture and values that govern human societies. Perception of the environment is associated with human recognition of the environment, and based on this, the result is the interaction of sensory perception and cognition. Visual values as the most important part of environmental quality contribute to the strengthening or weakening of the sense of place of individuals in their environment. This phenomenon has a major role in the identity and desirability of the place or, ultimately, society, which has a constructive effect on the economy of its place and its tourist destination. The axis between the Soltaniyeh dome and the tomb of Mullah Hasan Kashi is in the northern part of the city, and in the south the section passes through the farms. The southern part of the route has landscapes that look like vast plains. The different portions of this route create different images that, despite the heterogeneous texture of the city's margin, progress toward the city of Soltaniyeh. The best known landscapes are outside the city, and the most inappropriate sites are from outside the city to the city's fringes. Understanding the appropriate visual effects allows for the proper placement of the activities of the observatory, as it relates to the historical urban landscape.

25.7.2   Network of Roads

In the southern part of the city of Soltaniyeh, there is a border between the dome and the grave of the Hasan Taki, the direct route (Mullah Hassan Street), which connects the two buildings and is now connected with ordinary machinery and agricultural machinery, and is also used by pedestrians and livestock on the sides of this route and across the agricultural land.

In the soil paths to the hills of the valley and the lands of the east can be seen the comparison of the routes in the aerial images and existing maps of the city from the Safavid era, known for their historical dating. These pathways, which are based on the ability of each person to be built as a function of the site and historical background, can be used by planners to access other ancient historical sites and agricultural lands.

25.7.3   Farms and Gratings

A large part of the land around the city of Soltaniyeh is farmland, located in the south of the city, on the two sides of the Mullah Hassan Tile, which is a regular tourist destination in different months of the year. These lands are considered to be agricultural landscape as one of the main infrastructures forming the natural landscape of this city and are now used by farmers. Based on the findings of this study, by recognizing and strengthening the networking of farms, the use of these routes (to a limited extent) for those who tend to naturalize can improve the natural and agricultural landscape along the shores of the mainland. However, the mode, the corresponding paths, is not only the labeling of farmers' activity but can also be a guarantee of a risk to the lands. The tourism attraction is called the farm area.

25.7.4   Water Network

Soltaniyeh has a desirable position regarding the status of groundwater resources. Also, the existence of underground waters and the geological status of the area is one of the most important factors in the appearance of the massive and long grass of Soltaniyeh's Historical Grass. Understanding and improving the current status of the water resources of the region is important as a vital infrastructure for the survival of natural landscapes. Use of existing reservoirs will create sustainable methods in line with the traditional practices of the region in using water resources for agriculture. In the area of the periphery of the city and the landscape, proper understanding of the aqueducts of the aquifer in relationship to the farms is needed to prevent damage to this network that is not avoided by irresponsible design. By installing the blue water network and plotting the network of the farms, plots that are located on the separator responsible for water supply to the land are also identified. The surroundings of the farm are not involved.

25.7.5   Positive Positions for Spatialization and Deployment

The status of human factors in the area, activities in the course, places suitable for positioning, gradient of the earth, proximity to the paths, and the current use of the existing paths are identified. It is necessary for these temporary files to include a number of functions that, while guiding and improving the activities of the present visitors, will make the historic personality of the site possible and can be eliminated (Australia ICOMOS [3]:6).

It should also be noted that the location of spatial and landing applications, such as the location of the site, results from the nature of the activity, the amount of acceptable and suitable distance for the individual, the type of land ownership, and the activity in the region, as well as all the results of the review of the human and social activities of the contacts.

25.7.6   Historical Centers

The city of Soltaniyeh, on the other hand, has a number of other historic and valuable architectural monuments, located a little further than the dome. According to the docket of the Soltaniyeh Dome, these works include the citadel, the tower and its barracks, and the remnants of the interior of the citadel in the Bella dome area as parts of the buried Ilkhani, the ancient mosque of Allameh Helli in the citadel, the tomb of Molana hasan ibn Mahmud Kashani Amoli, known as Hasan Kashi, 2.5 km south of Soltaniyeh, the Chelpy Oghli Tomb in the southwest of Soltaniyeh Dome, the remains of the Ghazan Khan dome, the light trap on a large hill 1.5 km south of Soltaniyeh Dome, and the remains of workshops for pottery and bricks in the east of the mausoleum of Mullah Hasan Kashi, a group of Mahoori Mountains located 2.5 km southwest of the city Soltaniyeh known as the hangar, the far hill, the far emamzadeh, the Fenjanabad lands, the lands of the crewmen northeast of Soltaniyeh, the three-hill area near the caravans, and the hills of the castle and lands of Mustafa Khan in the middle of the Historical Grass of Soltaniyeh (ICHHTO [1]:22–26).

25.7.7   The World Heritage Site

Various studies have been conducted on the protection of the world's heritage and related issues. These ranges, depending on the type of effect, and its areas, have different public and private norms to protect the text. In UNESCO World Heritage Centre [7], because of the importance of the Gonbad Soltaniyeh, there are certain criteria for the observance of the statue of the site by the Cultural Heritage Organization to be considered. According to the approved plan of the country's archeological and cultural heritage in 2005, in the scope of vision and landscape, any developmental activity and physical organization should be carried out by the supervision and approval of the Cultural Heritage and Tourism Organization. These protective laws include general points about the scale, height, type of use, and constructive materials that alone are not sufficient, and thus provide many safeguards based on issues available on each site individually. Hence, to achieve the location of tourism spaces and their type in the limited range, the study and recognition of land ownership is of particular importance. Therefore, on the basis of the implementation of the hermitage of the province and of the spatially functional areas, the information is to be provided by the designer.

In this regard, it is necessary to carefully consider the height of the elements to be considered; under the southern part of the city, there are extensive agricultural fields, which, apart from several small tree communities, have no other vertical element.

25.7.8   Social and Cultural Behaviors and Values

As mentioned, according to the eighth recommendation, social and cultural behaviors and values should be considered as one of the components of the historic urban landscape. Sultaniyeh is a city with historic works and historic sites that have been linked to the daily routine of the city with its historic landscapes. The multitude of people affected by each location affects their affiliation. As more space is experienced, it will be more dependent on the general concepts and place meanings. In general, social and cultural behaviors and values have a range of components that are composed of several factors. Social value includes the qualities on the basis of which each place of the spiritual, national, or other cultural affinity has for a group of the majority or becomes a minority.

This range includes many valuable processes and funds that are considered to be the distinction of that community. In fact, "the historical urban perspective approach learns from local communication traditions and perceptions, while responding to the value of national and international communication."

In the study of the social behaviors and values of Soltaniyeh, the knowledge of the audience of this city is of inestimable importance. Cities with historical values include a wide range of individuals and groups each with their own activity patterns and needs. To categorize these groups, it is possible to separate Soltaniyeh's audience into three groups that include people living in the city, visitors, and tourists. The human and social activities of the audience are as follows.

25.7.9   Summarizing Through Layouts

Based on analytical-comparative analyzes, the most important components of the architectural monument of the historical city of Soltaniyeh are located in the three main areas: the natural environment, the artificial environment, and human-social works. Each of the components is one of the components considered as the constructive layers of the historical observation of Soltaniyeh were placed. It is necessary to consider each in the process of development and planning to improve the historical perspective of the city of Soltaniyeh, because based on the principles of the UNESCO Recommendation 2011, in terms of the historical urban landscape, these layers together, and not individually, form the entire urban landscape as an integrated identity and value. Therefore, in examination of the suburban departments in the southern parts of Soltaniyeh, it is also important that this layer be examined for the mutual effects that they create on each other. When the effect of the grapevine is evaluated in other respects, the role of each component becomes more intense and will be reinforced as the existing capabilities of the region. In this way, the factors and parts that exacerbate or weaken the effects of each other are identified, and the program for the expansion and development of this urban landscape is based on the identified features and the planned constraints.

25.8   Conclusion

The city of Soltaniyeh has unique characteristics that must be considered to know the components of its historic urban landscape. The removal of the ancient city of ancient times, amidst the urban environment and the everyday life of people, has a strong interaction with nature and the natural phenomena of selfishness. In this regard, the comparative analysis and analysis of components from the historical urban perspective in the UNESCO Recommendation 2011 and other conservation texts with a historical urban landscape of Soltaniyeh in the southern part of the city showed that three areas of the natural environment, the artificial environment of mankind, and the social processes and human beings in this city should be taken into consideration. Based on the separate studies provided in this essay, the most basic features of the city's southern harbor include community-based social activities shaped by visitors and locals, and the visual effects of the environment. The city is marked by its location on the south side of the city, the water network, the historic centers, the privacy protection rules on the effects of the World Heritage domes of Soltaniyeh and other ancient sites inside and outside the city, as well as the positions of space and activity. The exact recognition of these funds is as it lies. The architectural perspective of the historic city of Soltaniyeh, their mutual effects on each other, as well as improvement of the quality of the present situation, will provide a basis for improving its quality and preserving the survival of this valuable worldwide site.

The most important developmental needs of the Soltaniyeh historical monument are considering all the components of this landscape. To understand and analyze the historical perspective of the city of Soltaniyeh (the emphasis is on the southern parts of the city), it is necessary to consider the effect of the global registration of Soltaniyeh domes with the protection of their surroundings as a single unit, not individually. This requirement implies that mere consideration of the area as peripheral is not enough, and that all aspects of the artificial environment, the natural environment, and the social-human activities of this area are also to be understood and arranged in planning and development.

References

1. ICHHTO (2005) Inscription Dossier for Soltaniyeh. Tehran. Retrieved September 2012, from: http://whc.unesco.org/en/list/1188
2. World Commission on Environment and Development (1987)
3. Australia ICOMOS (1992) The illustrated Burra Charter. Australia ICOMOS, Sydney. Retrieved September 2012, from: http://australia.icomos.org/publications/charters
4. Thompson IH (2005) Ecology, community and delight. Taylor & Francis Group, London
5. UNESCO World Heritage Centre (2009) World heritage papers 25, world heritage and buffer zones. Netherlands and Paris: UNESCO World Heritage Center. Retrieved February 2012, from: whc.unesco.org/en/series/25/

6. UNESCO World Heritage Centre (2010) World heritage papers 27, managing historic cities. Netherlands and Paris: UNESCO World Heritage Center, Retrieved February 2012, from: whc. unesco.org/en/series/27/
7. UNESCO World Heritage Centre (2011) Recommendation on the historic urban land-scape. Paris. Retrieved December 2012, from: http://portal.unesco.org/en/ev.php-URL_ ID=48857&URL_DO=DO_TOPIC&URL_SEC-TION=201.html

Chapter 26
Influence of the Period of Measurements on Wind Potential Assessment for a Given Site

H. Nfaoui and A. Sayigh

Abstract The use of wind energy requires knowledge of the wind speed on the site concerned and its temporal variation over a long period, for example, 10 years. But, this requires the control and maintenance of the anemometer to minimize the error on the measurements of wind speed. Moreover, measurements, when they exist on the chosen site, constitute an enormous volume of data, difficult to use in their brut form. Moreover, in general, to install a wind farm, due to the lack of time, we are limited to a short period of measures to assess wind potential, 1 year, for example.

The study of statistical characteristics of hourly average wind speed (HAWS) for the Tangier site, based on 12 years of measurements (1978–1989), was carried out. It has shown that 9 years with four measurements per day (6 h, 9 h, 12 h, and 18 h) is necessary for an adequate study of wind speed, such as histogram, frequencies, and daily, seasonal, and annual variations and for the calculation of K and C parameters of Weibull hybrid distribution, in order to evaluate the wind potential for a given site.

For Tangier case, by limiting the potential assessment to 1 year, the overestimation and underestimation of wind potential can reach 75% for the windiest years and 68% for the least windy for the considered period (1978–1989). From which one may conclude the importance of the measurement step, the number of years, and the reliability of the measurements to study the statistical characteristics of wind speed and, consequently, to assess the wind potential for a given site before installing a wind farm

H. Nfaoui (✉)
Solar Energy & Environment Laboratory, Sciences Faculty,
Mohammed V University in Rabat, Rabat, Morocco

A. Sayigh
Renewable Energy, Brighton, UK

© Springer International Publishing AG, part of Springer Nature 2019   311
A. Sayigh (ed.), *Sustainable Building for a Cleaner Environment*,
Innovative Renewable Energy, https://doi.org/10.1007/978-3-319-94595-8_26

26.1   Introduction

Apart from the book on wind energy in Morocco published by Knidiri and Laâouina, research work on the evaluation of the Moroccan wind potential and its exploitation for large-scale electricity generation remains limited until the beginning of the twenty-first century, despite the existence of wind speed data up to 1948, of which the "Direction de la Météorologie Nationale" (DMN) is in charge. But, its exploitation to supply drinking water to the rural population and livestock, by using multi-wind turbines, started in the thirties of the twentieth century.

In order to evaluate and exploit wind energy, statistical and dynamic analysis of wind, related to the climatic data of the concerned region, consists essentially of determining the force of the wind, frequency, as well as the periods during which it blows. So in that effect, it can be measured at different times of the day, the year, and year over year.

On the other hand, the evaluation of wind energy necessitates a statistical assessment of the measurements of the wind speed available on the site over a long measurement period, taken at a small step, for example, an hour, or by using reconstruction models using a limited number of parameters, such as Weibull hybrid function, for example.

To contribute to the improvement of the wind atlas in Morocco, we will take up again the research work of Knidiri and Laâouina for the site of Tangier. The choice of this windy zone is due to the fact that, for the Tangier site, we have available 12 years of HAWS data. In addition, Tangier is located in a promising region for large-scale wind power electricity through the installation of wind farms.

Measuring wind speed every hour needs the use of measuring instruments (anemometer, integrator or acquisition chain, etc.). They operate all the time in addition to their permanent control and maintenance, in order to minimize errors due to breakdowns, as well as their calibration in order to increase the reliability of the measurements. This requires an investment and a skilled staff working continuously, which is not always easy to achieve. To facilitate taking measurements in an isolated site in order to evaluate its wind potential, it is practical and economical to optimize the taking of measurement by selecting only one to four measurements daily, so as to evaluate the wind potential with a good accuracy.

26.2   Statistical Assessment of Wind Speed

In order to evaluate and exploit wind energy, statistical and dynamic analysis of wind, related to the climatic data of the concerned region, consists essentially of determining the force of the wind, frequency as well as the periods during which it blows. So in that effect, it can be measured at different times of the day, the year, and year over year.

On the other hand, the evaluation of wind energy necessitates a statistical assessment of the measurements of the wind speed available on the site over a long measurement period, taken at a small step, for example, an hour, or by using reconstruction models using a limited number of parameters, such as Weibull hybrid function, for example.

26.2.1 Statistical Analysis of Wind Speed Data

A more accurate estimation of wind energy depends on the measurement step and the wind speed measurement period. In this section, we will review some of the compilation and calculation steps required by this method of data analysis.

26.2.1.1 Annual Mean and Seasonal Variation of Wind Speed

The average wind speed gives the order of magnitude of the wind force for a time period at the considered site. Moreover, we know that the winds are different according to the seasons. This seasonal variation is usually represented by a series of monthly average wind speeds. The latter are important when it comes to adapting wind energy to energy needs.

26.2.1.2 Daily Variation

Also, the wind varies between the day and the night. For example, in summer the wind can be strong during the day but relatively low at night, whereas in winter it can be practically constant day and night. Knowing this type of variation makes it possible to harmonize the wind power supplied and the energy needs.

26.2.1.3 Interannual Variation

The interannual variation is a temporal variation that corresponds to the difference observed from 1 year to another. It gives information on the periodicity and irregularities of the wind. This variation is defined quantitatively by the following formula (26.1):

$$I_i = \frac{\sigma_i}{V_i} \qquad (26.1)$$

where:

σ_i: standard deviation of monthly mean wind speeds for the ith month of the calendar, over a period of several years

v_i: overall average speed for the ith month.

26.2.1.4   Distribution of Wind Speed Frequencies

It is a form of data presentation that favors threshold phenomena rather than linearity of the wind conversion systems. The histogram represents a number of hours, for example, for which the wind speed is equal to 0 m/s, 1 m/s, 2 m/s, etc. (on the abscissa). It informs about the hour distribution in different classes. Such a presentation is useful to know the operation threshold of a wind turbine and to evaluate its generated energy.

26.2.2   Available Power in the Wind

The available power in the wind, which can be extracted by a wind turbine, is:

$$P = \frac{1}{2}\rho A v^3 \tag{26.2}$$

where: A: surface crossed by air (m^2)

P: air density (kg/m^3)

v: wind speed (m/s)

P varies proportionally to the cube of wind speed. Therefore, an assessment of the available wind energy at a given site requires the knowledge of the wind speed taken at smaller measuring steps, 1 h, for example, for a sufficiently long period.

26.2.3   Mathematical Modeling of the Distribution of Wind Speed Frequencies by Weibull Hybrid Function

In regions where the frequency, for wind speed is equal to zero, is relatively high (more than 15–20%), the Weibull distribution is not always suitable. For this reason, we define a new probability density [1, 2]:

$$\begin{aligned}
p_h(v) &= F_0 & \text{pour} \quad v = 0 \\
p_h(v) &= (1 - F_0)p(v) & \text{pour} \quad v > 0
\end{aligned} \tag{26.3}$$

The frequency F_0 comes directly from the wind data. A relatively long but highly accurate method, called maximum likelihood method, was used, with good results, for the calculation of parameters K and C. It requires the resolution of the following two equations by the iterative method [1, 2]:

$$k = \left[\left[\frac{\sum\limits_{i=1}^{N} V_i^k \ln(V_i)}{\sum\limits_{i=1}^{N} V_i^k} \right] - \left[\frac{\sum\limits_{i=1}^{N} n_i \ln(V_i)}{N} \right] \right]^{-1} \quad \text{and} \quad C = \left[\frac{\sum\limits_{i=1}^{N} n_i V_i^k}{N} \right]^{\frac{1}{k}} \quad (26.4)$$

N: total number of non-zero wind measurements
N_i: number of measurements in the ith interval
v_i: center wind speed of the ith interval

26.3   Wind Potential in Morocco: Bibliographic Study

Research work on the evaluation of the Moroccan wind potential and its exploitation for large-scale electricity generation remains limited until the beginning of the twenty-first century, despite the existence of wind speed data up to 1948, of which the "Direction de la Météorologie Nationale" (DMN) is in charge. But, its exploitation to supply drinking water to the rural population and livestock, by using of multi-wind turbines, started in the thirties of the twentieth century [1, 2, 5–7].

26.3.1   Available Wind Data

Since 1948, a dozen meteorological stations, several of them located in the airports, were equipped with anemometers and wind vanes to measure wind speed and its direction. By 1978, the number had risen to about thirty. They are managed by the DMN. For most of these stations, we have available hourly average wind speed (HAWS) and wind direction data, measured at a height of 10 m above sea level with the exception of Laâyoune where it is 15 m. The measurement data are archived on files, but their exploitation has remained very limited, apart from some research carried out at the universities and some institutions for research and development of renewable energies, such as the National Agency for the Development of Renewable Energies and Energy Efficiency (ADEREE), former Center for the Development of Renewable Energies (CDER) [1–3].

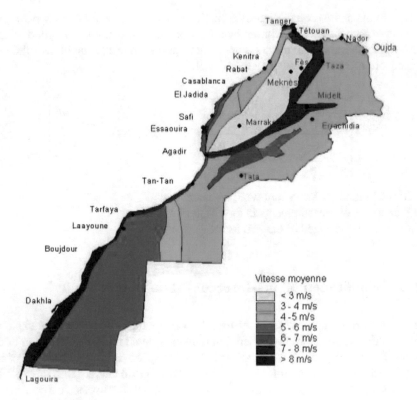

Fig. 26.1 Moroccan wind potential [6]

26.3.2 Quantitative Assessment of Wind Potential in Morocco

In 1986, CDER published the "Atlas Eolien du Maroc" on the basis of data from 17 synoptic stations of the DMN. These stations are generally installed in airports that are located in the less windy sites for the safety of the aerial navigation. This book concerns mainly the mapping of the annual mean wind regimes and the assessment of the wind potential, based on 5 years of wind speed measurements (1978–1982), limiting into 4 measurements per day, responding to the standards defined by the World Meteorological Organization (WMO) (Fig. 26.1) [1–3].

The series of wind speed data used for the establishment of the Moroccan Wind Atlas, published by Knidiri and Laâouina, are insufficient despite the interest of the subject being treated. The map of the average wind speed of the winds in Morocco is interesting, but does not take into account corrections of the local environment (topography, obstacles, etc.) as described, for example, in the *European Wind Atlas* [4].

In 1991, CDER launched an evaluation program of wind potential in Morocco that is more precise and complete, in order to locate the most high wind sites that could be profitable if wind farms were installed. In this context, CDER installed more than 40 measuring stations in all the windy regions of Morocco [3].

One of the most detailed research work carried out by Nfaoui [2] is concerned with the specific study of the statistical and dynamic characteristics of the Tangier wind potential. This study used the measurements carried out by the DMN at the Tangier synoptic station, located at the airport, considering a period of 12 years of HAWS data (1978–1989), that is to say 105,000 brut data (12 × 365 days) and 2 years of hourly data for wind direction (1988–1989). The data are archived on cards at DMN in Casablanca. Nfaoui seized and stored them on the computer support. Thus, it has set up a wind speed database for Tangier site.

26.4   Statistical Characteristics of Wind Potential in Tangier

The choice of this windy zone is due to the fact that, for the Tangier site, we have available 12 years of HAWS data. In addition, Tangier is located in a promising region for large-scale wind power electricity through the installation of wind farms.

The use of monthly and annual averages as well as the maximum wind speed is often satisfied. This presentation provides interesting information to wind energy users on wind speed fluctuations: the annual average wind speed gives an order of magnitude of the wind importance at the site under consideration and the average and maximum monthly speeds inform on seasonal and interannual variations. They also serve to estimate usable annual and monthly wind power. The maxima permit to know the critical speed of the wind turbine and then to predict the resistance of its rotor.

During the considered period (1978–1988), Table 26.1 shows July is the most windy month with a monthly average of 6.5 m/s and June is the least windy month,

Table 26.1 Monthly and annual averages and standard deviation of wind speed, V (m/s) (24 meas./24 h, 1978–1989), for Tangier [2]

Ann Mont	78	79	80	81	82	83	84	85	86	87	88	89	Aver	σ
J	5.02	4.05	3.95	6.07	4.48	6.69	5.07	6.29	6.78	7.73	4.44	5.97	5.54	1.16
F	3.67	5.57	3.62	5.70	7.10	6.07	5.70	7.27	8.07	5.84	6.11	7.77	6.04	1.35
M	4.05	4.48	5.09	4.49	6.48	6.68	6.93	7.04	5.72	5.80	4.28	5.51	5.55	1.03
A	4.50	4.43	6.00	5.21	5.95	6.81	5.70	6.50	7.15	7.45	4.77	5.95	5.87	0.96
M	4.42	6.16	5.01	5.11	6.34	4.78	6.14	5.67	7.72	5.81	5.22	6.07	5.70	0.85
J	2.97	5.92	4.44	6.28	5.56	4.95	6.98	4.23	6.33	7.47	5.68	5.54	5.53	1.19
Jt	4.27	6.39	5.55	8.74	4.60	3.38	7.57	6.08	10.5	7.24	7.50	6.30	6.5	1.9
A	4.59	5.31	5.36	5.85	6.87	5.18	7.78	6.15	5.71	5.20	5.38	4.50	5.66	0.89
S	5.68	5.89	7.24	5.48	6.61	6.56	4.83	6.51	7.28	6.32	10.2	4.37	6.41	1.41
O	4.99	5.44	6.65	5.21	5.91	6.18	8.91	8.37	5.35	4.44	5.81	7.02	6.19	1.2
N	5.21	6.05	5.56	8.46	6.14	5.24	6.96	6.44	6.23	6.44	5.88	6.24	6.24	0.8
D	2.41	5.48	6.13	6.88	5.36	6.91	6.12	7.02	5.15	5.17	5.76	5.44	5.65	1.18
Aver	4.31	5.43	5.38	6.12	5.95	5.79	6.56	6.46	6.82	6.24	5.92	5.89	5.90	0.6
σ	0.90	0.71	1.01	1.26	0.79	1.04	1.14	0.95	1.39	1.01	1.52	0.91	0.3	

$V = 5.5$ m/s. The maximum monthly average reached during July 1986 is of the order of 10.5 m/s, but the minimum value of the order of 2.41 m/s is recorded for December 1978. The most windy year is 1986 while the least windy year is 1978. The maximum speed that can be reached is of the order of 30 m/s [2].

26.4.1 Influence of the Frequency of Measurements and Number of Years on the Calculation of the Monthly Averages of Wind Speed

The study of the influence of the frequency of measurements and the sample size (length of the series of measurements) is a good test of representativeness of this sample from the standpoint of climatological norms.

26.4.1.1 Frequency of Measurements

Measuring wind speed every hour needs the use of measuring instruments (anemometer, integrator or acquisition chain, etc.). They operate all the time in addition to their permanent control and maintenance, in order to minimize errors due to breakdowns, as well as their calibration in order to increase the reliability of the measurements. This requires an investment and a skilled staff working continuously, which is not always easy to achieve. To facilitate taking measurements in an isolated site in order to evaluate its wind potential, it is practical and economical to optimize the taking of measurement by selecting only one to four measurements daily over a shorter period, so as to evaluate the wind potential with a good accuracy.

Figure 26.2 shows that there is a more pronounced approach between the curves corresponding to 4 and 24 measurements per day. So, it is possible to limit to 4 measurements per day (0 h, 6 h, 12 h, and 18 h) to estimate the monthly and annual averages with a good precision for Tangier. For realizing their book on wind potential in Morocco, Knidiri and Laâouina used, also, only 4 measurements per day [1].

26.4.1.2 Number of Years

The visual examination of Figs. 26.3 and 26.4 shows that as the number of years of data increases, the curves tend to approximate to each other, those corresponding to 9 and 10 years being virtually coincided. We conclude that we can limit measurements to 9 years (1978–1986) with four measurements per day to perform a statistical analysis of HAWS for Tangier, therefore, to assess its wind potential.

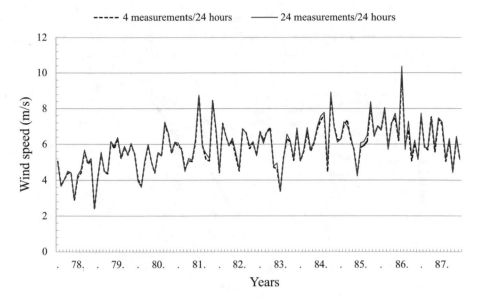

Fig. 26.2 Interannual variations of wind speed for Tangier, V(m/s) [2]

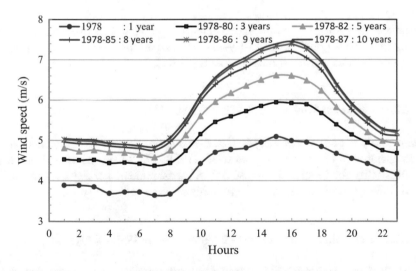

Fig. 26.3 Influence of number of years on daily variation of wind speed [2]

26.4.2 Validation of the Results Obtained

To confirm that only four measurements per day at 0 h, 6 h, 12 h, and 18 h are required over a period of at least 9 years, and in order to characterize the wind potential for Tangier and to reduce the enormous wind speed data, the results of the

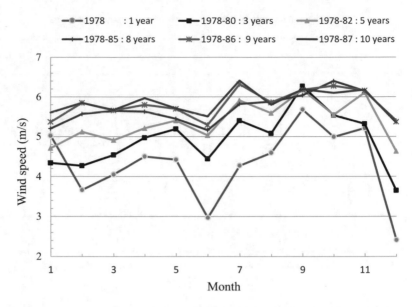

Fig. 26.4 Influence of number of years on monthly averages of wind speed [2]

statistical assessment of wind speed data are presented in different forms by comparing the original series (10 years, 24 measurements/day) and 9 years of measurements, using only 4 measurements per day.

26.4.2.1 Seasonal Variations of Wind Speed

The monthly variation of wind speed for 9-year series with 4 measurements per day is comparable to that of the original series (1978–1987) (Fig. 26.5). This confirms the possibility for Tangier site to use only 4 measurements per day over the period 1978–1986 (9 years) to estimate the monthly averages of wind speed.

26.4.2.2 Modeling of Wind Speed Frequencies by Weibull Hybrid

Figure 26.6 shows that Weibull hybrid distributions for the two considered series, 9 years with 4 measurements/day and 10 years with 24 measurements/day, coincide and adapt well to the observed data. As for estimating the monthly averages of wind speed, only 4 measurements per day over a 9-year period can be used to calculate K and C (Fig. 26.5).

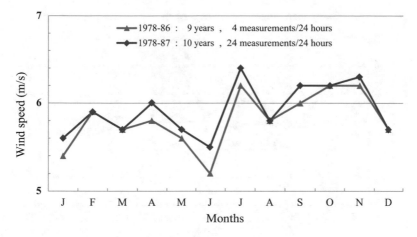

Fig. 26.5 Monthly variations of wind speed [2]

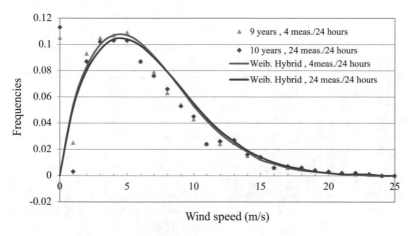

Fig. 26.6 Yearly frequencies of wind speed [2]

26.4.3   Influence of the Choice of the Most and the Least Windy Year of the Considered Series on the Estimation of Wind Potential

26.4.3.1   Seasonal Variations of Wind Speed

In general, wind speed varies with the seasons. This seasonal variation is usually represented by a series of average velocities (Fig. 26.7). The monthly averages for the considered period (12 years, 24 measurements/day) are close to the long-term average (5.90 m/s). On the other hand, the seasonal fluctuations corresponding to the most windy year and the least windy are more important. For the first, they vary

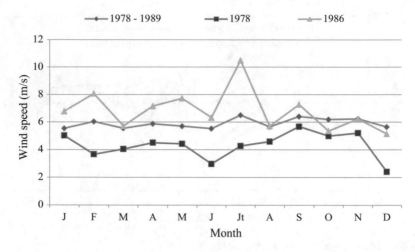

Fig. 26.7 Seasonal variations of wind speed

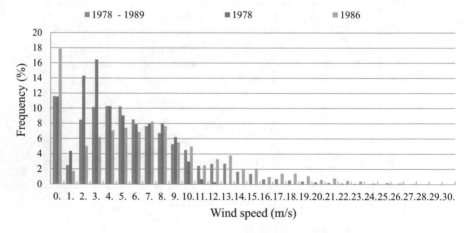

Fig. 26.8 Wind speed histograms

between 5 m/s and 10 m/s and, generally, are above the annual average, while for the other, it is the opposite and varies between 2 m/s and 5 m/s.

26.4.3.2 Wind Speed Histograms

The histogram (Fig. 26.8) represents a number of hours for which the wind speed is equal to 0 m/s, 1 m/s, 2 m/s, etc. (on the abscissa). It informs us about the distribution of hours in different classes. For the windy year, however, there is a maximum of 1565 h per year for class 0 m/s corresponding to $V = 0$ m/s (no wind), or about 18%, a figure that needs to be known, for example, when it is about sizing wind

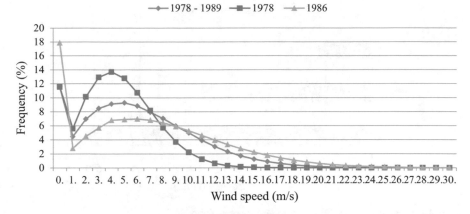

Fig. 26.9   Modeling of wind speed frequencies by Weibull hybrid

energy storage. For the less windy year, we notice that classes 3 and 4 (low wind) are higher corresponding to 1253 and 1441 h per year, respectively, that is to say the frequencies 14% and 16% on average. The classes with high values are not empty, for the windy year. They correspond to the strong wind that can reach 30 m/s for Tangier.

26.4.3.3   Modeling of Wind Speed Frequencies by Weibull Hybrid

Parameters K and C define the shape and the scale of Weibull distribution. A high value of K implies a narrow distribution, with wind speed concentrated around a value, while a low K value leads to widely dispersed wind speed. The scaling factor, C (m/s), adjusts the curve to the frequency histogram. Generally, its value is high for windy sites.

Figure 26.8 shows the adjustment of the observed distributions for the three datasets under consideration using the annual Weibull hybrid distribution. For the most windy year, the distribution of Weibull hybrid is more flattened and shifted toward high wind speed values; the scale factor ($C = 7.53$ m/s) is high compared to that of the least windy year ($C = 9.35$ m/s), but the form factors are comparable (Fig. 26.9).

26.4.3.4   Estimation of Wind Potential

Figure 26.10 shows that the long-term wind potential (366 W/m²) for the period considered (12 years, 24 measurements/day) is three times that corresponding to the least windy year but is only half that of the most windy year, which confirms the results obtained previously, that is to say the importance of taking a long series of measurements to estimate the wind potential. By limiting to 1 year, the overestimation and underestimation of the wind potential can reach 75% (more windy year)

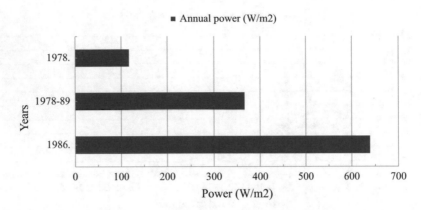

Fig. 26.10 Annual power (W/m²) (24 measurements/day)

and 68% (less windy one), respectively, for Tangiers, which is enormous. Hence the importance of the period, the step, and the reliability of measurements to study the characteristics of wind speed and consequently to assess the wind potential for a given site.

26.5 Conclusion

The series of wind speed data used for the establishment of the Moroccan Wind Atlas, published by Knidiri and Laâouina, are insufficient despite the interest of the subject being treated. The map of the average wind speed of the winds in Morocco is interesting, but does not take into account corrections of the local environment (topography, obstacles, etc.) as described, for example, in the *European Wind Atlas*.

The results of the statistical analysis of the data obtained for Tangier show that the minimum period of data that should be considered is 9 years with four measurements per day (6 h, 9 h, 12 h, and 18 h) in order to study adequately the statistical characteristics of the wind speed in Tangier and consequently evaluate its wind potential.

The influence of the choice of the most and the least windy year of the considered series on the estimation of wind potential confirms those obtained previously, that is to say the importance of taking a long series of measurements to estimate the long-term wind potential for Tangier. By limiting itself to 1 year, the overestimation and underestimation of wind potential can reach 75% for a windy year (1986) and 68% for a less windy year (1978), which is enormous.

From which one may conclude the importance of the period, the step, and the reliability of the measurements to study the statistical characteristics of wind speed and consequently to assess the wind potential for a given site.

References

1. Knidiri F, Laâouina A (1986) L'énergie éolienne au Maroc. CDER, Marrakech
2. Nfaoui H (2004) Caractéristiques du gisement éolien marocain et optimisation d'un système aérogénérateur/groupe électrogène pour l'électrification des villages isolés. Thèse de Doctorat d'Etat, Faculté des Sciences de Rabat
3. Enzili M, Nayssa A, Affani F (1995) Le gisement éolien marocain. le Centre Développement des Energies Renouvelables, Marrakech
4. Troen I, Lundtang E (1988) European wind atlas. Riso National Laboratory, Roskilde
5. Nfaoui H, Buret J, Sayigh A (1995) Wind characteristics and wind energy potential in Morocco. Sol Energy 63(1):51–60
6. Nfaoui H, Sayigh AAM (2012) Renewable energy, large project in Morocco 500MW – solar and wind potential in Ouarzazate Region, Morocco. In: Proceedings of WREF2012, Denver
7. Nfaoui H, Sayigh AAM (2014) Contribution of renewable energy to energy independence in Morocco: wind energy case. In: Proceedings of World Renewable Energy Congress 2014, London

Chapter 27
Integration Strategies of Luminescent Solar Concentrator Panels: A Case Study in Florence, Italy

Lucia Ceccherini Nelli and Giada Gallo Afflitto

Abstract The paper deals with few solutions for the integration of a luminescent solar concentrator (LSC) realized with color dye-sensitized solar cells, some of them produced by ENI Donegani Institute and analyzed by Politecnico di Milano. By this paper, we want to show the versatility of LSC panel either on the facade of a building or on urban lighting.

Luminescent solar concentrators have the capability to produce electricity on transparent surfaces [1], to be used in architecture, and to be integrated into the building envelope, such as in vertical walls [2]. These panels do not need to have south-oriented surfaces because LSC panels are perfect in the presence of diffused light and their performance does not decrease during this condition [3]. The visual effects of the dyed LSC integration are analyzed to find the potential use of such a component in the built environment.

A typical LSC panel consists of three elements: a layer containing fluorophores (fluorescent molecules), a waveguide plate in PMMA or similar, and lastly solar cells along the edges of the plate. Peculiarities of these panels are: a brilliant colour, transparency, lightweight system good for building integration, use of direct or diffuse light, no heat production, use of low-cost materials, and 10% efficiency, and they also glow during the night with their own colourful light.

It is also a feasible alternative to the classic PV solar panels. LSC panels can show identity value, sense of belonging, and iconography which is so much needed in an iconic building like university residence "M. Luzi," in Florence. By four project ideas, we demonstrated the versatile usage of these panels; furthermore, we made a comparison, for equal dimensions, between same efficiency LSC panels and Si-polycrystalline PV panels [4]. In a similar context, like the service industry, even an industrial building can evolve by changing the nature of a sad and empty suburb. LSC panels can be used for public urban lighting and also traffic lights.

L. Ceccherini Nelli (✉) · G. Gallo Afflitto
Department of Architecture DIDA – Centro ABITA, University of Florence, Florence, Italy
e-mail: lucia.ceccherininelli@unifi.it

© Springer International Publishing AG, part of Springer Nature 2019
A. Sayigh (ed.), *Sustainable Building for a Cleaner Environment*,
Innovative Renewable Energy, https://doi.org/10.1007/978-3-319-94595-8_27

The LSC panel integration in the building envelope can contribute greatly to produce electricity and characterize the envelope through transparency and color [5]. However, during the planning stage, it is necessary to focus on the internal and external context of the building. This is because LSC panels are characterized by very bright colors such as yellow and red (which have been commonly used up until nowadays). The usage of these colors can occur into a visual discomfort and dazzling light if they are not used correctly.

27.1   Objectives

Through the explanation of the LSCs (luminescent solar concentrators) and their positive or negative aspects, we want to prove that these are better than the classic PV panels. Sometimes specific local conditions or just common sense leads us to prefer the LSCs.

There are two different examples proposed in this paper: firstly, the M. Luzi student hall. It is situated just outside Florence city center, toward the north. In this case, if we used classic PV panels, we would have low-energy performances due to the local natural conditions. Using LSC panels could be a valid aesthetical alternative, probably more appropriate to represent a dorm iconography. Actually, LSC panels could give a significant improvement to the quality and performance of the energy.

Secondly, urban lighting in Florence. The aim is to improve the urban lighting through road signs integrated from LSC technology. The primary goal is to increase the urban and extra-urban nighttime visibility (during rain, fog, or snow).

Therefore, the objective is to establish a link between squares, parks, gardens, and roads. Urban and extra-urban lighting management is also an area that is seeing energy-saving innovations and develops three categories which will be discussed later on.

Every brought example serves to confirm that LSC panels are versatile and repeatable in different contexts despite low-energy performances. LSC panels have a very low or void cost, so these devices have become the object of deepened studies in different parts of the world.

27.2   Concentration Technology and LSCs

LSC panels belong to the family of the systems to concentration, which use optical systems as mirrors or lenses to focus the solar radiation into high-efficient PV cells.

The ability to produce energy from sunrise to sunset is the greatest characteristic of these devices. This solves the problem of the panel's disposition to a certain inclination and orientation.

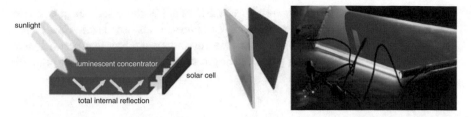

Fig. 27.1 LSC panel. A typical LSC panel during the night with its fluorescence

Initially, it seemed that the problem linked with the classical PV system by using concentration technology was solved (cost and bigger spaces). However, it soon became clear that these also had other problems.

The heat produced from such devices was difficult to reduce, and every mechanical issue of the solar pursuer would have been able to decrease the efficiency of the whole system [6].

In 1970 LSC panels were considered an evolution of existing concentration system for the lack of the solar pursuer and were not able to produce heat.

Generally, the LSC panel is composed of flat plate waveguide (in plexiglass, glass, or similar), which has high optical quality (Fig. 27.1). The flat plate is a matrix of this system and it is covered with a desired thickness of fluorophores. These are particular chromophores, which are colorful and are able to light up in particular situations. Presence of such particles makes the system photoactive; it means that the panel can convert the incidental photons on the surface into energy. Along the edges of the plate, there are PV cells that absorb the solar radiation, which is turned into electricity [7].

The most important characteristic of LSC panels is the ability to convert the solar radiation into more efficient wavelengths: ultraviolet radiation (UV) [8].

The matrix material of the LSC panel is transparent, colorful, and very light, so it is perfect to use in different architectural contexts.

A typical LSC panel is divided into only three parts: a thin layer containing organic molecules, a flat plate and transparent plate waveguide, and finally a small Si-monocrystalline PV cell along the edge of the plate.

For the University of Science and Technology in China, the ideal LSC panel must have the following characteristics:

1. Grid parity with form of energy currently used
2. Ample range of absorption, to absorb more efficiently light energy within the sunlight range
3. 100% quantum fluorescence yield
4. Large stock shift to minimize the overlap between absorption and emission spectrum
5. Simple in functioning
6. Able not to get overheated
7. Realizable in more colors and shapes
8. Low cost
9. Long-term stability (over 20 years)

The second important characteristic of LSC panel is the matrix in polymeric material like PMMA or glass. Part of the UV radiation is absorbed from LSC panel, so during summer time, the energy demand for indoor cooling decreases, hence saving on costs. The remaining part of sunlight range goes unchanged through the plate illuminating indoor areas.

27.3 Methodology

This paper presents a case study of the integration of the Luzi University student hall with LSC colored modules.

Before designing a PV installation, it is important to understand the local project and global conditions.

If the building was unfavorably directed and the shadow factor was too raised and it was not possible to integrate classic PV cells on the rooftops for aesthetic or urbanistic reasons, then it could be necessary to integrate LSC panels. That is exactly the case of M. Luzi student hall (Fig. 27.2).

This building is like a rectangular and white parallelepiped divided on its surface from vertical, narrow, and long elements. The dorm has six stories and it is surrounded by other buildings (2–3 stories), and there is a square around it.

If we integrated classic PV panels in front of this building, then we would have a low performance because of limited solar energy accident (the front of the building has northwestern exposure), because of shading factor and albedo existing in this place. Actually, surrounding buildings' shadows hinder the sunlight on its front. Because of this, we would have an increase of costs and times of return of the initial investment.

Fig. 27.2 Luzi University student house, Florence

In effect, the shading coefficient, calculated on "Solarius PV" ACCA software, is equal to 0.25, while reflectance values are equal to 0.26 (monthly average albedo value).

However using classic PV cells in Si-polycrystalline on M. Luzi dorm, we would have a return greater than using LSC panels. Nevertheless, we prefer these second categories because the first type produces an overheating on the skin of the building and because they are more appropriate to represent the iconography of a university dorm.

We have then drawn four different dispositions of panels on dorm surface and for every possible scenario; we have designed a daily and nighttime view (Fig. 27.3).

The second case study is about efforts to improve urban lighting and safety on the streets by integration of LSC technology to the roads system of signs.

Putting totems near benches in parks or along avenues can generate electric production usable in common services and contributes to improving nighttime visibility of these areas. In this way, it is possible to stay there after sunset (Fig. 27.4).

In urban areas, for example, in one of the most famous streets of Florence "Via dei Calzaiuoli," we have thought about making more visible advertising shop signs using LSC technology along the edges of roadside billboards. This could effectively be used inside open-air shopping malls, by illuminating, regulating, and identifying the shop signs (Fig. 27.5).

Fig. 27.3 Luzi University student house. Day and night views, some solutions

Fig. 27.4 Totem in green LSC panels, "Parco delle Cascine," Firenze

Fig. 27.5 LSC frame with red dye, "Via dei Calzaiuoli," Firenze

Scientific literature has agreed that improving the visibility of roadway can decrease the number of accidents [9]. So, we thought we'd integrate LSC panels in road signs like traffic lights, visible lane marking (means delineators on the border-line on the lane that are visible by the driver), danger sign, give way signs, and obligation or prohibition signs. Despite small energy power, this system is effective and functional in different situations and contexts (Fig. 27.6).

27.4   Results

Through the study of existing examples about LSC panels like Eni's bike sharing shelter in Rome (500 Wp from 60 mq of transparent photoactive yellow plates), and studies of Sergio Brovelli from University of Milano-Bicocca, it is deduced that these panels have about 10% efficiency.

Sergio Brovelli's team studies are based on the use of particular plates die chromophores, which can propagate sun light for long distances without energy dispersing (because concentrators are incorporated of particular colloidal crystals, nanomaterials). Therefore, fluorescence can propagate for long distances without

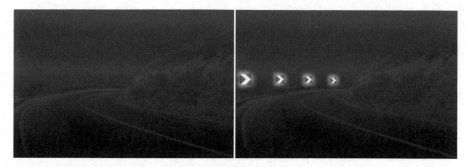

Fig. 27.6 Pictures show the difference roads in visibility with LSC panel integration

losses, and then it is possible to make LSC panels hundreds of centimeters in width. If directly used in buildings (windows for example), they do not do not increase costs.

For Eindhoven University, the color of LSC panel influences yield of panels; in effect a blue dye is more efficient than red dye (21.9% ± 1.6% versus 10.83% ± 1.4%), for example [10].

Using colorful (red, blue, yellow, violet, and green), LSC panels with 10% efficiency on M. Luzi dorm, we would have only 50 KW (by 770 modules for 1000 mq).

Similarly in "Parco delle Cascine" we calculated that every totem can generate around 12 watts, so if we put a totem along boulevards every 5 m (for 160 ha), we will get about 20 kW of electric production.

While if we put a frame in LSC panel on every shop sign in "Via dei Calzaiuoli" street, we would get 270 watts (10 watts for every frame of size 10 × 200 cm).

Lastly, for urban and extra-urban lighting case study, the energy power of every road sign depends on the greatness of signs. In effect, for the small sign (40 cm in diameter), we will have 6 watts; for the middle sign (60 cm in diameter), we will have about 14 watts; and for the large sign (90 cm in diameter), we will have 30 watts.

27.5   Conclusions

The present work allowed assessing daylighting performances of a series of innovative smart windows developed in Italy by Eni and the Politecnico of Milano, based on configurations realized with LSC plates characterized by different colors.

Nowadays, 10% performance for LSC panels is too low to persuade the public administration to finance such farsighted projects, but scientific research offers prototypes that are more and more powerful and durable every day (quantum dots, EuTT with organic binders like thenoyltrifluoroacetone, etc.).

In effect, if we had LSC panels with 30% performance, surely classic PV panels would disappear from the market because LSC panel does not produce heat, for low costs and weight and for great building integration.

References

1. Myong SY, Jeon SW (2016) Efficient outdoor performance of esthetic bifacial a-Si: H semitransparent PV modules. Appl Energy 164:312–320
2. Aste N, Tagliabue LC, Palladino P, Testa D (2015) Integration of a luminescent solar concentrator: effects on daylight, correlated color temperature, illuminance level and color rendering index. Sol Energy 114:174–182
3. Wang T, Zhang J, Ma W, Lou Y, Wang L, Hu Z, Wu W, Wang X, Zou G, Zhang Q (2011) Luminescent solar concentrator employing rare heart complex with zero self-absorption loss. Elsevier, Solar Energy
4. Meinardi F, Colombo A, Velizhanin KA, Simonutti R, Lorenzon M, Beverina L, Viswanatha R, Klimov VI, Brovelli S (2014) Large-area luminescent solar concentrators based on 'Stokes-shift-engineered', nanocrystals in a mass-polymerized PMMA matrix. Nat Photon 8(5):392–399. 13/04/2014
5. Aste N, Adhikari RS, Del Pero C (2012) An algorithm for designing dynamic solar shading system. Energy Procedia 30:1079–1089
6. Aste N, Tagliabue LC, Del Pero C, Testa D, Fusco R (2015) Performance analysis of a large-area luminescent solar concentrator module. Renewable Energy 76:330–337
7. Desmet L, Ras MAJ, De Boer DKG, Debije MG (2012) Monocrystalline silicon photovoltaic luminescent solar concentrator with 4.2% power conversion efficiency. Optical Society of America, Luglio
8. Aste N, Del Pero C, Tagliabue LC, Leonforte F, Testa D, Fusco R (2015) Performance monitoring and building integration assessment of innovative LSC components. Clean Electrical Power (ICCEP), 2015 international conference on IEEE
9. Zarcone R, Brocato M, Bernardoni P, Vincenzi D (2016) Building integrated a photovoltaic system for a solar infrastructure: Liv-lib' project, in SHC 2015, international conference on solar heating and cooling for buildings and industry. Energy Procedia 91:887–896. http://www.sciencedirect.com/science/article/pii/S1876610216303551
10. Li Y, Olsen J, Nunez- Ortega K, Dong WA (2016) Structurally modified perylene dye for efficient luminescent solar concentrators. Solar Energy 136:668–674

Chapter 28
Photovoltaic and Thermal Solar Concentrator Integrated into a Dynamic Shading Device

Giulia Chieli and Lucia Ceccherini Nelli

Abstract This paper presents the design of a photovoltaic/thermal solar concentrator (PV/ST) integrated into a system for external shading device suitable for different building typologies such as office facilities or residential houses.

The paper addresses the problem of designing efficient shading devices for buildings. The novelty of this project is a dynamic design for shading devices, new in the market of photovoltaic and concentrator systems, that changing orientation and configuration allows the best lighting comfort in buildings while producing energy.

This device consists of a series of linear low concentrator with monoaxial sun tracking and a semi-parabolic profile reflector providing the solar radiation concentration into a string of monocrystalline cell PV, with 5.5× concentration factor. The thermal energy production is achieved through the cooling system of the device whereby hot water is produced.

The contribution of this study is a new concept, and developments in solar thermal concentrator systems reveal new trends for the technology to be integrated into architecture, especially in dynamic solution for shading devices.

28.1 Introduction

The reference framework for this project is the performance of the building envelope. One of the most studied topics related to this theme is kinetic architecture [1]: an architecture able to change its configuration adapting to the environment in which it is situated, to the passage of time, and to different climatic conditions.

The dynamic part of a building is principally the façade or the façade component. The step forward in comparison with static solutions is both technological and stylistic because it allows to minimize or maximize the thermal loads of a building while opening new possibility in terms of design and aesthetic qualities.

G. Chieli (✉) · L. C. Nelli
Department of Architecture DIDA – Centro ABITA, University of Florence, Florence, Italy
e-mail: lucia.ceccherininelli@unifi.it

© Springer International Publishing AG, part of Springer Nature 2019
A. Sayigh (ed.), *Sustainable Building for a Cleaner Environment*,
Innovative Renewable Energy, https://doi.org/10.1007/978-3-319-94595-8_28

The dynamism in architecture can be reached with different technologies and materials. A popular case study is *The Arab World Institute* in Paris [2] in which the architect Jean Nouvel reinterprets the traditional theme of "mashrabiya," a wood grid widespread in Islamic architecture, transforming it in the characteristic element of the building able to regulate the light in the internal environment, thanks to the movement of a series of diaphragms. In the *Beijing National Aquatics Center* [3], the dynamism is obtained, thanks to the use of the polymer ETFE [4] that forms a pattern of bubble able to regulate the internal temperature by regulating the pumping of air inside the bladder formed by the two layers of ETFE. There are a lot of study cases to refer to, and all of them show how the best results, both technological and stylistic, are reached in projects characterized by a high iconic value and big investments of money. At the level of common constructions instead, it is hard to find examples that are so technologically advanced and also successful from an architectural point of view: the dynamism of a façade is normally consigned to the possibility to move, manually or electrically, the shading systems such as curtains or brise-soleil. If we look to more advanced solutions, it can be seen that the aesthetical appearance is quite overshadowed and the building appears more as a facility than as an architecture. Example of this is photovoltaic brise-soleil that haven't achieved a wide dissemination due to a design hard to integrate.

28.2   Research Objectives

The idea for this project emerged from the will to design a dynamic component, simple, versatile, and adaptable to different construction categories and contexts: a component that can be an expression of technological innovation without overlooking both the economic sustainability and architectural integration.

The element brise-soleil has been taking in consideration a starting point for the development of the project due to its features of adaptability in both new buildings and renovation projects. The first step was thinking a way to implement its technological potentiality without penalizing its construction simplicity. A common feature of the shading system available in the market is the option to tilt the shovels so to change the brightness conditions inside a building according to user needs: the handling can be manual, mechanical, or controlled by a sensor sensitive to external climate change so that the building's envelope can manage itself. A product already developed and that can be widely found in the market is the one of photovoltaic brise-soleil: this element can be realized applying a photovoltaic film in the shovel's surface or using small-scale PV panels, in place of shovels, so to form a shading system.

Doing a research about the state of the art of photovoltaic brise-soleil, it appears clear that their use is quite restrained. Despite all the leading companies producing shading devices have PV products in their catalogs, their spread in the building sector is not comparable to the spread of standard products although they present the advantage to produce energy. This lack of dissemination can be easily explained considering the PV brise-soleil also just from an aesthetic point of view: in comparison with standard product, the PV ones have bigger dimensions that reduce their versatility, often totally out of scale when applied in small/regular size constructions such as residential buildings. The replacement of shovels with PV panel negates the possibility to paint the elements so to create nice chromatic effects characterizing the façade. In this way, the building ends to look like an energy production facility without any aesthetic value.

The integration of PV in architecture is a thematic that a lot of architects had already tried to face, sometimes with very good results, but more often it happens that the use of PV in buildings is just intended to the production of energy or, at most, to show a special attention to environmental issues.

This project tries to fill this gap in the design of PV shading device by using the technology of the concentrated solar system.

28.3 Methodology

28.3.1 Concentrated Solar Technology

One part of the research in the field of PV and solar thermal collector is focused on concentrated technology: this system, thanks to mirrors or lens, concentrates sun rays in a small area where the active material is located. The active material is represented by photovoltaic cells or pipes in which a heat-transfer liquid flows.

The components of a solar-concentrated photovoltaic system are:

_Receiver
_Reflector
_Tracking system
_Inverter

The *receiver* is located in correspondence of the focal line of the system, where all the sun's rays are concentrated. In this place is situated the PV material. Because of the high concentration of solar energy, the temperature goes up very fast so that a heat sink system is required. The heat sink prevents the PV cells from degradation and loss of efficiency (about -0.5% every degree exceeding the operating temperature of $40°$). The dissipation system can be active when cooling is achieved by using the flow of a low-temperature liquid or passive in case a heat sink with straight fin is applied.

The *reflector* is an optical system that can be "imaging optics" if consisting of a curved mirror and "non-imaging optics" when using lenses as the Fresnel lens. Both the solutions use the principle for which when a surface **A** is hit by the sun's rays, it is possible to converge the rays onto a smaller surface **a** through a reflector so that the incident energy in **a** results the same of the incident energy in **A**. The ratio **A/a** is called *concentration factor C* and is measured in "suns" (x). The concentration factor subdivides concentration system in low-concentration photovoltaic (LCPV) system and high-concentration photovoltaic (HCPV) system, when suns are higher than 300×. Thanks to this principle it is possible to reduce the quantity of PV material used and thereby the costs of the system. As known, in a standard PV panel, the price of the photovoltaic material affects the final price for the 70%. Monocrystalline silicon cells are used for LCPV, and they still are the one with greater efficiency. In HCPV system are instead preferred multi-junction cells because of their ability to endure deterioration when subject to high temperature.

The *tracking system* is a fundamental device, while concentration system can only converge direct solar radiation. The system's optical axis has to be aligned with the source of light so that the system can give the greater amount of energy. Tracking systems can be single- or dual-axis trackers: dual-axis trackers adapt themselves both to azimuth and altitude, while in case of single-axis tracker, the choice of the right axis must be done considering the features of the installation place. A device that follow the path of the sun in north-south direction has the maximum performance during the central hours of the day and collects more energy in summer, vice versa a tracker that follows the sun in east-west direction catch collect energy steadily during the year.

Like in any other PV system, an *inverter* is required to change the direct current into alternating current.

At this moment concentrated technologies are mainly used in big industrial facilities to produce heat or electrical power. Big plants, however, still have some points of criticism not completely solved and currently under study. The main problem is related to the noncompetitive price of multi-junction cells: originally used in the space sector, these cells have now a widespread use in HCPV since the high temperature reached with this system prevents the use of monocrystalline cells, easily subject to deterioration. Multi-junction cells use compounds of semiconductor elements as an alternative to silicon, such as gallium arsenide or indium phosphide. Their efficiency is between 25% and 50%, they have better resistance to high temperature and less degradability, but they present the big inconvenience of a price from 3 to 5 times higher than a monocrystalline cell due to the complicated production process and the reduced availability of the alternative elements. Other factors that affect the price of HCPV systems are the heat sink system that is always required and the continuous maintenance needed. These systems are based on the law of reflection and refraction; for this reason, mirrors and lens have to be kept clean so to avoid loss of efficiency. The problems experienced in HCPV considerably decrease with small size and small concentration systems: smaller concentration factors correspond to lower temperature so that is possible to use monocrystalline cells; moreover, heat can be dissipated using heat-transfer fluids as in a classic solar thermal collector. Furthermore, because of the smaller size, costs for maintenance are significantly reduced.

For all these reasons, the use of a low-concentration system has been taken into consideration for the development of the project presented in this paper. On the current market, small size concentration systems are very rare although it is possible to find some more examples considering research projects. The aim of the products on the market is mainly energy production through renewable sources, but less attention is paid to their appearance so that they look more like facilities than an architectural element. The topic of integration has a bigger importance in research projects although they present also bigger complexity regarding shape, gears, and technology so that their prices are not competitive and the use of normal brise-soleil combined with classic PV panel is still preferred.

28.4   Results

The guidelines from this project have been delineated starting from the study of the state of the art of concentration systems already existing in the market and in research projects. The basic idea was to design a prototype that, thanks to the link with a sensor, could modify its configuration, opening and closing itself, according to necessities so to significantly change its appearance as well the environment's perception from the point of view of a user located inside a building where the prototype is applied.

One of the most important goals was to keep the whole system very simple so to ensure a large-scale applicability. In this sense was chosen to focus on a linear element, instead of punctiform, so to minimize and optimize the number of gears required for the handling system. The device is composed of two specular semi-parabolic parts able to revolve around their focal line, with an opening-closing movement, and to move together revolving upward and downward so to follow the sun apparent motion. The first step was the geometrical study of a parabolic shape suitable as a profile for the component. In the selected shape, the focal point results in internal respect the two semi-parabolic profiles so that the two wings forming the parabola can protect the most delicate part of the component represented by the PV cells (Fig. 28.1).

The mechanisms responsible for the handling system are very simple: both the monoaxially tracking and the opening-closing movements of the parabolic wings are provided by the anchorage of the component to a central linear pivot. Three cogwheels are mounted around the linear pivot: one is welded to the pivot itself and regulates the integral movement of the whole device, and the other two cogwheels are, respectively, welded to one wing of the device and regulate opening and closing

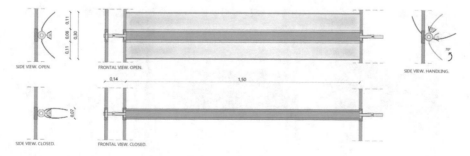

Fig. 28.1 Section and elevation view of the component

of the brise-soleil. The motion is transmitted to the cogwheel by three screw pumps controlled by two small engines able to react to the impulse of a climatic sensor (or eventually controlled by internal users). The handling gears are hidden behind two aluminum carters through which the whole shading system is anchored to the building's façade. In the internal part of the carters, there are also notches in which the pivots are situated so to fasten their position and stability. The device has been designed to work as a brise-soleil producing energy in its open configuration by standing always perpendicular to the sun's rays, following the sun apparent motion and directing the sun rays, through its mirror part, to the focal line of the system. In case of absence of sun, during nighttime or in every situation in which the maximal brightness is desirable, the two parabolic wings can be closed so that the device's dimension results reduced for more than one-third in comparison with its dimension in the open configuration. This option not only solves a real need but also gives to the device a changing appearance so that the whole façade can change according to clime and passage of time in conformity with the dictates of dynamic architecture. The internal part of the two aluminum wings is coated with a reflective film (reflectivity from 85% to 95% depending on the product) suitable for outdoor use, thanks to its weather-resistant property. The external part can be basically coated at will, according to the aesthetic effect desired. More simply and economically, it can be painted in every color and nuance (Fig. 28.2).

Corresponding with the focal line of the parabola, an extruded profile with a trapezoidal shape is mounted in which the PV cells and the electrical network required for the transport of the produced energy find place. As for every concentrator system, a cooling device is necessary due to the high temperature reached along the focal line. In case of low-concentration system, an adequate cooling can be achieved by the flowing of a heat-transfer liquid along the focal line, in this case inside the trapezoidal profile. In this way, the liquid not only works as cooler but can be also be used to produce domestic hot water from a renewable source that can be used in the building itself. To provide this dual use, a simple distribution system is required, the same used in standard solar thermal collector usually located on the roof of buildings. The choice of this cooling system not only allows the shading device to produce both electrical and thermal energy but also does not complicate

Fig. 28.2 Axonometric view of the component

its functioning because the technology used is common and run-in in the field of the classic thermal collector. Inside the trapezoidal profile, properly isolated so to not dissipate the heat, flows a duct with a 1.5 cm diameter in which runs the heat-transfer liquid. This duct is connected, through specific hoses, to a column pipe with a diameter of 3 cm, which is adequate to ensure hot water's distribution. These pipes, located on each side of the parabolic element, are situated inside the aluminum carter that hides and protects also the screw pump and all the handling system's gears. The conduct inside the trapezoidal profile has also the function of structural pivot being connecting to the second pivot, a little bigger, where the two parabolic wings are mounted and are free to rotate. This bigger pivot allows also the simultaneous rotation of the shading system so to follow the apparent motion of the sun. The trapezoidal profile is designed so that the PV cell is mounted in its oblique sides having a surface of 3×150 cm = 450 cm^2 each. The surface corresponding to the bigger parallel side, the one visible in the front view, is completely free so that can be painted or coated according to the will planners.

Once the concentrator dimensions are defined, the concentration factor C has been calculated as follows:

- C = Collecting surface / PV surface

  - Collecting surface = 0.11×1.5 m = 0.165 m^2
  - PV surface = 0.02×1.5 = 0.03 m^2

- $C = 0.165/0.03 = \mathbf{5.5 \times}$
- C is referred to one of the two specular parabolic wings.

The linear devices are assembled together so to form a shading module. The module's dimensions are 1.5 m width $\times$ 3 m height so that one module is high enough to cover the standard floor's height. The width has been determined considering both aesthetical and technical reasons: the largest dimension would have influenced too much the total weight of the module so that more complicated handling systems would have been required. On the other hand with a smaller dimension, the building's façade would have appeared too fragmented. A width equal to 1.5 m results moreover suitable for the building of different dimensions and proportions. Every module is assembled with seven linear shading elements, placed at a calculated distance so that they can't shade each other, and the global efficiency of the system is not reduced. The distance between linear elements allows also internal user to have always the perception of the outside environment when the elements are in the open working configuration (Fig. 28.3).

28.4.1 The Distance Between the Linear Shading Elements

The choice of a suitable distance between the shading elements affects both the aesthetic and the energy production: the shading of part of them can indeed be the cause of a drastic loss of efficiency if not attentively calculated. Empiric studies

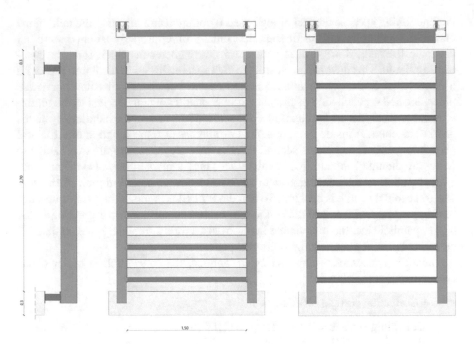

Fig. 28.3 Elevation of the Module in the two positions: open and close

carried out about horizontal prototype [5] facing south, 30 cm height, have shown that the loss of efficiency is less than 98% when the elements are situated at a distance of 20 cm. Being this project firstly a shading device, a distance of 20 cm between the linear element has been rejected because of the big quantity of light allowed inside the building. On the other hand, also the option of having no distance between the elements has been rejected due to the big loss of efficiency related to energy production. A distance of 10 cm has been chosen why the mutual shading within the element is not present in winter time, when there is a bigger need of electrical and thermal energy, while during summer time efficiency losses are around 50% (Fig. 28.4).

28.4.2 Productivity of a Linear Shading Element (Relating to a Horizontal Element Facing South)

Since the impossibility to carry out test directly on this project, due to the absence of a prototype, data related to the energy production has been calculated by interpolating experimental data derived from tests conducted by a team from the Engineering Department of the Florence University [6]. These tests concern a concentrator prototype which features are comparable with one of the projects presented in this paper (Fig. 28.5).

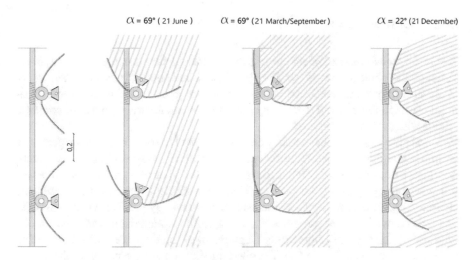

Fig. 28.4 Position of the receiver of the solar concentrator in relation to different inclinations of sunlight

Fig. 28.5 Few photomontages of the integration of the solar concentrator in façade

Engineering prototype		Project	
Colletting Surface	0.25 m	Colletting Surface	0.22 m
Length	3 m	Length	1.5 m
Distance between elements	0.2 m	Distance between elements	0.1 m
Annual Production	600 Thermal kWh	Annual Production	265 Thermal kWh
Annual Production	120 Electric kWh	Annual Production	53 Electric kWh

28.5   Conclusions

The link between technological innovation and architecture is nowadays very strong although their integration is not always efficient and successfully designed, especially in case of common constructions. The main goal of this project is to experiment new technologies not in special cases but in standard architecture for common clients with ordinary investments.

During the design process, attention has been paid to try to keep the project very simple although innovative. Simplicity not only affects the spreading of a new technological solution positively but also can reduce the costs of a product, and this was the other big purpose of the project. Until this moment, there has not been the chance to produce a prototype, so the estimation of energy production and costs has been done, with proper proportions, through the comparison with the results obtained by the team of Engineering Department of the Florence University testing a similar and comparable device. Despite this, during the project phase, the device has been simplified in comparison with the University's prototype, both concerning the materials and handling gears used, so to be sure that the final product could be more affordable in the market.

The final goal was to design a technological device that was easy to integrate into a lot of building typologies, in the same way as standard brise-soleil is used but providing a new value to a classical shading solution. It was very important for the first design phase that the concentrator device would not appear as a facility system but that could interact with architecture representing instead an added value also from an aesthetical point of view. The possibility to connotate the single linear element with a different color, in both the open configuration and the closed one, facilitates the dialogue with the building and the environment around. The contained dimensions, similar to the dimensions of a lot of shading system already existing in the market, assure that the device does not impose its presence overriding the architecture.

The choice to use the concentration technology avoids the problem of the photovoltaic integration in building's façade, a topic that has not been well assimilated yet in the architectural field. Moreover by focusing on the use of a low-concentration system, the critical issues concerning high prizes for materials and maintenance, easy deterioration, and heat disposing have been prevented.

Notes

1. Also called dynamic architecture
2. Jean Nouvel, *The Arab World Institute,* Paris,1981–1987
3. PTW Architects, CSCEC, CCDI and Arup, *Beijing National Aquatics Center,* Beijing, 2004–2007
4. Ethylene tetrafluoroethylene (ETFE) fluorine-based plastic

References

1. AA. VV (2008) Manuale pratico di edilizia sostenibile. Esselibri, Napoli
2. Cappelletti A, Catelani M, Ciani L, Kazimierczuk MK, Reatti A (2016) Practical issues and characterization of a photovoltaic/thermal linear focus 20x solar concentrator. IEEE Trans Instrum Meas:65:2464–2475
3. Catelani M, Ciani L, Kazimierczuk MK, Reatti A (2016) Matlab PV solar concentrator performance prediction based on triple junction solar cell model Measurement. J Int Meas Confed 88:310–317
4. Reatti A, Kazimierczuk MK, Catelani M, Ciani L (2015) Monitoring and field data acquisition system for hybrid static concentrator plant. Measurement. 98:384–392
5. Cappelletti A, Reatti A, Martelli F (2015) Numerical and experimental analysis of a CPV/T receiver suitable for low solar concentration factors. Energy Procedia 82:724–729
6. ENEA (2014) ENEA Atlante italiano della radiazione solare [Online]. Available: http://www.solaritaly.enea.it/

Chapter 29
A University Master's Course and Training Programme for Energy Managers and Expert in Environmental Design in Italy

Marco Sala, Lucia Ceccherini Nelli, and Alessandra Donato

Abstract This paper presents the learning activities of the second level Master course, ABITA – Architecture Bioecology and Technological Innovation for the Environment – that involves energy expert training in the field of energy efficiency.

The ABITA master's course training programme is offered in Florence in Italy and covers nearly all the energy-relevant issues that can arise in public and private companies and sectors. The final output of the practical training is to achieve an elevated professionalism in the study of environmental design and energy management in buildings.

The master's includes studies on low-energy architecture and energy efficiency measures, integration of renewable energies in buildings, building information modelling, dynamic software for energy simulations and energetic diagnosis. Master ABITA training provides a solid basis for increasing the knowledge and skills of energy managers and is developed with an emphasis on practical experiences related to the knowledge through case studies, measurement and verification of energy-efficient solutions in buildings, in the industry and in the cities.

Highlights

- Sustainable design and energy analysis through BIM
- Training programme for energy experts and managers in green buildings
- Topics and energy-saving opportunities evaluated for young architects and engineers
- Lessons learned and future challenges

M. Sala · L. C. Nelli (✉)· A. Donato
Department of Architecture DIDA – Centro ABITA, University of Florence, Florence, Italy
e-mail: marco.sala@unifi.it; lucia.ceccherininelli@unifi.it; alessandra.donato@unifi.it

© Springer International Publishing AG, part of Springer Nature 2019
A. Sayigh (ed.), *Sustainable Building for a Cleaner Environment*,
Innovative Renewable Energy, https://doi.org/10.1007/978-3-319-94595-8_29

29.1   Introduction

The master's degree programmes offered by the Department of Architecture of the University of Florence are postgraduate courses designed with the purpose of developing proficiency in terms of knowledge, skills and new approaches. The courses blend theoretical knowledge with practical applied activities and respond to specific requirements of the labour market.

The courses are intended for students but also for professionals wishing for further education and advanced training. Second-level master's courses are accessible only to students that have a second-cycle degree or equivalent.

The main purpose of the master's course is to design the energy consumption of building technologies, components and structure at the conceptual design stage, so it could be very helpful for designers when making decisions related to the selection of the most suitable design alternative and for choice materials that will be used in an energy-efficient building. Building information modelling (BIM) has the capability to help users assess different design alternatives and select vital energy strategies and systems at the conceptual design stage of proposed projects. Furthermore, by using BIM tools, designers are able to select the right type of materials early during the design stage and to make energy-related decisions that have a great impact on the whole building life cycle.

The main objective of master ABITA is to propose an integrated methodology that links BIM and energy analysis tools with green building certification systems. This methodology will be applied at the early design stage of a project's life.

Around 70% of the Italian master ABITA training participants confirmed that in addition to their energy concepts, they had also realized additional energy efficiency projects after they finished the training course.

Based on the experience, gained during the past 15 years master's training programme's implementation, the graduated energy experts have really appreciated their practical training education and with the support of professional coaches.

The future challenges of the education and training programme are related to follow-up activities, the development of interactive tools and the curriculum's customization to meet the constantly growing needs of energy experts from industry.

29.2   Objectives

The development of low-carbon, resource-efficient buildings and cities is becoming extremely important against the ever-increasing consumption of natural resources. The ABITA second-level master's course provides students with opportunities to develop the advanced knowledge and skills required to find innovative solutions to these needs and become experts in energy management in buildings. The ABITA master's course training programme is offered by the University of Florence and covers nearly all the energy-relevant issues that can arise in public and private companies and sectors. The final output of the practical training is to achieve an elevated professionalism in the study of environmental design and energy management in buildings.

The master's includes studies on:

- Low-energy architecture and energy efficiency measures
- Integration of renewable energies in buildings
- BIM – building information modelling
- Dynamic software for energy simulations
- Energetic audit and diagnosis – green audit
- Smart cities – green urban planning
- Water and waste reuse and economic evaluation management

The training MASTER ABITA provides a solid basis for increasing the:

- Knowledge and skills of energy managers and expert in energy
- Practical experiences through case studies, measurement and verification of energy-efficient solutions in buildings, in the industry and in the cities

The master's course is divided into frontal lectures, e-learning activities, workshops, study trips and meeting with enterprises:

- *Frontal lectures* are divided in four UNITs: Each unit is articulated in many frontal lectures and hours of project work.
- *E-learning activities*: The online training materials, as well as online tutoring for project work, will be implemented, thanks to the MOODLE e-learning platform of the University of Florence.

29.3 Methodology

The course includes different teaching activities: theory classes, seminars, design workshops, speeches, construction labs, external construction lab and visits to case studies. The first module will be dedicated at introducing students to the basic skills of indoor comfort, sustainable architecture and the basic skills of BIM and principles of environmental strategies.

From the second module and through the next 11 months, students will work on a challenging thesis project that will be presented to a public audience at the end of the master's. During this phase, the master's course will stimulate students on the topics of green energy design strategies, smart technologies for the built environment, design rating systems and certifications, renewable energy systems, installation, Italian regulation and smart urban infrastructures. All the input coming from these activities will be embedded in the students' design work to develop a compelling thesis project.

The core modules are four:

- *M1* Sustainable architecture and building deep renovation
- *M2* Energy management and integrated design for NZEB – mass modelling for conceptual design (BIM) – energy performance evaluations
- *M3* Integrated design for NZEB – nearly zero-energy building – energy modelling and simulation of buildings
- *M4* Building the future: green building and smart cities
- *ML* Project work training – workshops

29.3.1   Module 1

Core module 1 is on the teaching of sustainable architecture and its principles with the main purpose to:

- Analyse the energy consumption of building technologies, components and structure at the conceptual design stage
- Help designers relate to the selection of the most suitable design alternative and for choice materials

Smart buildings and sustainable design address the issue of sustainability in terms of design strategies that optimize environmental, energy and social behaviours of the urban environment. The course is aimed at the inclusion of interactive tools to connect technology with architecture, with the aim of achieving an energy efficient environmental design. BIM tools help designers to develop sustainable architecture in accordance with the challenges and problems of post-industrial polluted cities, combining design and technical skills with theoretic, critical and cultural knowledge.

The master's programme aim is to learn the principles and methodology of environmental design and to develop critical thinking skills to challenge established practices, in order to improve technologies for:

- Green building strategies
- Energy efficiency
- Environmentally conscious design
- Experimenting technologies
- Procedures and tools
- Renewables
- Climatic and environmental control
- Reduction of the use of natural sources and energy consumption
- Comfort behaviour
- Environmental and energy quality
- Developed eco-efficiency
- Greenhouses: strategies and evaluation on energy efficiency
- Natural ventilation and cooling systems

- Automation, management systems
- Passive house – NZEB
- Green building efficiency

The students are continuously trained in interdisciplinary cooperation in order to implement the integrated design method in their professional practice.

Students will hold the knowledge and the practical tools to better understand existing buildings for retrofit and to design new ones – positively driving change in this field and moving towards a truly environmentally conscious architecture.

The master's course starts generally in November and ends in April or May. Lessons will take place 3 days a week and at the end of the course will follow an internship. The final thesis is expected within the next year in April.

29.3.2   Module 2

The main contents of the *core module 2* are:

- Integration of renewables in architecture
- EGE – regulations
- Energy audit
- Tools

29.3.3   EGE: Expert in Energy Management

The MS2 provides knowledge and skills required of a professional energy manager for the qualification according to the standard UNI CEI 11339:2009 for the industrial and civil sector.

Students will be involved in practical workshops on the use of tools and on the development of analytical methods, which will be directly applied to a design studio project on the evaluation of case studies.

In this module, students will learn about climate and microclimate analysis and fieldwork methods for the measurement of environmental and energy parameters, thermal comfort surveys and post-occupancy evaluations.

The EGE qualification allows students to demonstrably achieve the highest level of competence in energy management.

At the end of the course, students will be able to attend the exam for certification according to EGE TUV e UNI EN ISO 50000.

The specialization of the expert in energy is to be an independent energy consultant that offers a service to help to cut the energy costs, reduce carbon footprint and invest in renewable energy production.

The expert in energy can identify and help owners of productive buildings or offices to prioritize actions to cut their consumption and advice on how best to invest in energy efficiency and renewable energy systems.

Businesses	Homeowners
Energy surveys	Make homes more comfortable, cut bills and carbon footprint
Energy auditing	Home energy survey
Energy management	Thermal imaging survey
Lighting surveys	New-build consultancy
Sustainable build consultancy	
Training	
Project support	
Renewables consultancy	

29.3.4   BIM and Design Builder Tools

Workshops and design application of parametric massive design, dynamic thermal modelling and daylighting are the contents of the core module 2.

The software used (BIM): climate data analysis to building form, daylighting and thermal modelling, in order to maximize the impact of shadows and solar control systems [3].

These will be directly applied to a design studio project running in parallel to the workshops. The energy analysis of the project will be run with the software DesignBuilder.

29.4   Building Information Modelling (BIM)

- Helps users to select energy strategies and systems at the conceptual design stage of proposed projects.
- Are able to select the right type of materials and make energy-related decisions that have a great impact on the whole building life cycle.
- Integrated methodology that links BIM and energy analysis tools with green building certification systems. This methodology will be applied at the early design stage of a project's life [4].

29.4.1   Green BIM

In particular, green BIM implements the process of green designs; sustainable design methods can be used to analyse the impacts of green buildings, including all aspects of lighting, energy efficiency, sustainability of materials and other building performances. The construction of green technology should be optimized with green building parameters and standards.

Green BIM can control all these aspects:

- The use of natural ventilation, natural lighting and shading effective measures
- The use of solar energy
- Rainwater recycling and waste recycling
- The outdoor use of permeable ground
- The use of green, sustainable and energy-efficient materials
- Focusing on ecological maintenance
- The application software featuring of energy-efficient computing
- The use of natural ventilation and performance analysis

BIM (building information modelling) is an effective tool for the integration of natural and technical systems in architectural design [1].

Green BIM allows to develop integrated multidiscipline building models. The result is better performing buildings and infrastructure because complex information is combined to allow decisions to be made early in the design stage.

Energy model uses green BIM to model a project's energy performance and helps to identify choices to optimize the building's life cycle energy efficiency during the early design phase when changes can be made without incurring high costs [2].

During the design phase, the project teams can ensure that relevant building codes or baselines are technically and cost-effectively verified. The models can also be used, during the operation of maintenance of the building to validate its energy consumption.

High-quality information modelling is able to achieve cost-effective NZEB buildings and districts in the shortest time and managed all planning creating economic benefit for the investments (Fig. 29.1).

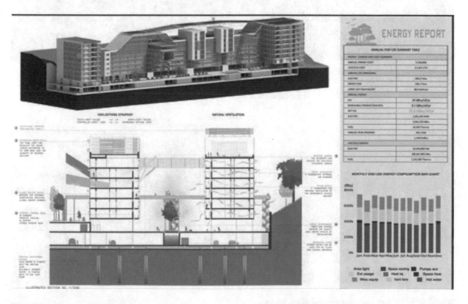

Fig. 29.1 Few master's students' drawings realized with BIM programmes

29.5 Core Module 3

- Energy modelling and building simulation
- Environmental and energy quality
- Integrated design for NZEB nearly zero-energy building

The *tools* used generally during the master's course are:

- Revit (Autodesk BIM) Ecotect
- DesignBuilder (dynamic simulation)
- EnergyPlus (energy simulation)
- PVsyst (photovoltaic)
- TerMus (energy certification ACCA)
- Solarius PV (photovoltaic ACCA)
- DOCET (energy certification ENEA)
- Relux (lighting design)
- Daylight Visualizer (lighting design VELUX) (Figs. 29.2, 29.3 and 29.4)

Fig. 29.2 Thermal analysis realized with thermal camera

Fig. 29.3 Smart city approach with BIM tools

Fig. 29.4 VELUX tool for daylighting

29.6 Core Module 4

By 2050, a great part of the population will live in new urban cities that do not exist yet. Those smart cities need to create better life conditions and to reinvent citizenship for every stakeholder, young/elder, families/professionals and tourists/dwellers, and by giving meaning to the best technologies.

The smart city is supported by information and communication technology (ICT) for learning, adaptation and innovation. ICT means the enhancing of competitiveness of cities through the development of the digital economy, and this means to understand the various challenges and opportunities with which cities are faced:

1. The evolution of cities: social, political and spatial planning models.
2. City services: utilities (water, energy and communications), public street lighting, roadways and traffic, public transport, signage, environmental quality, cleaning of public spaces, waste and sewage management, maintenance, security, civil protection, government, education, public health, social services, town planning, public housing, planning of economic and tourist activities, user and consumer protection and cemeteries.
3. Interactions: housing, education, work, culture, religious services, sport, trade, industry (processing/digital/cultural industries, etc.), professional services, tourism, mobility, logistics, associative and civic activity, crime, etc.
4. Quality of life in the city.
5. Migration flows and their impact on cities.

6. Human diversity in cities.
7. Physical and social virtual networks.
8. Efficiencies and inefficiencies in cities; challenges and opportunities.
9. The impact of ICT on the social fabric, on the management of cities and their innovation potential.
10. City resilience.
11. Green infrastructures aim to guarantee protection, recovery and the enhancement of territorial development, become an integral part of territorial planning by improving the environmental quality and sustainability of cities. Green infrastructures reduces and treats rainwater by providing environmental, social and economic benefits.

Core module ML: **Project work training**

Professional opportunities Good Professional opportunities can be achieved by graduates that can progress or start career in national or international architecture and engineering studios, as designers or specialists in sustainable strategies and innovative technologies for low energy architecture. The highly technological profile of the master's course, including the use of BIM and dynamic building technologies and tools, opens the door to public or private research centres, experimental labs and institutions related to product design and technology innovation.

This module includes lectures based on:

- Personal presentation
- Personal impact
- CV writing lesson plan
- Finding employment
- Interview tips and skills lesson plan
- Creativity and idea generation

29.7 Study Tours

Tours form a central part of the MASTER ABITA experience; the travel consists of a *week-long study tour* in Europe, in countries such as Denmark, Germany, England, Spain, Austria or others or in a neighbouring country, combined with a seminar in Florence. While on tour, theories learned in the classroom come to life by meeting with professionals and experts in the green building and NZEB field which contribute to furthering student understanding of course topics. Study tours are hands-on and experiential, combining theory with practice, and expose students to additional cultural perspectives. They can have the opportunity to visit sites and experts in their field of interest that they may not otherwise have access to. *Study tours are integrated into specific core courses, generally towards April.*

29.8 Internship

Internship will be locally conducted in relative institutions such as universities' specialization departments, state institutions, international organizations, design companies, etc.

The master ABITA Internship is an initiative that helps students in making connections between their study of architecture and the real world of practice by undertaking supervised work placements in architectural firms. The internship exposes students to the work environment and gaining work experience in the industry. It provides opportunities for students to be involved in the conceptual stages of real architectural projects and put into practice what they have learned in their architectural studies. Internships offer students exposure to actual working life, an experiential, an overview on career choices and the chance to build valuable business networks. Students can serve as a valuable resource for organizations as a bridge to understand employment needs and fill immediate needs for labour sources. During the internship stint, master's course students have the opportunity to develop their technical competence for possible employment after graduation.

29.9 Thesis

Upon attending five modules and two workshops and passing them successfully, students are allowed to submit their thesis plan to the master's coordinator.

During summer time, the thesis topics will be approved and a tutor will be assigned to each student. This policy will allow the students to start writing their thesis in advance, and hence their graduation can be achieved successfully by the end of December (2018) or by the end of the third semester (April 2019).

29.10 Conclusion

Based on the experience gained during the past 15 years of the master's training programme's implementation, the energy experts who graduated have really appreciated their practical training activity, and job placement support given by the trainer professional coaches.

The future challenges of the education and training programme are related to follow-up activities, the development of interactive tools and the curriculum's customization to meet the constantly growing needs of energy experts from industry.

Career opportunities are the real result of our master's course; the professional application of the knowledge and skills acquired during this master's programme can be used within two main areas:

- Professionals already working as local council officers or elected members or as specialists or managers in businesses which supply services to cities
- Professionals who wish to use the project element to launch their own entrepreneurial or economic venture or community activities

References

1. Bonenberg W, Wei X (2015) Green BIM in sustainable infrastructure. In: 6th international conference on applied human factors and ergonomics (AHFE 2015) and the affiliated conferences, AHFE; Krygiel E, Nies B (2008) Green BIM: successful sustainable design with building information modeling. Wiley Publishing, Indianapolis, pp 75–86

2. Bernstein HM, Jones SA, Russo MA (2010) Green BIM: how building information modeling is contributing to green design and construction, smart market report. McGraw-Hill Construction, Bedford
3. Krygiel E, Vandezande J (2014) Mastering autodesk revit architecture 2015: autodesk official press. Sybex, Indianapolis
4. Gu N, Wang X (2012) Computational design methods, and technologies: applications in CAD, CAM and CAE education. IGI Global, Hershey

Chapter 30
A Project for the NZero-Foundation in the South of Italy

Lucia Ceccherini Nelli, Vincenzo Donato, and Danilo Rinaldi

Abstract NZero-Foundation is a complex future-oriented student housing.

It has been designed thinking about the characteristics of the place that surrounds it, the specific climatic properties of the Mediterranean region, and the needs and the behavior of the final users.

The model of the building is created using a BIM software, which permitted to simulate the performances of the envelope, annual energy consumption, life cycle cost, daylight, and shading of the facades and, most importantly, permitted to use a unique tool for the different phases of the design management.

The NZero-Foundation is a green building; the energy consumptions are near to zero, maintaining reasonable realization costs. To reach the target, it adopts different strategies to reduce heat losses during winter (buffer spaces, roof gardens, low-transmittance glasses and walls, led lights) and during summer (brise-soleil, natural shading, natural daylighting, high-efficiency mechanical ventilation systems). The complex collects rainwater, coming from roofs and non-permeable soils, for domestic and watering uses buildings, recovers heat from geothermal, and produces energy from photovoltaic integrated into the rooftop obtaining a balanced system between energy performances and an affordable realization cost.

L. C. Nelli (✉) · D. Rinaldi
Department of Architecture DIDA – Centro ABITA, University of Florence, Florence, Italy
e-mail: lucia.ceccherininelli@unifi.it

V. Donato
DISEG – Department of Structural, Geotechnical and Building Engineering, Torino, Italy
e-mail: vincenzo.donato@polito.it

© Springer International Publishing AG, part of Springer Nature 2019   361
A. Sayigh (ed.), *Sustainable Building for a Cleaner Environment*,
Innovative Renewable Energy, https://doi.org/10.1007/978-3-319-94595-8_30

30.1 Introduction

The concept of green architecture, also known as sustainable architecture, is the philosophy and science of buildings designed and constructed in accordance with environmentally friendly principles.

The purpose of the study is to manage the entire design process through a single software, from the preliminary conceptual phase until the energy simulation phase, allowing the designer to correct in real time the strategic decisions and avoiding multiple software problems.

The use of a holistic BIM methodology permits to create a visual high-detailed model, useful to choose the shapes of the architectural object and, at the same time, to calculate the performance of the system, which is essential for a responsible shared strategy. The difficulty of the study is to manage a very great volume in this calculation, keeping in mind that the project refers to a social housing district and that passive strategy can't be easily computed in a conceptual energy building model.

30.2 Project Objectives

The target is to demonstrate with a conceptual model that even a very large-scale building, mostly reserved for public use, can reach nearly zero-energy building (nZEB) standards. NZEB building is defined in Article 2 of the EPBD Directive 2010/31/EU as *building that has a very high energy performance*. The nearly zero amount of energy required should be covered to a very significant extent by energy from renewable sources, produced on-site or nearby [1]. In our case the transmittance requirements that are specific for different climatic zones in the south of Italy (south area is mostly assigned to B and C) refer to D.M. 26/06/2015 that defines the limits of the opaque ($U = 0.38$ W/mq + K) or transparent envelope ($U = 2.40$ W/mq*K) in addition to roof and floor transmittances ($U = 0.36$ W/mq*K). The metric that is used to compare the energy consumption for different buildings by accounting conditioned floor area is EUI, energy use intensity, that is, defined as annual energy consumption divided by conditioned floor area and is expressed in the units of kBtu/sf/yr. or kWh/sm/yr. Our specific objective is to reach at least 16 kBtu/sf/yr. = 52 kWh/sm/yr., which is a good goal for a social housing complex that can accommodate over 2600 students and contain many services for the entire community in 89.000 square meters of conditioned floor area.

In addition to the technical requirements, the very important target is the implementation of a passive strategy which contributes to maintain high level of comfort using basic knowledge of bioclimatic architecture, essential for creating a healthy green design with a limited budget.

A rendered view created with internal program engine

30.3 Methodology

In order to achieve the stipulated aim, the study presented in this paper traces the following steps:

1. Describing the benefits of applying adopted passive strategies that depend on site climatic considerations
2. Exposing the technical properties of the building
3. Defining the criteria assumptions for the creation of a simplified conceptual model
4. Describing the results of the simulation in order to validate the strategic assumptions

30.3.1 Passive Strategies

The passive bioclimatic strategies are the following and allow to reduce the energy requirements of the entire system without requiring additional costs:

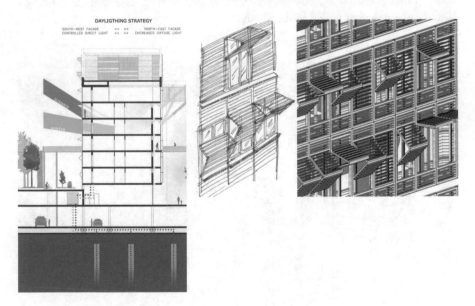

Illustrated south-west section

- *Wind control:*
  Take advantage of fresh winds during late hours of summer to increase cross ventilation and reduce mechanical intervention as much as possible.

- *Environment:*
  Benefit from existing vegetation buffers to reduce noise and cold winds;
      create gardens at different levels to multiply green socialization spaces and increase healthy places;
  recover rainwater for domestic use.

- *Sun shading:*
  A double-skinned wall surrounds the building, reduces sun penetration during summer months, and lets the light and heat in during winter months.

  In particular, a "parametric family" has been expressly created to speed up the creation process of the shading facade, which allows to regulate the right angle of the brise-soleil fixture according to the final user needs.

- *Buffer spaces:*
  The passive solar design helps to preheats air entering from adjacent rooms and to maintain thermal comfort indoor.

  Adjustable windows permit the users to control natural ventilation according to their own needs.

  Intermediate space permits a flexible use, ideal for drain clothing, summer lecture room, or outdoor doghouse.

- *Distribution balcony:*
  Facing the internal court makes it possible to create two-sided overlooking apartments and generates vertical stack effect through ventilation grids.

- *Roof garden:*
  Create high-insulated rooftops.

  Generate exclusive shared spaces ideal for socialization and with high solar exposition.

- *Efficient heating:*
  Mechanical air systems with heat recovery reduce heat losses.

  The vertical geothermal system linked with heat pumps permits high-efficient heating system.

- *Envelope:*
  Low U-value (transmittance) walls and windows reduce heat losses during cold season and minimize mechanical intervention during summer.

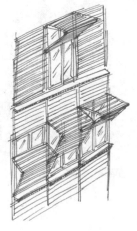

From the sketch to the "parametric family"

30.3.2 Project Assumptions

The site is characterized by high level of relative humidity, influenced by the sea breeze, and the diurnal average temperature has a range between 9 °C and 25 °C.

The "block" object of study, that is reduced compared to the total project area, has a floor area of 23.000 square meters, for an estimated number of 970 people. It was necessary to isolate a single block since every block is independent of each other.

The building envelope is crucial for high performance and has very low-transmittance properties:

Wall R-value: made in R-20 wood frame wall, equivalent to $U = 0.28$ Watt/mq*K.

Roof R-value: R-30 wood frame floor with overlying garden roof, equivalent to $U = 0.18$ Watt/mq*K.

Window U-value: triple low emission glass, equivalent to $U = 1.45$ Watt/mq*K. Window to wall ratio: 28%.

The heating and cooling system and the water system are based on ground source heat pumps having 4.2 COP of efficiency, and the ventilation strategy is based on natural ventilation and supply fan flow. LED lighting has an average power density of 0.28 W/sf, controlled by photosensors [2].

The renewable energy solutions adopted are 90 meters depth geothermal for heat recovery, polycrystalline and thin-film photovoltaic for the production of about 600 kWatt/h/yr. over a surface of 2200 square meters.

Block object of study

Photorealistic simulation

30.3.3   Energy Model

Because of the complexity of the architectural shapes, it was necessary to create a conceptual massing elements model, easily manageable even if less accurate for the calculation.

This step, as the ideation one, has been performed with a single software, Revit Architecture, just to maintain the initial purpose of the study: manage the entire process adopting only one tool, from the design to the energy calculation.

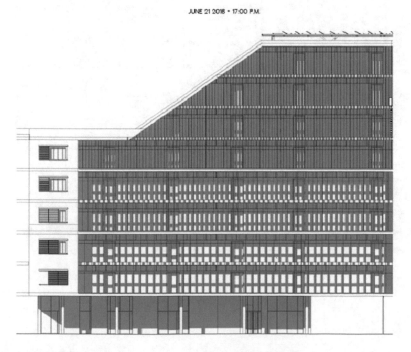

Conceptual model created to calculate energy performance of the block Block object of study

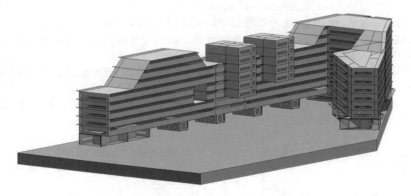

Solar simulation during summer solstice – west-south/west elevation

30.3.4   Environmental Impact and Results

After the modeling step, next one is to direct the information, contained inside the conceptual model, to the cloud tool GBS by Autodesk.

Green Building Studio is a cloud-based service that allows you to run building performance simulations to optimize energy efficiency and to work toward carbon neutrality earlier in the design process.

The results of the calculation, having acceptable approximation because of the simplicity of the model, are the following:

The energy use intensity (EUI) is about 24 kBtu/sf/yr., from which we can subtract the photovoltaic production value of about 8.1 kBtu/sf/yr., that brings the final value to 15.9 kBtu/sf/yr., equivalent to about 51 kWh/sm/yr.

The estimated annual CO_2 emissions are 494 tons for electric and 258 for on-site fuel.

The estimated life cycle energy demand is about 43.540.000 kW for electric and 1.335.000 therms for fuel.

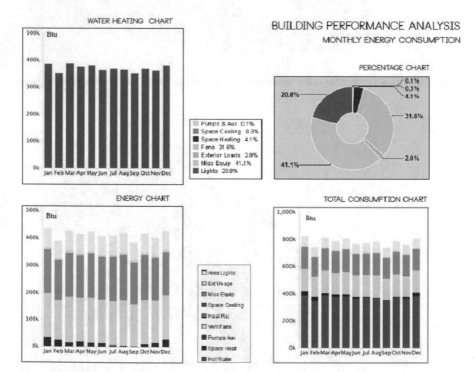

30.4   Conclusions

Green architecture produces environmental, economic, and social benefits. Environmentally, green architecture helps reduce pollution, conserve natural resources, and prevent environmental degradation. Economically, it reduces the amount of money that the building's operators have to spend on water and energy and improves the productivity of those using the facility. Socially, green buildings are meant to serve the population by inserting beautiful artificial objects within the context, both natural and anthropic [3]. To achieve this aim, the designer needs specific tools and preparation, essential to face the unforeseen and propose the solution. A holistic BIM-oriented methodology helps the operator to make decisions in time relatively short, almost in real time, which allows generating a better architecture without neglecting aesthetics nor functionality.

[4] Tools for the energy analysis are used by architects and engineers to design energy-efficient buildings. Energy was chosen. The energy consumption analysis should be defined at the conceptual design stage; this could be very helpful for designers when making decisions related to the selection of the most suitable design alternative that will lead to an energy-efficient building. Building information modeling (BIM) has the capability to help users assess different design alternatives and select vital energy strategies and systems at the conceptual design stage of proposed projects. Furthermore, by using BIM tools, designers are able to select the right type of materials early during the design stage and to make energy-related decisions that have a great impact on the whole building life cycle.

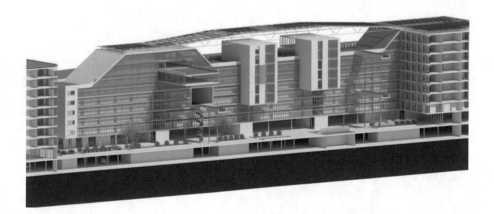

References

1. Ecofys (2013) Towards nearly zero-energy buildings. Definition of common principles under the EPBD. Final report for Europen Commission
2. Piano R, Piano C, AA VV (2012) Almanacco dell'Architetto, Costruire L'architettura. Proctor Edizioni Spa, Bologna
3. Trombadore A (2015) Mediterranean smart cities, Collana MED smart cities. Altralinea Edizioni s.r.l., Firenze
4. Jalaei F, Jrade A (2014) Integrating BIM with green building certification system, energy analysis, and cost estimating tools to conceptually design sustainable buildings. J Inf Technol Constr 19:140–149. https://doi.org/10.1061/9780784413517.015

Chapter 31
Planning Without Waste

Adolfo F. L. Baratta, Laura Calcagnini, Fabrizio Finucci,
and Antonio Magarò

Abstract This paper aims to assess the benefits of a Design for Environment (DfE) at the design stage, according to the goal of "not harming" first. The promotion of conscious solutions for waste reduction and proper waste and resources management must necessarily employ easily available resources, low environmental and economic impact, adopt recycled materials and products, and use low-complexity solutions also due to a limited number of materials. Communication is a relevant item too, because effective information and data transmission between operators reduces errors and, consequently, waste generation. The DfE principles can be applied for buildings that last as long as possible by intervening with maintenance technologies, or alternatively, with a limited lifetime but planning their end of life.

In cither case, easy to maintain or easily to dismantle, environmental benefits can be obtained.

Decisions taken during the design phase determine the kind of waste that will be produced and affect the way it is collected and disposed of.

In DfE, a waste elimination or contraction planning, as well as a proper management of waste and resources designing, is necessary:

- Use easily available resources, with low environmental and economic impact, and carefully evaluate the transport phase
- Mainly adopt recycled, recovery or, at least, recyclable items
- Use low-complexity solutions because the involvement of a growing number of operators generates more waste because of implemented design changes
- Communicate effectively because the transmission of information and data between operators reduces the possibility of errors and, consequently, waste generation

From general point of view, it is possible to apply the aforementioned principles through two distinct strategies, resulting in projects that include:

- The adoption of reliable and environmentally friendly materials and solutions, capable of making buildings that last as long as possible by intervening with maintenance technologies

A. F. L. Baratta (✉) · L. Calcagnini · F. Finucci · A. Magarò
Roma Tre University, Department of Architecture, Rome, Italy
e-mail: adolfo.baratta@uniroma3.it

© Springer International Publishing AG, part of Springer Nature 2019
A. Sayigh (ed.), *Sustainable Building for a Cleaner Environment*,
Innovative Renewable Energy, https://doi.org/10.1007/978-3-319-94595-8_31

- The construction of buildings that provide a limited period of lifetime only when they are functionally and technologically appropriate, planning their dismantling and recovery

The availability of second raw materials is crucial for the industrial sectors, especially for emerging realities such as Asian ones that have redesigned the entire raw and second raw materials market. The objective must be the decrease of waste production, which also depends on how a building is conceived. Everything must help to make a more conscious approach to the responsibility of the designers. At design stage, it is possible to facilitate the use of economically and environmentally friendly materials or the use of reversible systems such as the dry-assembled ones.

Integrating the environmental topic into design, especially the one related to waste management, is the result of a continuous process fueled by innovative materials and a better understanding of existing maintenance practices. This process, which is by definition infinite and needs pragmatic eco-tools, describes scenarios that allow to imagine a variety of possible developments.

All this is needed to pass definitively "from the incivility of waste, the culture of dissipation, the consuming and disposing of consumption, the squander of non-renewable waste, to the civilization of recycling, reuse, highest conservation of raw materials, especially if not renewable, in conclusion to the reduction of waste."

31.1 Resources and Waste

A building can be considered a set of materials scientifically organized according to a precise logic, but it is also the result of a process that causes environmental changes because of the resources, mostly nonrenewable, that are used. These changes come from both the acquisition and transformation of raw materials and the waste production at different stages of the production cycle.

According to the European Commission, "every material object placed on the market is bound, sooner or later, to become waste. Futhermore every production process produces some waste" [1]; therefore, the actions taken at every stage of the design and construction process may originate, directly or indirectly, waste.

Furthermore, the waste treatment is a high-degree complex process due to waste quantity and heterogeneity, as well as the number of operators involved.

The size of the problem can be better understood, thanks to a picturesque image used over 20 years ago by Nicola Sinopoli: if you were to load the waste from the building sector onto lorries, the column of them, bumper-to-bumper, would go from the South Pole to the North Pole, and if a minimum distance were kept, it would equal the circumference of the equator [2]. Two decades after Sinopoli gave us this alarming picture, the situation has certainly not improved.

In addition to the problem of the amount of waste, some assessments about its quality are necessary: waste in the construction sector is extremely heterogeneous, also because of the disruptive development that the chemical industry has applied to the building sector.

The widespread use of synthetic materials for new building components, obtained through technologies borrowed from other applications or industrial sectors, and the use of newly formulated materials (especially in the field of waterproofing, thermal and acoustic insulation, surface treatments, and sealing and protection) are practices that prevent the association of construction debris with chemical-physical inertia and harmlessness for the environment.

Given the undeniable benefits of proper waste management, the real solution to the problem is to not produce waste, especially to avoid producing any kind that is harmful, so as not to have to deal with its disposal.

By pursuing this goal, attention is shifted to the cause and not to the effect, with obvious advantages for the community. In this respect, only a few research studies are concerned with identifying solutions to eliminate waste production during the phase of planning and design [3, 4]; most focus on waste recovery and recycling.

31.2   Strategies and Instruments

This contribution is based on the principle that waste is a result of a design error [5]. With this in mind, any waste containment strategy must necessarily start from the design process, therefore identifying the phases when waste is generated and providing solutions to contain or, rather, eliminate its production.

This principle allows "first of all not to harm" by foreseeing, in search of economically advantageous solutions, a waste reduction at the source: instead of looking for a solution for when the waste has already been generated, it can be done with solutions that mitigate its production at stages where the waste has not yet been produced.

Moreover, this precautionary principle is the fifteenth principle of the responsibilities and rights of the countries participating in the United Nations Conference on Environment and Development in Rio de Janeiro in 1992: "in order to protect the environment, the precautionary approach shall be widely applied by States according to their capabilities" [6].

The real goal must not be to properly dispose the ever-increasing amount of waste products, but the diminution of their production so that they do not give rise to pollution in the territory.

It is not enough to believe in correct waste management and recycling practices in building by simply choosing most of the products among those suitable for disposal in diversified landfills; rather the design process should focus on criteria such as reuse, bearing in mind that you cannot reuse something unless you decide to do so in advance and unless you design products and projects through precise reuse methods [7].

In the field of building design, there are many limits but also many opportunities for developing waste containment strategies.

While designers are now sharing the need to adopt energy-efficient solutions, the appropriate choice of materials solutions is not yet perceived as a priority.

This approach leads to the choice of excellent thermal performance products without assessing their impact on the environment once dismantled. In fact the current energy strategies are based on the research for more productive materials to create not only buildings that do not consume energy but also buildings that could even produce it.

Although the design of zero-energy buildings has found support in stringent European legislation, it seems the path toward the regulation of zero-waste buildings still needs to be made despite the intention stated in 2014 EU communication "Towards a circular economy: a program for a zero-waste Europe."

Decisions made during the design phase determine the kind of waste that will be produced and affect the way it is collected and disposed of.

With reference to the environmental topic, a commonly used term, mainly in the United States, is Design for the Environment (DfE): a design that can reduce environmental impact without compromising function and quality [8].

In DfE, which is committed to waste elimination or waste contraction, as well as to proper waste and resource management, it is necessary to:

- Use easily available resources, with low environmental and economic impact, and carefully evaluate the transport phase.
- Mainly adopt recycled, salvaged, or, at least, recyclable items.
- Use low-complexity solutions because the involvement of a growing number of operators generates more waste because of implemented design changes. The use of a limited number of materials helps to reduce the solution's complexity because it scales down separation difficulties in decommissioning and favors the determination of homogeneous waste categories.
- Communicate effectively because the information and data transmission between operators reduces the possibility of errors and, consequently, waste generation. Conversely, incorrect communication or ineffective coordination implies a potentially greater waste production.

From general point of view, it is possible to apply the aforementioned principles through two distinct strategies, resulting in projects that include:

- The adoption of reliable and environmentally friendly materials and solutions, capable of making buildings that last as long as possible by intervening with maintenance technologies
- The construction of buildings that provide a limited period of existence only when they are functionally and technologically appropriate and the planning of their dismantling and recovery

A building can be designed to be easily maintained or dismantled.

In the first case, the use of highly durability materials, which can be repaired without replacement, can be a good preventive criterion for waste recycling strategies. In the second case, architecture becomes an open, editable work in its formal, functional, and technological ways, thanks to solutions that allow for subsequent recovery operations and which make nondestructive, but conservative, interventions possible.

Starting from the stated sustainability objectives, however, the two strategies have opposite positions in determining the whole building life cycle: in the first case, the building process ends with the management activity leaving the material's life cycle unknown; in the second case, the useful life is temporally limited, and the demolition phase is added as a programmed need [9].

31.3   Some Design and Process Experiences

In order to define suitable instruments for conceiving a zero-waste project, we proposed some process and design experiences aimed at building zero-waste architecture or, at least, making as much effort as possible toward this goal.

In 2016, Hiroshi Nakamura & NAP designed a commercial building in Japan: the *Kamikatz Public House* (Fig. 31.1), which in the same year won the WAN (World Architecture News) Sustainable Buildings Award.

The building has been made with recycled materials and creatively reused items. The project is the pride of Kamikatsu, a small Japanese city committed to zero-waste community: residents divide their waste into 34 categories and their material recycling rate is about 80%.

The building, committed to the principle of zero-waste design, has an 8-meter high glass facade, made with frames from the nearby abandoned houses. The entire building is an abacus of reuse and recycle solutions, including floor tiles and chandeliers made with used bottles and wallpaper made out of recycled newspapers.

Both in the construction phase and in its useful life, the building saves energy and resources by significantly reducing emissions through reuse and recycling,

Fig. 31.1 Hiroshi Nakamura & NAP Kamikatz, Public House, Kamikatsu (J) 2016. The building was made of recycled and reused materials and products

Fig. 31.2 Encore Heureux Circular Pavilion, Paris 2016. The temporary building was built with recycled products and subsequently dismantled with the complete recovery of material resources. (Photo by Cyros Corut courtesy of Encore Heureux Studio)

enabling many principles of the circular economy [10], based on the idea that one person's waste can be a resource for another.

A similar project is the *Circular Pavilion* at the Hôtel de Ville in Paris (Fig. 31.2), a temporary structure designed by the Encore Heureux Architecture Studio.

The pavilion had been built in just over 30 days, and for 90 days, it stood in front of the City Hall as the object of the exhibition "Matière Grise" at the Arsenal Pavilion. The *Circular Pavilion* is an experimental construction made with reuse materials; it takes its name not from its form but from the regard to the principles of the circular economy.

By virtue of this principle, the facade is covered with 180 wooden doors recovered from the refurbishment of a residential building; the rock wool panels for thermal insulation comes from the disposal of a supermarket roof; the floor is made up of wood panels previously used for other exhibitions; the furniture is gathered from several Parisian merchants; the lighting comes from obsolete street lamps; the windows come from the surplus of material from other Parisian construction sites.

The pavilion demonstrates the potential of reusing in architecture not only because of the construction choices (the pavilion is made almost exclusively of reuse materials) but also because it has been created as a place for exchanging ideas.

The construction site was opened to visitors who were able to follow the assembly steps; the pavilion had been dismantled in the early days of 2016 and replaced as a permanent structure in the 15th *arrondissement* of Paris as sports associations' clubhouse [11].

The *Upcycling House* in Nyborg, Denmark (Fig. 31.3), clearly captures the name from the goal of realizing a reuse, recycling, and recovery process aimed at transforming second raw material into components with higher quality.

Like the examples above, this is a virtuous design contribution in terms of saving energy and raw materials saving. Made to be affordable for all Danish households,

Fig. 31.3 Lendager Architekten, Upcycling House, Nyborg (DK) 2015. The building is completely made of recycled and recyclable materials and items

the building demonstrates how the upcycling principles are also economically advantageous. The 140 m² building is a one-family home designed according to low energy consumption criteria, well oriented to ensure maximum utilization of passive and active systems for water recycling and the use of renewable energy sources for electricity generation [12].

The life cycle assessment (LCA) was carried out on all design materials, showing 86% reduction of emissions compared to a home built with traditional materials. The load-bearing structure is realized by reusing two naval containers, while thermal insulation has been guaranteed through the use of paper from old newspapers. The same was done with other coating materials and finishes, all from recycled materials (such as glass used for the finishing tiles and aluminum cans for roof plates).

The Norman Foster's Hong Kong Bank was designed several years earlier, between 1979 and 1986. It's a bigger project if compared with the examples above (Fig. 31.4). The project (29-, 36-, and 44-story towers) shows a design idea committed to a very high degree of prefabrication. Although it cannot include any choice in recycled materials, the building is potentially removable and transferable, if the political conditions were to change.

The incredible value of this project, inspired by the ability to relocate a construction created without the conception of being temporary (as in the case of the *Circular Pavilion*), finds successive applications in most of the constructions of international events such as the Expo or the Olympic Games. Two important experiences emerged in the Olympic Games of London 2012.

Fig. 31.4 Norman Foster, Hong Kong Bank, Hong Kong (RC) 1986. The project was designed to be dismantled into pieces, loaded on boats, and restored in another location (if China were to occupy Hong Kong)

The first one is a three buildings complex designed by Magma Architecture (Fig. 31.5), designed to be dismantled and reconstructed elsewhere; at the end of the Olympic Games, they were disassembled and reassembled at Glasgow, for the 2014 Commonwealth Games. Thanks to the temporary nature of the event, the buildings are meant to have an impressive architectural image that can be remembered by visitors and by the local community.

The three buildings' cladding is made of a PVC membrane characterized by large colored circular elements (each building has a different color: red, pink, and blue) which, in addition to a significant aesthetic value, have an important functional value too: they both regulate the ventilation inside and form tension nodes under the coating membrane, collaborating with the steel structure.

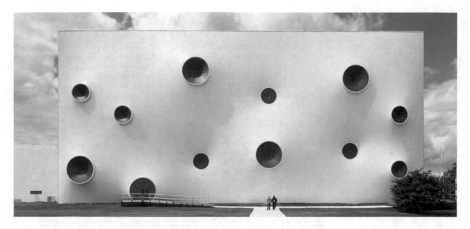

Fig. 31.5 Magma Architecture, Shooting Boat, London (UK) 2012. This structure is one of the symbol structures of the Olympic Games: the building has been dismantled and rebuilt elsewhere

In these experiences, though virtuous, there is no design solution for the end of life, namely, the prediction of waste that may be generated in the case of maintenance work or in a possible demolition phase: in fact, if some waste products are resumed during construction, once the new cycle is over, they become waste again, though with a useful life that already entails a significant saving of resources.

The US Environmental Protection Agency (EPA) estimates that the incidence of waste generation during construction and demolition (C&D) phases was 90% in 2014 [13].

This data highlights how necessary it is to introduce waste containment and disposal in the design phase and how sporadic the experiences are. The slogan coined by EPA "Thinking beyond waste," which concludes the above mentioned report, synthesizes the mindset that is needed.

In the London 2012 context, the Olympic Delivery Authority (ODA), in line with the "Zero Waste Games Vision" [14], has implemented a process strategy with the aim of tracing as much as possible the hierarchy of waste production and avoiding its generation.

As part of designing out waste (DOW), ODA applied the principles of the "waste management hierarchy" outlined in European Directive 2008/98/EC: the design approach to waste follows a hierarchy that provides recycling and reuse actions as a last possible approach, whereas reducing the use of materials is a priority. By associating the need to determine the source, reusability, and quantification of any potential waste that has been spared or generable to this hierarchy [15], the London Organising Committee of the Olympic and Paralympic Games (LOCOG) established a methodological approach for the design phases of the Olympic Park buildings. LOGOC asked to apply a set of design strategies to build without waste [16][1]; a

---

[1] Le strategie erano: *Design for reuse and recovery*; *Design for offsite construction*; *Design for materials optimization*; *Design for waste efficient procurement*; *Design for deconstruction and flexibility*.

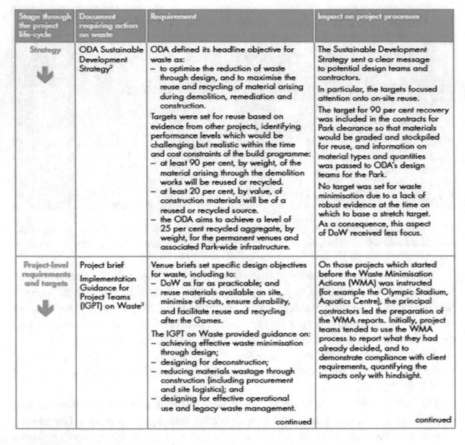

Stage through the project life-cycle	Document requiring action on waste	Requirement	Impact on project processes
Strategy	ODA Sustainable Development Strategy[2]	ODA defined its headline objective for waste as: – to optimise the reduction of waste through design, and to maximise the reuse and recycling of material arising during demolition, remediation and construction. Targets were set for reuse based on evidence from other projects, identifying performance levels which would be challenging but realistic within the time and cost constraints of the build programme: – at least 90 per cent, by weight, of the material arising through the demolition works will be reused or recycled. – at least 20 per cent, by value, of construction materials will be of a reused or recycled source. – the ODA aims to achieve a level of 25 per cent recycled aggregate, by weight, for the permanent venues and associated Park-wide infrastructure.	The Sustainable Development Strategy sent a clear message to potential design teams and contractors. In particular, the targets focused attention onto on-site reuse. The target for 90 per cent recovery was included in the contracts for Park clearance so that materials would be graded and stockpiled for reuse, and information on material types and quantities was passed to ODA's design teams for the Park. No target was set for waste minimisation due to a lack of robust evidence at the time on which to base a stretch target. As a consequence, this aspect of DoW received less focus.
Project-level requirements and targets	Project brief Implementation Guidance for Project Teams (IGPT) on Waste[3]	Venue briefs set specific design objectives for waste, including to: – DoW as far as practicable; and – reuse materials available on site, minimise off-cuts, ensure durability, and facilitate reuse and recycling after the Games. The IGPT on Waste provided guidance on: – achieving effective waste minimisation through design; – designing for deconstruction; – reducing materials wastage through construction (including procurement and site logistics); and – designing for effective operational use and legacy waste management. continued	On those projects which started before the Waste Minimisation Actions (WMA) was instructed (for example the Olympic Stadium, Aquatics Centre), the principal contractors led the preparation of the WMA reports. Initially, project teams tended to use the WMA process to report what they had already decided, and to demonstrate compliance with client requirements, quantifying the impacts only with hindsight. continued

Fig. 31.6 Extract of DOW's design actions for London 2012 [17]

programmatic document provided strategies and objectives for the different design phases (Fig. 31.6).

31.4 Which Future

The waste production during C&D phase and during life cycle management have complex causes. Most of the material resources used are not renewable or at least cannot be used again in the original state they were before industrial transformations. The environmental costs of this situation are there for all to see, with great damage to biodiversity, ecosystems, landscape (natural and anthropic), and the related contaminations of different environmental matrices. The waste origin should be sought in every unavoidable action to produce a building construction, and it is therefore essential to evaluate how each passage of the constructive process interacts with others.

The growth of the international secondary raw materials market, governed by the Basel Convention adopted in 1989 and coming into force in 1992, has two main economic drivers: (1) the relocation of productive capacities in countries with lower production costs and scarce resources and (2) the generation of secondary raw material flows in countries other than production ones. In fact, the availability of secondary raw materials is now crucial for several industrial sectors, especially for emerging realities such as the Asian ones that have redesigned the entire business sector. Suffice it to think that Chinese paper imports have gone from 1.6 to 22.5 million tons in a decade (1997–2007), over the same decade, plastic waste increased from 0.5 to 6.9 million tons, ferrous scrap from 1.8 to 10.1 million tons, and aluminum scrap from 300,000 to over 2 million tons [18]. The objective, however, must be the contraction of waste production, which depends, above all, on how a building is conceived. Everything must contribute to making it possible for a more aware approach of the designers in the creative act since the design phase, using materials easy to find, economically and environmentally friendly, or using reversible systems such as those assembled in dry conditions. The growing user expectations can be satisfied by delivering durable, innovative, and sustainable solutions that incorporate the latest generation technology that can ensure the material's performance upgrade. It can be a lasting innovation with solutions that allow modular updates of single parts. In addition, the key issues of coordination and communication have an immediate impact on the waste containment. Integrating the environmental dimension into design, especially the one related to waste management, is the result of a continuous and gradual process, fueled by innovative materials and processes, as well as a better understanding of existing maintenance practices. This process is, by definition, infinite and needs pragmatic eco-tools.

There are several reasons for planning without waste. In particular, it is important to point out that there are many instruments (such as the environmental certification protocols, the LEED and the ITACA Protocol) that emphasize the use of recycled materials in design. What really needs to be done is project design in a forward-looking manner, through recycles that reckon the availability to reuse entire components rather than mere materials. There are functional elements that could be fully reused, for example, fixed kitchens, typically constructed so as to require their actual destruction at the end of the building's life or following a refurbishment; these could be designed as easily removable modular components for a different function. Planting (cables, piping, etc.) could have the same potential. In order not to generate waste, coatings require a more complex analysis. The interior and exterior window frames can be designed to not go to a storage area for possible recycling at the end of life but could be reusable, preventing easy removal by designing standard (and reversible) fixing methods so as to keep them in standardized positions and sizes. It should be remembered that whenever C&D phases are called into question, a low level of reuse is determined [7]. All this to pass definitively "from the /incivility of waste, from the culture of dissipation, of the consuming and disposing, of the waste of nonrenewable resources, to the civilization of recycling, reusing and of the maximum preservation of raw materials, in conclusion at the waste reduction" [19].

References

1. Commissione Europea (2004) Verso una strategia tematica di prevenzione e riciclo dei rifiuti. Gazzetta Ufficiale 76:11
2. Sinopoli N (1994) L'edilizia che avanza, vol 133. Costruire, Segesta editore, Milano, pp 93–96
3. Keys A, Baldwin AN, Austin SA (2000) Designing to encourage waste minimisation in the construction industry. In: CIBSE national conference, Dublin, 20–22 settembre 2000, pp 1–11
4. Liu Z, Osmani M, Demian P, Baldwin A (2011) The potential use of BIM to aid construction waste minimisation. In: International conferences information and knowledge management in building, CIB, Sophia
5. Köhler M (2014) Trash oder treasure. Recycling in der Architektur. In: Competition, vol 8. Verlages GmbH, Berlin (D), pp 34–37
6. www.europa.eu/legislation_summaries/development/sectoral_development_policies/l28102_it. Last visited on 25.03.2018
7. http://zerowasteinstitute.org/. Last visited on 25.03.2018
8. Telenko A, Seepersad C, Webber ME (2008) A compilation of design for environment principles and guidelines. In:,International design engineering technical conferences & computers and information in engineering conference, New York, 3–6 agosto 2008, pp 1–13
9. Bologna R (ed) (2002) La reversibilità del costruire. L'abitazione transitoria in una prospettiva sostenibile. Maggioli editore, Rimini
10. http://inhabitat.com/zero-waste-japanese-town-builds-a-unique-building-from-abandoned-materials/. Last visited on 25.03.2018
11. Calcagnini L (2016) Non aprite quella porta. In: Modulo, vol 401. BeMa Editrice, Milano, pp 78–79
12. www.archdaily.com/458245/upcycle-house-lendager-arkitekter. Last visited on 25.03.2018
13. EPA. Advancing sustainable materials management: 2014 fact sheet assessing trends in material generation, recycling, composting, combustion with energy recovery and landfilling in the United States, www.epa.gov/sites/production/files/2016-11/documents/2014_smmfact-sheet_508.pdf. Last visited on 25.03.2018
14. LOCOG (2012) London Organising Committee of the Olympic Games and Paralympic Games Limited. London 2012 Zero Waste Games Vision, London
15. Van Houten RS, De Lange NA (2016) A zero-waste approach in the desing of buildings. Introducing a new way of approaching sustainability in buildings with a conceptual industrial building design as an illustrative example. Graduation Thesis, Delft University of Technology, Delf
16. Altamura P (2015) Costruire a zero rifiuti. Strategie e strumenti per l'upcycling dei materiali di scarto in edilizia. Edizioni Franco Angeli, Milano
17. Moon D, Holton I (2011) Learning legacy. Lessons learned from the London 2012 Games construction project, London
18. Bianchi D (ed) (2008) Il riciclo ecoefficiente. Performance e scenari economici, ambientali ed energetici. Edizioni Ambiente, Milano
19. Boato M (1995) Dall'inciviltà dei rifiuti alla civiltà del riuso, Ambiente Risorse e Salute, vol 33. Centro Studi l'Uomo e l'Ambiente, Padova, pp 38–40

Chapter 32
Production of ZnO Cauliflowers Using the Spray Pyrolysis Method

Shadia J. Ikhmayies

Abstract Zinc oxide (ZnO) is a semiconductor material with direct wide bandgap energy (3.37 eV) and a large exciton binding energy (60meV) at room temperature. It currently attracts worldwide intense interests because of its importance in fundamental studies, and its numerous applications, especially as optoelectronic materials, UV lasers, light-emitting diodes, solar cells, nanogenerators, gas sensors, photodetectors, and photocatalyst. ZnO micro$-$/nano-cauliflower arrays are of potential use for electrochemical sensing, dye-sensitized solar cells, an electron transporting layer, and photocatalytic applications.

In this work, ZnO cauliflowers were produced as thin films on glass substrates at a substrate temperature of $350 \pm 5°C$ using the low-cost spray pyrolysis (SP) method. Zinc chloride ($ZnCl_2$) was used to prepare the precursor solution. The produced films were characterized using X-ray diffraction and scanning electron microscopy and X-ray energy-dispersive spectroscopy (SEM- EDS). X-ray diffractogram revealed that the films have hexagonal structure with preferential orientation along the (002) line. EDS analysis showed that the films are nonstoichiometric and they contain chlorine. ImageJ software was used to estimate the size distributions of the cauliflowers.

32.1 Introduction

Zinc oxide (ZnO) is a II-VI compound semiconductor with direct wide bandgap energy (3.37 eV) and a large exciton binding energy (60 meV) at room temperature [1]. ZnO is currently attracting intense interests because of its numerous applications in optoelectronic materials [2, 3], UV lasers [4], light-emitting diodes [5], solar cells [6], nanogenerators [7], gas sensors [8], photodetectors [9], and photocatalysts [10]. Zinc oxide (ZnO) micro$-$/nanostructures have attracted increasing attention as gas sensors, photocatalysts, and an electron transporting layer in dye-sensitized

S. J. Ikhmayies (✉)
Al Isra University, Faculty of Science, Department of Physics, Amman, Jordan

© Springer International Publishing AG, part of Springer Nature 2019
A. Sayigh (ed.), *Sustainable Building for a Cleaner Environment*,
Innovative Renewable Energy, https://doi.org/10.1007/978-3-319-94595-8_32

solar cells (DSCs) due to the fast electron injection efficiency and superior electron transport ability compared to conventional sintered TiO_2 [11].

There are several experimental routes to produce ZnO micro−/nanostructures. These include sol-gel process [1], pulsed laser ablation [12, 13], DC thermal plasma [14], chemical vapor deposition [5], electrodeposition [11], RF magnetron sputtering [15], chemical bath deposition (CBD) [16], and spray pyrolysis (SP) [17–26] techniques. The spray pyrolysis (SP) method is chosen for the production of ZnO films in this work because of its simplicity, low cost, and ability to change properties of the films by controlling several deposition parameters.

Several authors prepared and characterized ZnO micro- and nano-cauliflowers for different purposes and applications [1, 11, 16, 17, 27, 28]. Rajabi et al. [17] produced and characterized cauliflower like ZnO nanostructures using the SP method on ZnO nanorods which were grown on p-type silicon substrates. Yamaguchi et al. [11] produced ZnO nano-cauliflower array dye-sensitized solar cells using a two-step electrodeposition method, and Wang et al. [13] prepared them by chemical bath deposition on a conducting glass substrate and used them as electrode in dye-sensitized solar cells. Also Wang et al. [27] prepared ZnO cauliflowers on a template of silicon nanoporous pillar array (Si-NPA) for humidity sensing. In this work, ZnO cauliflowers are synthesized on glass substrates by the low-cost SP method using $ZnCl_2$ as a precursor. Such morphology is of potential use as a forecontact and/or a window layer in thin-film solar cells, a photoanode in dye-sensitized solar cells. In addition it has several applications in optoelectronic device industry, photocatalysis, and gas sensors.

32.2   Materials and Methods

Cauliflower like ZnO thin films were produced using the spray pyrolysis (SP) technique on glass at a substrate temperature of about 350 °C. The substrates have dimensions of $10 \times 10 \times 1\ mm^3$, were precleaned by soap, and then rinsed in distilled water to remove the impurities from their surfaces, and finally they were dried by lens paper. Zinc chloride ($ZnCl_2$) of purity 99% was used as the precursor, where 2.5 g of $ZnCl_2$ are dissolved in 60 ml of distilled water to get a 0.03 M solution. The solution has been sprayed intermittently on the hot substrate with a deposition time of 10 s followed by a period of 1–2 min without spray. The time without spraying is required to avoid strong cooling of the substrate, to give enough time for crystal growth, and to give enough time for chemical reactions to be completed. After finishing the deposition process, the heater was turned off, and the samples were left on the heater to cool to room temperature gradually.

The microstructure of the films is determined using X-ray diffraction (XRD), where a Shimadzu XRD-7000 diffractometer utilizing X-Ray Cu K_α radiation ($\lambda = 1.54\ A°$) is used. The step size and scan speed are 0.02° and 2°/min, respectively. The measurements were taken in the continuous 2θ mode in the range 2–70° using a current of 30 mA and voltage of 40KV. An FEI scanning electron microscope

(SEM) (Inspect F50) supported by X-ray energy-dispersive spectroscopy (EDS) is used for the observation of surface morphologies and identifying composition of the films. The thickness is estimated by the same SEM microscope and found to be around 1.277 μm.

32.3   Results and Discussion

Figure 32.1 shows the X-ray diffractogram of one of the as-prepared ZnO films of cauliflower morphology. All diffraction peaks are characteristic of the hexagonal (wurtzite) phase of ZnO, and they are indexed with the corresponding Miller indices. The highest intense peak at 33.88 corresponds to reflections from the (002) plane, which has the smallest energy of formation. This means that preferential orientation of crystal growth is along the c-axis perpendicular to the substrates. No other diffraction peaks arising from metallic Zn or $ZnCl_2$ are present in the XRD pattern, which indicates the high phase purity of the synthesized samples.

Figure 32.2 displays the SEM images of three as-deposited ZnO films. The images show the surface morphologies of the films, which appear as micro−/nano-cauliflowers of diameter estimated using ImageJ software and are found in the range 185–2036 nm. The hexagonal structure of ZnO grains is apparent in all images, and their diameters are also estimated using ImageJ software and found to be in the range 39–260 nm. As the figure shows, the films are highly porous, which ensures a large internal surface area (large surface-to-volume ratio of ZnO cauliflowers). The large surface area is advantageous in the case of thin-film solar cells where it enables more absorption of light, and in the case of dye-sensitized solar cells, it enables more

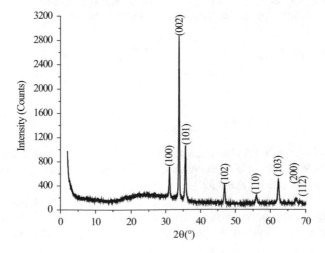

Fig. 32.1   XRD diffractogram of one of the as-deposited ZnO films

Fig. 32.2 SEM images of three spontaneously deposited ZnO thin films of thickness about 1 μm and cauliflower morphology. (**a**) First film, (**b**) second film, and (**c**) third film

Table 32.1 Results of EDS analysis

Element	Concentration (at.%)	Error (%)
O	54.29	5.0
Cl	2.13	0.2
Zn	43.58	2.5

adsorption of dye, while in the case of photocatalysts, it enables more adsorption of sensitizers which increases the photocatalytic reactions sites and promotes the electron-hole separation (Table 32.1).

Figure 32.3 displays the EDS spectrum of one of the films, where the vertical axis displays the number of X-ray counts, while the horizontal axis displays energy in KeV. The figure confirms the presence of elemental zinc and oxygen signals of

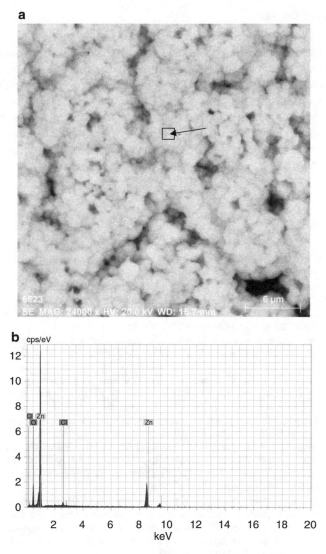

Fig. 32.3 Energy-dispersive spectroscopy (EDS) data. (**a**) Position A at which the measurement is recorded. (**b**) Spectrum of one of ZnO films of cauliflower structure

the ZnO film. From the EDS report, the elements Zn and O are in an atomic ratio of 43.58 at.% and 54.29 at.%, respectively. Taking into account the precision limits of the EDS especially for light elements such as oxygen, these values are comparable to the atomic ratio in the compound ZnO which is 1. The presence of chlorine in the films is expected because of the use of $ZnCl_2$ in the precursor solution. Chlorine improves the n-type conductivity of the films which is an advantage when the films are used as a fore contact and/or a window layer in thin film solar cells.

32.4 Conclusions

Thin films of ZnO micro−/nano-cauliflowers were fabricated on glass substrates using the low-cost SP method. X-ray diffractogram showed that the films have hexagonal structure and preferential orientation along the c-axis perpendicular to the substrates. It is found that the concentrations of Zn and O in the films are comparable to the compound ratio within the accuracy limits of EDS system. The cauliflowers have diameters in the range 185–2036 nm, and they consist of hexagonal grains with diameters in the range 39–260 nm. These results are attractive for a number of potential applications such as photocatalysis, solar cells, and gas sensors.

References

1. Nipane D, Thakare SR, Khati NT (2013) Synthesis of novel ZnO having cauliflower morphology for photocatalytic degradation study. J Catal, Article ID 940345, 8 p
2. Lu F, Cai W, Zhang Y (2008) ZnO hierarchical micro/nanoarchitectures: solvothermal synthesis and structurally enhanced photocatalytic performance. Adv Funct Mater 18(7):1047–1056
3. Qu J, Luo C, Cong Q (2011) Synthesis of multi-walled carbon nanotubes/ZnO nanocomposites using absorbent cotton. Nano-Micro Lett 3:115–120
4. Chu S, Wang G, Zhou W, Lin Y, Chernyak L, Zhao J, Kong J, Li L, Ren J, Liu J (2011) Electrically pumped waveguide lasing from ZnO nanowires. Nat Nanotechnol 6(8): 506–510
5. Na JH, Kitamura M, Arita M, Arakawa Y (2009) Hybrid pn junction light-emitting diodes based on sputtered ZnO and organic semiconductors. Appl Phys Lett 95(25):253303
6. Sudhagar P, Kumar RS, Jung JH, Cho W, Sathyamoorthy R, Won J, Kang YS (2011) Facile synthesis of highly branched jacks-like ZnO nanorods and their applications in dye-sensitized solar cells. Mater Res Bull 46(9):1473–1479
7. Wang ZL, Yang R, Zhou J, Qin Y, Xu C, Hu Y, Xu S (2010) Lateral nanowire/nanobelt based nanogenerators, piezotronics and piezo-phototronics. Mater Sci Eng R 70(3–6):320–329
8. Xu J, Han J, Zhang Y, Sun Y, Xie B (2008) Studies on alcohol sensing mechanism of ZnO based gas sensors. Sensors Actuators B 132(1):334–339
9. Lu CY, Chang SJ, Chang SP, Lee CT, Kuo CF, Chang HM (2006) Ultraviolet photodetectors with ZnO nanowires prepared on ZnO:Ga/glass templates. Appl Phys Lett 89(15):153101
10. Cho S, Kim S, Jang JW, Jung SH, Oh E, Lee BR, Lee KH (2009) Large-scale fabrication of sub-20 nm diameter ZnO nanorod arrays at room temperature and their photocatalytic activity. J Phys Chem C 113(24):10452–10458
11. Yamaguchi I, Watanabe M, Shinagawa T, Chigane M, Inaba M, Tasaka A, Izaki M (2009) ZnO nano-cauliflower array dye-sensitized solar cells. ECS Trans 16(36):3–10
12. He C, Sasaki T, Shimizu Y, Koshizaki N (2008) Synthesis of ZnO nanoparticles by pulsed laser ablation in aqueous media and their self assembly towards spindle-like ZnO aggregates. Appl Surf Sci 254:2196–2202
13. Fan XM, Lian JS, Guo ZX, Lu HJ (2005) Microstructure and photoluminescence properties of ZnO thin films grown by PLD on Si(1 1 1) substrates. Appl Surf Sci 239:176–181
14. Ko TS, Yang S, Hsu HC, Chu CP, Lin HF, Liao SC, Lu TC, Kuo HC, Hsieh WF, Wang SC (2006) ZnO nanopowders fabricated by dc thermal plasma synthesis. Mater Sci Eng B 134:54–58
15. Zhang DH, Xue ZY, Wang QP (2002) The mechanisms of blue emission from ZnO films deposited on glass substrate by r.f. magnetron sputtering. J Phys D Appl Phys 35:2837–2840

16. Wang Y, Cui X, Zhang Y, Gao X, Sun Y (2013) Preparation of cauliflower-like ZnO films by chemical bath deposition: photovoltaic performance and equivalent circuit of dye-sensitized solar cells. J Mat Sci Tech 29(2):123–127
17. Rajabi M, Dariani RS, Iraji A, Zahedi F (2013) Optoelectronic properties of cauliflower like ZnO-ZnO nanorod/p-Si heterostructure. Solid State Electron 90:33–37
18. Ikhmayies SJ (2016) Synthesis of ZnO microrods by the spray pyrolysis technique. J Electron Mater 45(8):3964–3969
19. Ikhmayies SJ, Abu El-Haija NM, Ahmad-Bitar RN (2014) A comparison between different ohmic contacts for ZnO thin films. J Semicond 36(3):033005-1-6
20. Ikhmayies SJ, Naseem M, El-Haija A, Ahmad-Bitar RN (2010) Electrical and optical properties of ZnO:Al thin film prepared by the spray pyrolysis technique. Phys Scr 81(1):art. no.015703
21. Ikhmayies SJ, Abu El-Haija NM, Ahmad-Bitar RN (2010) Characterization of undoped spray-deposited ZnO thin films of photovoltaic applications. FDMP: Fluid Dyn Mat Process 6(2):165–178
22. Ikhmayies SJ, Abu El-Haija NM, Ahmad-Bitar RN (2010) The influence of annealing in nitrogen atmosphere on the electrical, optical and structural properties of spray- deposited ZnO thin films. FDMP: Fluid Dyn Mat Process 6(2):219–232
23. Ikhmayies SJ, Zbib MB (2017) Spray pyrolysis synthesis of ZnO micro/nanorods on glass substrate. J Electron Mater 46:1–6
24. Ikhmayies SJ (2017) Formation of three dimensional ZnO micro flowers from self assembled ZnO micro discs. Metall Mater Trans A 48(8):3625–3629
25. Ikhmayies SJ, Zbib MB (2017) Synthesis of ZnO hexagonal micro discs on glass substrates using the spray pyrolysis technique. J Electron Mater 46(7):3982–3986
26. Juwhari HK, Ikhmayies SJ, Lahlouh B (2017) Room temperature photoluminescence of spray-deposited ZnO thin films on glass substrates. Int J Hydrogen Energy 42:17741–17747
27. Wang LL, Wang HY, Wang WC, Li K, Wang XC, Li XJ (2013) Capacitive humidity sensing properties of ZnO cauliflowers grown on silicon nanoporous pillar array. Sensors Actuators B Chem 177:740–744
28. Gordon T, Grinblat J, Margel S (2013) Preparation of "cauliflower-like". ZnO Micron-Sized Part Mat 6:5234–5246

Chapter 33
Evaluating Deep Retrofit Strategies for Buildings in Urban Waterfronts

Nicola Strazza, Piero Sdrigotti, Carlo Antonio Stival, and Raul Berto

Abstract The renovation and requalification of existing building stock and the exploitation of renewable energy sources are considered key actions in the European Energy Roadmap to 2050.

In this scope, a noticeable potential application is represented by urban constrained built heritage, whose rehabilitation can be considered both as a strategy for its conservation and an enhancement of competitiveness for a sustainable city.

The proposed research evaluates the potential of scalable retrofit strategies in existing buildings through the exploitation of on-site renewable energy sources available in urban waterfront. The methodology starts with the collection of data concerning environmental and climate conditions, predictable building energy demand for heating and cooling services, and timescales of available on-site renewable energy sources.

The approach has been tested by the development of a detailed building energy model (BEM) for a constrained building in the urban waterfront of Trieste, in Northeastern Italy; the building construction dates back to the early 1980s and it is located in a valuable historical and cultural context. According to the preliminary rehabilitation proposal by Trieste Municipality, a baseline energetic model has been carried out to evaluate and optimize retrofit strategies for building envelope and energy systems, including renewable energy source exploitation.

The study considers particularly hydrothermal energy stored in sea basin that could play a significant role in urban waterfronts contexts. The main results show that the most effective adaptation strategies are characterized by key factors such as envelope thermal inertia in reducing heating and cooling demand in middle seasons, thermal effects of roof greening, and the combination of plants that exploit several

N. Strazza (✉) · P. Sdrigotti · R. Berto
Department of Engineering and Architecture, University of Trieste, Trieste(IT), TS, Italy
e-mail: nicola.strazza@phd.units.it; rberto@units.it; http://www.dia.units.it;

C. A. Stival
Department of Civil, Environmental and Architectural Engineering,
University of Padova, Padova(IT), Italy
e-mail: carloantonio.stival@dicea.unipd.it; http://www.dicea.unipd.it

© Springer International Publishing AG, part of Springer Nature 2019   391
A. Sayigh (ed.), *Sustainable Building for a Cleaner Environment*,
Innovative Renewable Energy, https://doi.org/10.1007/978-3-319-94595-8_33

discontinuous renewable sources and accumulation systems, such as photovoltaic systems and sea hydrothermal energy.

This paper looks at a strategic vision that focuses the need to manage plant systems through advanced commissioning or by developing an integrated smart thermal grid.

This research is embedded in a pilot project concerning the development of a low-temperature thermal grid in the waterfront of the old town of Trieste.

33.1 Introduction

In the scope of functional and technologic retrofit of existing buildings stock in Europe, various limitations restrict suitable actions for the improvement of its energy efficiency, both in operational and management phase. Moreover, existing building heritage, if subjected to architectural and landscaping constraints or carrying historical and cultural values to be preserved, is not directly considered in the scope of the regulatory framework described by European Directives.

Since the issue of Directive 2012/27/EU on energy efficiency, the improvement of energy performance in existing buildings is considered a high-priority goal in European energy topics' agenda; thus, a higher efficiency and renewable energy source integration represent suitable actions in existing building refurbishment and, at the same time, an intervention range to reach Europe 2020 strategy and European Energy Roadmap objectives.

The primary purpose in this scope is to identify performance levels objectively achievable through intervention actions compatible with restoration principles, considering energy improvement as a strategy to preserve the building itself. In historic buildings, the conservation of construction characters should match with a suitable exploitation of their passive energy behavior; thus, different performance levels for new constructions and existing heritage should be adopted, balancing appropriate measures for landscape or architectural integration [1].

In Italy, a significant part of listed architectural stock includes historical or otherwise constrained buildings that have been excluded to energy efficiency applications until the issue of Legislative Decree No. 63/2013 [2–5]. Now energy efficiency requirements for historic buildings can be prescribed, excluding a mandatory application only if compliance would involve an unacceptable alteration in historical, artistic, and landscape characters or appearance [6]. Thus, energy efficiency is considered as a tool capable of preserving protected assets for future generations in the best possible conditions, rather than an intervention that may conflict with conservation requirements; the development of appropriate technologies for the renovation and correct maintenance of existing buildings heritage now represents a challenge for a sector of fundamental importance for future society [1].

The improvement of energy performance of constrained buildings represents a challenge to test and, subsequently, to direct urban policies for exploitation of renewable energy sources available in valuable contexts. Thus, the implementation of energy technical systems in listed building heritage, deriving from on-site renewable energy sources, is seen as a sustainable purpose.

33.2   Background and State of the Art

The combined goal of competitiveness, sustainability in urban processes, and quality of urban public spaces is considered a challenge for medium and small urban cities in Europe, characterized by remarkable cultural and historical heritage and limited potentialities in their urban development. Strategies for urban renovation must match interventions on built environment with network technologies development, introducing systems based on renewable energy sources and related to urban microclimate conditions.

In Northern Europe, seawater exploitation for air-conditioning of building stock has been successfully tested in large-scale applications. Since 1986, a 180 MW power plant in Värtan Ropsten provides about 60% of thermal energy requirement of Stockholm City customers; 26% of the whole amount of energy requirement is given by filtered seawater [7]. A seawater hydrothermal energy is suitable for integration in a smart thermal grid for heating and cooling provision; heat recovered by thermoelectric gas-fired and waste-fired power plant can be combined in a large heating and cooling facility. This plan concept has been applied in Katri Vala system in Helsinki and in Fornebu neighborhood in Oslo [8, 9].

In Italy, exploitation of seawater hydrothermal energy has been carried out in small and medium contexts. In the Aquarium of Genova, the conditioning system performed in 2003 consists of three refrigeration units for the cooling of common spaces and seals' tank, with a reuse of disposed condensation water for tropical fish tank conditioning. District conditioning project for San Benigno neighborhood in Genova consists in a plant distributing low-enthalpy energy for heating and cooling services. A soft-water secondary ring provides thermal energy to each central system by sampling and filtering seawater.

A hydrothermal system based on seawater source heat pump, in favorable climate conditions, provides steady and continuous energy source for heating and cooling, generating remarkable saves during transition seasons, in which a low flow rate is required [10]. Under the right circumstances, in warm fall or during spring, it is possible to apply a direct surface water cooling bypassing heat pump set [11].

The developed research brings into focus the retrofit of an existing building, not completed yet, part of the waterfront in the urban area of Trieste, in Northeastern Italy. This urban system, developed along city waterfront in which building typologies different by construction techniques and age concur, is characterized by high energy consumption, derived quite completely by fossil sources. The planned concept for seawater energy exploitation derives from the immediate availability of the resource: a heat pump-based plant or a heat exchange system makes possible the use of sea basin water in a hypothetic energy distribution network.

Since 2005, a technical plant that includes seawater exploitation provides air-conditioning services for "Salone degli Incanti" and Aquarium, an eclectic historical building in the old town of Trieste. A larger-scale installation has been recently performed in Sistiana "Porto Piccolo," an environmental requalification that consists in a new residential and touristic village for overall 450 dwellings, 32 commercial units, and a marina. The energy plant consists in a main pumping and distribution facility that ensures seawater circulation in a closed loop, with an integration provided by a solar thermal collector field 200 m² wide and a prearrangement for groundwater exploitation as cold sink.

Given these noticeable existing applications, the study considers a functional renovation and the implementation of solutions capable of maximizing the potential of renewable energy sources exploitation, highlighting the challenges arising for Trieste urban waterfront. The nature of intervention object, owned by Trieste Municipality, enhances the need for the restitution of this architectural heritage to a new intended use as expositive area, creating renewed social spaces for the community. Intervention actions can reverberate their potential to a wider urban scale, leading to the redevelopment of the 'Sacchetta' urban area, on the southern side of Trieste waterfront. The application of renewable energy sources, after energy potential harnessing and recovery of waste streams on-site, is considered an important issue to be targeted in built-up areas [12, 13].

Afterwards, the target concerning renewable energy sources has been reformulated to identify technologies for their maximum exploitation and strategies to manage building energy demand. The target has not been limited to parametric analysis of main factors influencing the overall performance of the building, including its technological installations, but identifies strategies for potential reinvestment of advantages carried out by these actions on the surrounding district.

33.3   Methodology

33.3.1   Approach

The proposed methodological approach firstly involves an optimization process of intervention actions: the main purpose is the improvement of the building overall energy performance, following applicable restoration criteria such as compatibility, durability, and respect of the original fabric [1]. On a second, the analysis defines specific strategies for the optimization of generation subsystem in operational phase at a more detailed level, regarding the specificity of each season climate conditions [14].

In the city of Trieste, the peculiarities of northern Mediterranean climate have been noticed: outdoor climate conditions determine remarkable thermal loads arising both in heating and cooling periods, combined with modest daily temperature ranges. In this context, the behavior of massive construction, characterized by noticeable thermal inertia performances, plays a major role expressing a bioclimatic design. Thermal inertia leads to the damping of thermal load peaks, necessary to ensure adequate comfort conditions in indoor spaces and to reduce energy consumption during a whole year.

Within the scope of functional renovation of existing buildings, new technological plants and facilities concept must accord with both sustainability principles and restoration criteria, notwithstanding an actual gap in relating HVAC systems configuration and renovation intervention in earlier contexts. To evaluate each possible solution to maximize energy efficiency, a preliminary evaluation should consider building-integrated photovoltaic technology, including the exploitation of thermal energy surplus and district heating and cooling systems, coupled with cogeneration and trigeneration plants [15, 16].

Other efficient solutions, in the scope of a sustainable refurbishment of building envelope, should be further considered, such as the extensive greening of the rooftop, potentially capable of a remarkable reduction of cooling loads, and implementation of solar passive systems basing on natural lighting, through an appropriate design of transparent envelope with a moveable solar shading system [17, 18].

Moreover, it is considered an essential feature to identify all renewable sources connoting the installation site. In addition to solar thermal and PV systems, Trieste waterfront potentially provides hydrothermal energy contained in Trieste Gulf seawater basin. It is possible to state that hydrothermal energy is a solution evaluable for exploitation both on technical and cost-effectiveness sides [19].

The study has been performed starting with a critical analysis of case study situation, evaluating the building status quo and Trieste Municipality preliminary project contents and the relationships between intervention site, climatic context, and landscaping mandatory constraints. The definition of occupation profiles and comfort requirements let the individuation of analysis features according to specific energy efficiency targets. An optimization process has driven the performance evaluation through a dynamic energy simulation toward a configuration capable of minimizing off-site and network energy requirements; results and discussion highlight the potentiality for an integration at a larger scale in a smart thermal grid that could include similar buildings in the area by the exploitation of thermal and electric surpluses [20].

33.3.2   Urban Context and Case Study Description

The described case study is the building named "Ex Meccanografico": its completion and functional recovery is one of the main priorities for the Trieste Municipality in order to start the planned renovation of the waterfront, which is one of the most important key elements in nineteenth-century urban development.

Fig. 33.1 Juxtaposition between "Ex Meccanografico" building (on the right) and early twentieth-century Campo Marzio station (on the left). The main façade has northern exposition, facing the waterfront in "Sacchetta" sea basin

In this area, the intrinsic historical value lies in the whole waterfront network and related visual perception. The position of this building, nearby historic waterfront and old town center, concurs to fix the need for an efficient public use of this urban area, partially devoted to exposition and cultural functions (Fig. 33.1). Building construction began in the early 1980s on the initiative of the owner Italian State Railways, but its completion never occurred despite the investment made by Trieste Municipality for its acquisition in 2005. Consequently, its structure has been damaged due to neglect and weathering. Possible guidelines for intervention are limited by landscape restrictions, in association with visual perception of the valuable skyline along the waterfront in which twentieth-century buildings are located.

The "Ex Meccanografico" is a three-story building showing a rectangular plan approximately 1000 m² wide; the total floor surface is 3122 m², while the overall volume is 12,722 m³. The structural system refers to reinforced concrete technology, partly in shear walls 20 cm thick and partly in columns defining three-dimensional frames; the floors, currently devoid of any finishing layer, are mainly made by "Predalles" slab technological solution, except for first beam and pot floor, with consistent longitudinal deep beams in concrete. Perimeter walls are completed by hollow brick unfinished masonry. Currently, the building lacks of finishing layers, window glazing and moldings; no indoor technical system (HVAC, DHW production, indoor lighting) is currently installed.

Trieste Municipality has proposed in 2014 a renovation design project. This document can be considered close to a preliminary design though not meeting all requirements connoting this design step [21]. The building would accomplish expositive use thanks to a shape and an internal structure capable of indoor spaces availability and adaptability to different display area types.

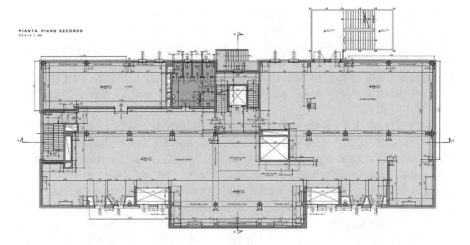

PIANTA PIANO SECONDO
SCALA 1:100

Fig. 33.2 Standard floor for expositive area according to preliminary project. (Iammarino & Ricci, 2014. Courtesy of Trieste Municipality)

With reference to this preliminary project, the building is supposed to be subdivided into three main areas: reception at ground floor; expositive area at 1st and 2nd floor, connoting a standard space distribution; and restaurant at 3rd floor, whose perimeter shape steps back if compared to lower levels. The completion of building aims to maximum flexibility goal; in this way, future changes in the internal volumetric distribution will be achievable (Fig. 33.2). In addition to preliminary baseline design, this research considers additional criteria such as CO_2 footprint minimization and the fulfillment of specific energy requirements providing adequate performance levels.

33.3.3   Definition of Renovation Technological Contents

With the aim to optimize energy retrofit of the building, the existing structure has been subjected to identify main design actions. Intervention scenarios consider a light-intensive green roof installation, the variation of thermal insulation thickness, and the variation of window-to-wall ratio on the southern façade, opposite to the waterfront. The optimization of these parameters lets a mitigation of outdoor temperature ranges effects. Green roof is designed with a composite vegetation layer 10 cm high, sedum carpet and rockery type plants characterized, at full growth, by a leaf area index = 2.7, an albedo $\rho = 0.22$, and a stomatal resistance $R_s = 100$ m^{-1} s; the lightweight growing media is designed in shattered pumice, 20 cm thick, giving an average U-value of 0.22 W m^{-2} K^{-1}. Vertical envelope consists in reinforced concrete shear walls. Notwithstanding visual constraint on main façade, it is possible to insulate opaque envelope with a 20 cm thick external layer in wooden fibers, providing a U-value of 0.22 W m^{-2} K^{-1}. Wooden fibers panels, externally protected by a lime plastering, provide an additional performance in heat storage capability. Windows will consist of a triple 3-mm-thick glazing with double low-e coating ($\varepsilon = 0.1$), and argon fulfilled spaces between glazing, wooden frame, and integrated blinds with high-reflectivity slats operating at 24 °C indoor air temperature set-point, for an overall shading coefficient SC = 0.47 and a U-value of 0.76 W m^{-2} K^{-1}.

The study carried out on building envelope refurbishment is directly integrated in the design of energy production system, a combination of a photovoltaic power plant with a thermal energy production facility. Because of its proximity, the energy contained in sea basin is available not only for the "Ex Meccanografico" building but also, in a wider vision, for a remarkable part of built heritage overlooking urban waterfront. This combined system is considered appropriate to satisfy building energy demands in summer, during which the maximum production from PV plant is available, while in winter an integrative production system will be required. Building generation subsystem should consider an energy integration that can provide support in the satisfaction of the winter demand peaks, both in terms of thermal energy for heating and domestic hot water production. As above stated, the combination of strategies about outdoor climate conditions and building components in view of the intended use of the building areas determines energy needs related to technological services that will be provided (Table 33.1).

Internal average operative conditions are assumed according to comfort requirements expressed in EN 15251, category II: 20 °C set-point temperature in heating period, 26 °C in cooling mode, and a supply air flow rate equal to 8·10^{-3} m^3 sec^{-1} per person. The optimal configuration of building technical systems

Table 33.1 Building envelope and occupation parameters set in the energy model for each thermal zone

Thermal zone	Area [m^2]	Volume [m^3]	Gross wall area [m^2]	Window glazing area [m^2]	Lighting installation [W m^{-2}]	People crowding rate [m^2 per person]	Plug and process [W m^{-2}]
3rd floor – restaurant	223.61	670.84	188.28	38.58	5.00	2.22	10.50
2nd floor – expositive space	916.92	2750.77	458.56	59.89	3.75	5.00	4.00
1st floor – expositive space	916.92	2750.77	458.56	59.89	3.75	5.00	4.00
Ground floor – reception	870.61	2698.90	456.59	93.39	3.75	5.00	4.00
Total/average values	2928.07	8871.28	1562.00	251.74	3.84	4.56	4.50

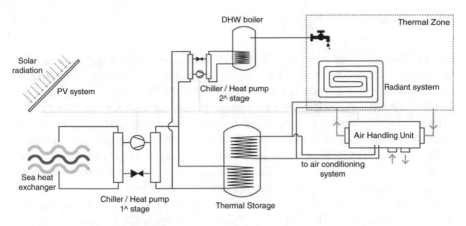

Fig. 33.3 Plant concept scheme for heating, cooling, and domestic hot water production. Exploited renewable energy sources consist in solar radiation and hydrothermal energy stored in basin seawater

considers heat recovery system with an efficiency close to 70%. HVAC systems are characterized by low-temperature radiant systems, with a moisture control in addition. PV collector field in monocrystalline silicon cells is 620 m² wide, characterized by an azimuth of 33° from south toward west and a 20° tilt angle on horizontal; conversion rates are assumed in 15% for PV modules and 90% for inverters. Hydrothermal seawater-to-water reversible chiller has a 70 kW peak power (Fig. 33.3).

An energy model of the building has been developed to operate a dynamic state simulation and to represent the most likely behavior of the building in actual outdoor climate conditions and occupational stresses related to its intended use. For this purpose, EnergyPlus v. 8.3 has been set as simulation engine. The climate data used for simulations derive from "IGDG climate data" (Gianni De Giorgio) created by Prof. L. Mazzarella; it is based on the period 1951–1970 that gives hourly weather conditions in Trieste in a reference year. In order to globally manage energy and architectural variables relevant to the building, a BIM model has been developed to produce a multidisciplinary analysis [22].

In operational phase, heat exchanger maintenance represents a remarkable cost burdened to control biofouling phenomena, namely, the undesired development, deposition, and growth of biofilms. Extracellular polymeric substance matrixes can adhere to heat exchanger surfaces, determining an additional insulation layer on the exchange surface and inhibiting convective heat transfer [23]. A regular maintenance of heat exchangers, deriving from an optimization of cleaning cycle's costs, prevents remarkable variations of seawater supply design temperatures and a serious decrease of heat exchanger efficiency [24]. Thermal treatment, mechanical cleaning, and disinfection/clarification of seawater could prevent piping from withdrawal and discharge of seawater from macrofouling phenomena caused by mussels, clams, and other large sea organisms.

33.4 Results

Once defined the plant concept, pointedly seawater-to-water reversible chiller plant section, sea basin conditions have been at first brought into focus. The operative temperature ranges for the reversible chiller are set at 8÷10 °C in winter and 22÷25 °C in summer. Basing on sea temperature profiles and seawater flow speed in Trieste Gulf, appropriate water source heat pumps have been identified. In building heating and cooling systems, using low-temperature emission subsystems, a COP = 4.3 is achievable; for DHW production a dedicated second stage is provided, with a COP = 3.5 is estimated. In this analysis, precautionary inputs refer to average seawater temperatures, consequently influencing COP value.

Although considering biofouling phenomena with an additional thermal resistance (9E-5 m^2 K W^{-1}) applied to heat exchanger, the overall system heat exchange with seawater is estimated in 148 MWh transferred from building to sea basin water during cooling season and 5,6 MWh taken from sea basin water to building during heating season. A great contribution to energy efficiency is given by mechanical ventilation in addition with heat recovery facility and by the high air-tightness performance provided by building envelope. During summer, the maximum cooling load amounts in thermal zones correspond to the maximum electricity production by photovoltaic plant (Fig. 33.4).

Simulation results show that "Ex Meccanografico", in preliminary design intended use, is an internally dominated building: in some utilization scenario, thermal loads to be removed from each thermal zone can be evaluated as about 25 times heating loads in the same context (Fig. 33.5). This is a favorable situation for the maximum exploitation of photovoltaic plant: regarding utilization scenario and

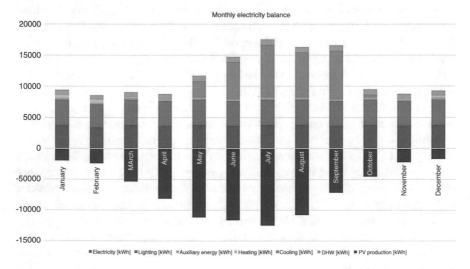

Fig. 33.4 Monthly energy balance in electric consumption and photovoltaic production, considering energy services provided in the building, once renovated

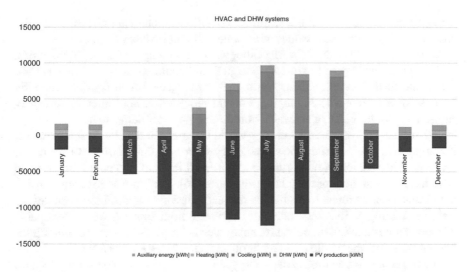

Fig. 33.5 Monthly energy balance, considering domestic hot water production and air-conditioning services

considering active all electrical loads, the consumption of self-produced energy is about 80% of the whole photovoltaic plant production.

Sensitive analysis demonstrates that the introduction of an electrical accumulation system could increase the self-use for a maximum percentage of +5% within the utilization scenario. By considering the building intended use, mainly occurring during the central part of the day, most of photovoltaic energy production is consumed by building services without any storage systems.

In addition to this solar system, an inertial thermal facility should completely cover up, with self-produced energy, the whole energy demand for heating and cooling services provided by heat pump.

A greater efficiency improvement should be given by a building automation system, which can predict the occupancy profiles and the weather, thus maximizing the inertial energy contribution, free energy contained in the thermal storage system, or other passive strategies implemented in the building.

Green roof contribution is greater in indoor comfort conditions control than in energy consumption reduction, anyway increasing positive CO_2 footprint in building lifecycle. If compared to a traditional covering, green roof offers a reduction from 1% to 5% for each indicator of environmental impact categories, the highest in eutrophication and abiotic depletion [25].

33.5   Conclusions

The future perspective of the research aims to analyze feasibility, effectiveness, and sustainability of urban rehabilitation strategies. This purpose will consist in network technologies development with renewable energy sources implementation and fruitful exploitation of local microclimate conditions, though not affecting peculiar architectural and historical values of existing building heritage.

In this perspective, the "Ex Meccanografico" building in Trieste, in view of a high-level availability of flexible spaces, in future could consent of a different use provision. In this case, analysis result may be subjected to a sensible variation because of a different technological content required. A new intended use may suggest a non-eligibility of seawater stored energy as heat pump source.

Whenever sea hydrothermal energy would be considered as the best alternative, if compared to other on-site renewable energy sources in further different scenarios, a vision of integration of this resource into a smart thermal grid could represent an advantageous situation for the whole urban area [26]. The innovative feature carried out in this research refers to evaluation and exploitation of hydrothermal energy by a single building, with a wide range of intervention reliability. It would be capable of integration into a low-temperature smart thermal grid, as improving solution for the renovation of the whole building heritage facing Trieste waterfront.

In this perspective, hydrothermal energy represents a great interest option, because of the immediate availability of a source carrying out remarkable advantages as a simple distribution network layout, compatibility with low-enthalpy energy carriers, a know-how technology content available also for small installations. Moreover, the involvement of Trieste Municipality will give to this case study a pilot project role-play. However, the output model should be considered in order to compare other RES technologies, particularly a low-enthalpy geothermal plant.

At urban planning level, the surplus production given by each thermal power plants could be transferred into the energy grid ensuring a remarkable global performance improvement, a proposal for the buildings involvement in a "diffused generation" view. A wider application context enables more efficient paths for energy delivery and storage, toward buildings characterized by a different pattern of energy requirements that may solve the overall energy balance at neighborhood level.

References

1. De Santoli L (2015) Reprint of guidelines on energy efficiency of cultural heritage. Energ Buildings 95:2–8
2. Commission of the European Communities (2005) Green Paper on energy efficiency or Doing more with less
3. Italian Government (2004) Legislative Decree n. 42/2004 of the Italian Government. Code of cultural heritage and landscape
4. Italian Government (2005) Ministerial Decree n. 192/2005 of the Italian Government. Implementation of Directive 2002/91/CE on energy efficiency in buildings
5. Italian Government (2015) Interministerial Decree of the Italian Government 26 June 2015. Application of energy performances calculation methodologies and definition of mandatory prescriptions and requirements of buildings
6. Italian Government (2013) Decree-Law of the Italian Government 4 June 2013, No. 63. Urgent provisions for the transposition of Directive 2010/31/UE of the European Parliament and of the

Council of 19 May 2010 on the energy performance of buildings for the definition of infringement procedures started by European Commission, and other provisions on social cohesion

7. Friotherm AG, Ropsten V (2005) The largest sea water heat pump facility worldwide, with 6 Unitop® 50FY and 180 MW total capacity. Friotherm AG, Winterthur

8. Riipinen M (2013) District heating and cooling in Helsinki. In: Proceedings of the international CHP/DHC workshop, Paris, 12–13 Feb 2013

9. Friotherm AG (2005) Oslo – Fornebu. Sustainable development with a district heating/cooling system using a Unitop® 28/22CY. Friotherm AG, Winterthur

10. Cao ZK, Han H, Gu B, Zhang L, Hu ST (2009) Application of seawater source heat pump. J Energy Inst 82:76–81

11. Mitchell MS, Spitler JD (2012) Open-loop direct surface water cooling and surface water heat pump systems – a review. HVAC&R Res J 19(2):125–140

12. Schumacher P (2015) Energy efficiency for Eu historic districts sustainability. Smart management and integration of renewable and energy efficiency solutions. BAU München, München, 19–24th Jan 2015

13. Vieites E, Vassileva I, Arias JE (2015) European initiatives towards improving the energy efficiency in existing and historic buildings. Energy Procedia 75:1679–1685

14. Pisello AL, Petrozzi A, Castaldo VL, Cotana F (2016) On an innovative integrated technique for energy refurbishment of historical buildings: thermal-energy, economic and environmental analysis of a case study. Appl Energy 162:1313–1322

15. Moschella A, Salemi A, Lo Faro A, Sanfilippo G, Detommaso M, Privitera A (2013) Historic buildings in Mediterranean area and solar thermal technologies: architectural integration vs preservation criteria. Energy Procedia 42:416–425

16. Allegrini J, Orehounig K, Mavromatidis G, Ruesch F, Dorer V, Evins R (2015) A review of modelling approaches and tools for the simulation of district-scale energy systems. Renew Sust Energ Rev 52:1391–1404

17. Gagliano A, Detommaso M, Nocera F, Evola G (2015) A multi-criteria methodology for comparing the energy and environmental behaviour of cool, green and traditional roofs. Build Environ 34:71–81

18. Athienitis A, O'Brien W (eds) (2015) Modeling, design, and optimization of net-zero energy buildings. Wiley, New York

19. Keirstread J, Nilay S (eds) (2013) Urban energy systems: an integrated approach. Routledge, London

20. Gang W, Wang S (2016) District cooling systems: technology integration, system optimization, challenges and opportunities for applications. Renew Sust Energ Rev 53:253–264

21. Iammarino L, Ricci P (Trieste Municipality, 2014) Preliminary project for Ex Meccanografico building renovation

22. Ceranic B, Latham D, Dean A (2015) Sustainable design and building information modelling: case study of energy plus house, Hieron's Wood, Derbyshire UK. Energy Procedia 83:434–443

23. Flemming HC (2002) Biofouling in water systems – cases, causes and countermeasures. Appl Microbiol Biotechnol 59(6):629–640

24. Ezgi C, Ozbalta N (2012) Optimization of heat exchanger cleaning cycle on a ship. J Nav Sci Eng 8:133–146

25. Saiz S, Kennedy C, Bass B, Pressnail K (2006) Comparative life cycle assessment of standard and green roofs. Environ Sci Technol 40:4312–4316

26. Schmidt RR, Fevrier N, Dumas P (2013) Smart cities and communities. Key to innovation integrated solution. Smart Thermal Grids, version

Chapter 34
Enhancing the Thermophysical Properties of Rammed Earth by Stabilizing with Corn Husk Ash

Amina Lawal Batagarawa, Joshua Ayodeji Abodunrin, and Musa Lawal Sagada

Abstract Conventional building materials such as sandcrete and concrete blocks have high thermal conductivity and low specific heat capacity which increase thermal discomfort in comparison with rammed earth in buildings. However, rammed earth is susceptible to cracking as a result of shrinkage due the heating effect of the sun. The aim of this study is to investigate the potential of corn husk ash as a stabilizer for the production of rammed earth blocks to improve thermal comfort in buildings. Three different levels of stabilization (0%, 10% and 20%) using corn husk ash are adopted for this study, beyond which the strength of the material is known to fail. The samples are moulded and subjected to thermal conductivity, specific heat capacity and density tests. In general there is a significant improvement in the thermal conductivity of stabilized rammed earth blocks. From the thermal conductivity test results, thermal conductivities of 0.996 w/k.m^{-1}, 0.637 w/k.m^{-1} and 0.489 w/k.m^{-1} were obtained for samples admixed with 0%, 10% and 20% corn husk ash, respectively. The 0%, 10% and 20% samples had heat capacities of 962.1 j/kg.K, 984.9 j/kg.K and 993.4 j/kg.K and density increased with the increase in the amount of ash from 862.3 kg/m^3 to 942.5 kg/m^3 and 959.5 kg/m^3, respectively. Overall the 20% sample performs best when compared to the 0% and 10%. Stabilizing rammed earth with corn husk ash can improve the thermal properties of rammed earth blocks, making them suitable for use as a building material in enhancing thermal comfort.

34.1 Introduction

As evident in developing countries such as Nigeria, it is challenging to fulfil the immense requirements for shelter with conventional construction techniques and building materials such as sandcrete blocks, concrete, aluminium and steel which are noted for their high energy consumption during production and associated

A. L. Batagarawa (✉) · J. A. Abodunrin · M. L. Sagada
Department of Architecture, Ahmadu Bello University, Zaria, Kaduna, Nigeria
e-mail: aminab_@hotmail.com

© Springer International Publishing AG, part of Springer Nature 2019
A. Sayigh (ed.), *Sustainable Building for a Cleaner Environment*,
Innovative Renewable Energy, https://doi.org/10.1007/978-3-319-94595-8_34

negative environmental impacts [1]. This leads to a need to source for, modify and make use of locally available materials for construction. Stulz and Mukerji [2] reports that earth is a natural resource which is one of the oldest and versatile that is commonly used throughout the world as building material. It is cheap, has excellent heat insulation capacity and is strong in compression. Rammed earth has been played down for use in the building construction industry because of its low strength, thermal conductivity inconsistencies and need for heavy maintenance due to cracking in shrinkage mode [3]. Earth can be stabilized so as to increase compressive strength, reduce shrinking and swelling, reduce or exclude water absorption, reduce cracking and reduce expansion and contraction using fibres.

This has led to the stabilization of soil blocks with sand and clay, lime and pozzolanas, Portland cement, gypsum, bitumen, sodium silicate, cow dung or horse urine, plant juices, resins, molasses, whey, animal products such as hair, termite hills and plant products such as cob and husk of stale crops such as rice and corn to improve its properties even though most of these stabilizers are costly and unsustainable. Stabilization of rammed earth is the process of modifying the soil properties in relation to its strength, texture, voids and water-resistant properties, to obtain permanent properties compatible with a particular application [4]. Stabilizing rammed earth leads to irreversible change in the physical properties of soil depending on the quality of building design, materials employed and economic aspects of the project or on issues of durability. Previous studies on rammed earth have shown that for experimental purposes, the stabilizers are added in 10% increment by weight and beyond 20% increment of ash; the strength of stabilized earth has been known to fail [5, 6].

Corn husks along with corn stalk and leaves are usually disposed either by burning or tilling into the soil in developing countries like Nigeria [7]. The ash of the corn husk is believed to possess substantial amount of siliceous compounds making it a pozzolanic material. This property makes it feasible to function as a cementitious material by improving the binding forces between the soil particles [8]. On a wider scale, the stabilization of rammed earth with corn husk ash can also curtail the pollution of the environment and reduce the cost of building while, most importantly, enhancing the durability of rammed earth and enhancing thermal comfort.

Thermal conductivity of a material is a very important factor in determining its insulation and therefore the thermal comfort which it brings. Insulation reduces unwanted heat loss or gain and can decrease the energy demands of heating and cooling systems [9]. Thermally effective materials require low thermal conductivity, high specific heat capacity and high density. Other properties that affect the thermal performance of building construction materials are U-value, admittance, solar absorption, visible transmittance, thermal decrement, thermal lag and emissivity [10]. However, to calculate these properties, specific heat capacity and density of the samples are required [10].

The aim of this study is to test the conductivity, specific heat capacity and density of rammed earth stabilized with differing percentages of corn husk ash.

34.2   Methodology

A laboratory test is conducted on rammed earth that is stabilized with 0%, 10% and 20% corn husk ash. The earth used in this study is sourced from the Department of Architecture, Ahmadu Bello University, Zaria, Nigeria, while the corn husk ash used is sourced by burning in open flame the corn husk gotten from a local farm where corn was recently harvested in Samaru also in Zaria (Figs. 34.1 and 34.2). Three different levels of stabilization (0%, 10% and 20%) using corn husk ash were adopted for this study. After measuring in the right proportions, the samples were thoroughly mixed and homogenized. They were then placed in a mould where they were compressed with a 20 tonne hydraulic press and then allowed to dry.

Fig. 34.1 Rammed earth sourced from the Department of Architecture. (Source: Authors' Fieldwork)

Fig. 34.2 Corn husk ash. (Source: Authors' Fieldwork)

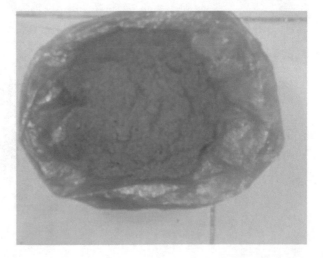

34.2.1 Measurement of Thermal Conductivity

The dried moulded samples as shown in Figs. 34.3, 34.4, and 34.5 are taken to the Searle's apparatus so as to measure their thermal conductivities. The different samples are put into the apparatus (Fig. 34.6) and the different temperature readings taken and are used in calculating the thermal conductivity.

$$k = \frac{x.m.s\ (\theta 3 - \theta 4)}{A(\theta 1 - \theta 2).t} \tag{34.1}$$

Equation 34.1 is used in calculating the thermal conductivity,

where k is thermal conductivity
x is length of rod between the holes
θ is change in temperature
A is cross-sectional area
t is time in supplying heat

Fig. 34.3 0% CHA

Fig. 34.4 10% CHA

Fig. 34.5 20% CHA

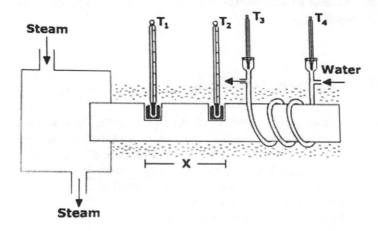

Fig. 34.6 Diagram showing a Searle's apparatus

34.2.2 Measurement of Specific Heat Capacity

There are several methods such as calorimetry method, electrical method, method of mixing, etc., but the electrical method is considered in this study. In this method, a calorimeter made of a wooden box is used. The inside wall is surrounded with a thin coksheet layer. The outside wall is also surrounded with a thin coksheet layer and a thin thickness of glass wool layer. The inside wall is covered with a mild steel plate. A hole is made on the top face of the box to insert the cable into the box. A sample was then inserted into the calorimeter and a heater is put into one hole and the temperature sensor in the other hole. The cable of the temperature sensor and the heater is drawn out through the hole of the calorimeter. Electricity is then supplied, and different parameters such as current (I), voltage (V), time (t), initial temperature (θ_0), final temperature (θ_1), etc. are measured. Finally the specific heat capacity of the samples is calculated using the specific heat capacity formula [11]:

Table 34.1 Showing the different thermal conductivities of the differing percentages of ash

S/no.	Sample ID	Thermal conductivity (W/m.k^{-1})
1	0% ash	0.996083
2	10% ash	0.637082
3	20% ash	0.488776

$$c = \frac{(q2 - q1)}{(m2 - m1)(\theta0 - \theta1)} \qquad (34.2)$$

Equation 34.2 is used in calculating the specific heat capacity, where C is the specific heat capacity.

34.2.3 Measurement of Density

The densities of the samples were calculated from a separate mass and volume measurement; the volumes were obtained from the measurement of the dimensions of the samples. The weights of the samples were calculated and the mass of an unknown metal cylinder recorded. The volumes of the samples were calculated by measuring in metres the height (h) and diameter (d) of the cylindrical samples and then applying this formula: Volume (m^3) = $h \times 0.785d^2$.

The density of a sample is determined from Eq. 34.3:

$$\rho = \frac{m}{v} \qquad (34.3)$$

where

ρ is density
M is mass of sample
V is volume of sample

34.3 Results

The following were the results obtained from the laboratory procedure and calculations.

34.3.1   Thermal Conductivity

Table 34.1 and Fig. 34.7 show the thermal conductivities calculated.

34.3.2   Specific Heat Capacity

Table 34.2 and Fig. 34.8 show the results from the specific heat capacity test.

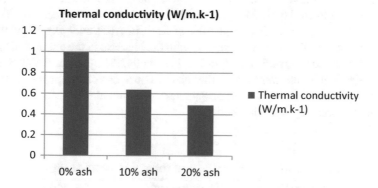

Fig. 34.7 Showing the thermal conductivities of the different samples

Table 34.2 Showing results from the specific heat capacity test

S/no.	Sample ID	Sample mass	Temp. of mix. (°C)	Heat cap. (Jkg^{-1}K^{-1})
1.	0% ash	18.97	38.5	962.1
2.	10% ash	19.90	39.5	984.9
3.	20% ash	18.85	40.0	993.4

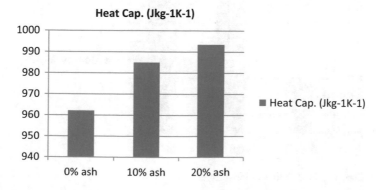

Fig. 34.8 Showing the specific heat capacities of the different samples

34.3.3 Density

Table 34.3 and Fig. 34.9 show the densities of the different samples as calculated.

34.4 Conclusion

In general there was a significant improvement in the thermal conductivity of stabilized rammed earth blocks. From the thermal conductivity test results, thermal conductivities of 0.996 w/k.m⁻¹, 0.637 w/k.m⁻¹ and 0.489 w/k.m⁻¹ were obtained from samples admixed with 0%, 10% and 20% corn husk ash, respectively. The 0%, 10% and 20% samples had heat capacities of 962.1 j/kg.K, 984.9 j/kg.K and 993.4 j/kg.K, respectively. The density increased with the increase in the amount of ash; the 0%, 10% and 20% samples had 862.3 kg/m³, 942.5 kg/m³ and 959.5 kg/m³, respectively. Stabilizing rammed earth with corn husk ash can improve the thermal properties of rammed earth blocks, making them suitable for use as a building material in enhancing thermal comfort.

Table 34.3 Showing the different densities of the differing percentages of ash

S/no.	Sample ID	Density (g/cm³)	Density (kg/m³)
1	0% ash	0.8623	862.3
2	10% ash	0.9425	942.5
3	20% ash	0.9595	959.5

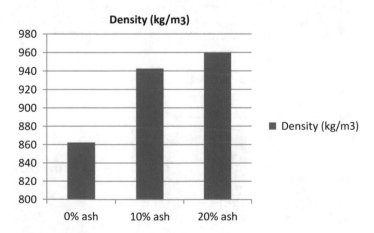

Fig. 34.9 Showing the densities of the different samples

References

1. Minke G (2006) Building with earth: design and technology of a sustainable architecture. Birkhauser Publishers for Architecture, Boston
2. Stulz R, Mukerji K (1993) Appropriate building materials. SKAT Publications and IT Publications, London
3. Kamang EE (1998) Strength properties of compressed earth blocks with earthworm cast as stabilizer. J Environ Sci 1:65–70
4. Rigassi V (1985) Compressed earth blocks: manual of production. Gesellschaft fur Tecchnische Zusammenaebeit GTZ, Deutsche
5. Fatih T, Umit A (2001) Utilization of fly ash in manufacturing of building bricks. International utilization symposium. Center for Applied Energy Research, Kentucky
6. Okunade EA (2008) The effect of wood ash and saw dust admixtures on the engineering properties of a burnt laterite-clay brick. J Appl Sci 8:1042–1048
7. Nazir M, Abeyruwan H, Mouroof M (2012) Waste ash pozzolans. Retrieved from reactivity and suitability for use in concrete: http:www.ricehuskashpozzolanic-materialforpdf
8. Kevern JT, Wang K (2010) Investigation of corn husk ash as a supplementary cementitious material in concrete. Second international conference on sustainable construction materials and technologies. Coventry University, Coventry
9. Bahobail MA (2012) The mud additives and their effect on thermal conductivity of adobe briks. J Eng Sci 40:21–34
10. Autodesk Ecotect Analysis (2011) Building energy software tool directory. US Department of Energy
11. Al Farouk AS, Alamin M, Hasan R, Haque T, Alam S (2013) Measuring specific heat of normal strength concrete and the comparison of the specific heat with different types of concrete. Int J Adv Struct Geotech Eng 2:69–76

Chapter 35
Thermal Monitoring of Low-Income Housings Built with Autoclaved Aerated Concrete in a Hot-Dry Climate

Ramona Romero-Moreno, Gonzalo Bojórquez-Morales, Aníbal Luna-León, and César Hernández

Abstract The construction of housing in series for low-income families located in cities with hot-dry climate like Mexicali requires the use of constructive systems in the envelope, which cushion the effect of ambient temperatures of 45 degrees Celsius during the summer period. The conventional constructive system used was concrete block walls and beam and polystyrene vault. This system causes environmental conditions outside the ranges of thermal comfort in summer time. The use of nonconventional constructional systems for the housing in series is an option that needs to be evaluated, as is the autoclaved aerated concrete (AAC-based building envelope systems); we have data of thermal properties of the manufacturer, but we do not count with experimental data of a house built with the AAC system in a zone of warm climate, to know its thermal performance in conditions of high ambient temperature. This article shows the results of the thermal monitoring of a proposed bioclimatic model of affordable housing built in Mexicali, based on the AAC construction system. Measurements were made from September 21 to October 6, 2016; the variables monitored on site were indoor ambient temperature, relative humidity, and black globe temperature; HOBO type sensors were used. The outdoor temperature was taken from the weather station of the Autonomous University of Baja California. The frequency of measurements was every 15 min; averaging data per hour was obtained. The results showed significant differences between internal and external environmental conditions and that environmental conditions are without the range of thermal comfort for the inhabitants of Mexicali, as well as lower demand for air conditioning. Therefore, the use of walls and roofs based on autoclaved aerated concrete was a viable thermal and energetic option for the construction of housing in series in areas of hot-dry climate.

R. Romero-Moreno (✉) · G. Bojórquez-Morales · A. Luna-León · C. Hernández
Universidad Autónoma de Baja California, Mexicali, Mexico
e-mail: ramonaromero@uabc.edu.mx; gonzalobojorquez@uabc.edu.mx; anibal@uabc.edu.mx

© Springer International Publishing AG, part of Springer Nature 2019
A. Sayigh (ed.), *Sustainable Building for a Cleaner Environment*,
Innovative Renewable Energy, https://doi.org/10.1007/978-3-319-94595-8_35

35.1 Introduction

The energy efficiency of buildings nowadays is one of the subjects that matter the most because of the gases that come with their greenhouse effect, partly, because of the electric consumption from the environmental conditioning of the spaces. In Mexico, in 2016, the residential sector consumes 27% of the total electric consumption [1]. In hot-dry climates, the 60% is from the use of air conditioning. In the case of extreme hot-dry climate such as Mexicali, the electric consumption from the residential sector goes up to 43.9% of the total electric consumption [2].

On the other hand, housing politics, that in their moment attended the need for dwellings, standardized housing models that were repeated in places with different climate conditions (Fig. 35.1).

At the end of 2001, the Economic Dwelling National Program promoted the construction of houses that would be available for low-income families; the cost of the dwelling was of 117.06 minimum wages [3]. From the urban point of view, this program propitiated the construction of dwellings in the periphery of the cities, because a lot of urban ground available at a low cost was there. This caused problems of urban mobility, as they were far from their works, schools, and healthcare centers, among others; from the architectonic point of view, dwellings were built initially of 32 square meters (a bedroom) and then of 38 square meters (two bedrooms). Studies showed that in this type of dwellings, the inhabitant average was of four to five people [4].

The public politics in housing have promoted the access to a worthy dwelling that at the same time can contribute when it comes to caring for the environment. This has been established in the Housing Law [5] and regulated in the Housing Code [6]. According to the National Housing Plan 2014–2018, the public politics are actually promoting the redensification of the cities and further sustainability actions applied to the housing sector [7].

Fig. 35.1 A series of one single-family detached dwelling, México

In Mexico, 48.3% of its territory corresponds to arid and semiarid zones [8]; in these, the climate impact interacts with the surrounding of the dwelling, and if it is not thermally efficient, the interior of the house will have out of comfort conditions during the warm and cold season, affecting directly the conditions of thermal habitability of the inhabitants. This is why it is necessary to improve the actual constructive systems or the research of new options, as a first step, to improve the energy efficiency of the dwellings to provide thermal comfort conditions.

The autoclaved aerated concrete (AAC-based building envelope systems) is a product applied frequently to European construction. In Mexico, it has been applied to the industrial, touristic, commercial, and particular housing sectors. However, it has only been applied to the city of Hermosillo, Sonora, within the Nationally Appropriate Mitigation Actions (NAMA) program [9].

35.1.1   AAC-Based Building Envelope System

AAC is an ultralight concrete formed by spherical air cells, homogenous and independent. Some properties are shown in Table 35.1 [10].

In Mexico, as a constructive system, it is not conventional for the mass construction dwellings; even if it shows a better performance when it comes to workforce, the envelope of the housings with this system are lighter compared to concrete block constructive systems (used in mass construction housings). Moreover, there are not published studies about thermal-energetic performance of mass construction dwellings built with ACC in the country, so this research is relevant due to the thermal monitoring of a housing prototype built in extreme hot dry climate conditions, as Mexicali.

35.1.2   Application Research on the AAC Constructive System

Internationally There are many studies related with ACC, among which stands up Jerman et al. [11]. They studied the hygrothermal behavior of material, using blocks with different densities and humidity absorption levels at different temperatures,

Table 35.1 Mechanical and thermal properties of autoclaved aerated concrete AAC-4 and AAC-6

Properties	AAC-4	AAC-6
Nominal density (kg/m³)	500	600
Bulk density (kg/m³)	509.91	579.01
Design weight (kg/m³)	600	720
Compression resistance (kg/cm²)	40.8	61.2
Thermal conductivity (W/m °C)	0.123	0.169

Source: Elaboration as from technical sheets of the products, Xella Mexicana, http://www.hebel.mx/es/content/resumen_t_cnico_1454.php

and identified that thermal conductivity increased to six times the humidity level and 18% when submitted to 25 °C - 40 °C (Table 35.2).

Studies done with thermal simulation analyzed the effect of adding a thermal isolation layer to a AAC wall; the hygrothermal behavior with different humidity conditions was annually evaluated and found that the most appropriate materials were those that contained hydrophilic mineral wool or calcium silicate [12]. Studies have been made regarding to the fabrication of AAC with the use of copper slag as a lime substitute [13] and the use of industrial waste to produce AAC concrete blocks [14].

National Scope The National Organism for Normalization and Certification of Edification and Construction (ONNCCE, for its acronym in Spanish) has validated AAC [15] (Table 35.3).

According to the rules of energy efficiency for housings [16], and to the adjustments made in 2014, the recommended "U" heat transference coefficient for slabs is 0.625 W/m^2 °C and for walls 0.714 W/m^2 °C [17]. So that according to those numbers, it was found that for AAC-6 slab (thickness 0.20 m) was of 0.807 W/m^2 °C, AAC-6 slab (thickness 0.15 m) was of 1.121 W/m^2 °C, and walls AAC-4 (thickness 0.15 m) was of de 0.698 W/m^2 °C [17].

Local Scope Studies made from a sample of 0.03078 m x 0.1098 m x 0.0593 m, and 0.1 kg weight are as follows: the calculated density was 499.07 kg/m3 ± 0.5%, thermal conductivity 0.1262 W/m°C ± 5.9%, and specific heat 749.31 J/kg°C ± 3.5% [18].

Table 35.2 Thermal conductivity, according to temperature

Material	Temperature (°C)					
	2	10	15	25	30	40
P4-500	0.0941	0.106	0.1088	0.1218	0.1288	0.1438

Source: elaboration from Jerman [11], 356
Note: P4-500 refers to the density of the material used for this study

Table 35.3 Thermophysical properties of AAC, NOM-ONNCCE #2260

Thermophysical properties	Unidades	Autoclaved aerated concrete (AAC)		
		AAC-2	AAC-4	AAC-6
Bulk density	kg/m^3	407.25	517.00	598.70
Thermal conductivity	W/m °C	0.104	0.130	0.135
Permeability to water vapor	ng/Pa*s*m	0.343	0.308	0.289
Humidity absorption	% mass	5.46	5.29	5.18
	% volume	2.11	2.71	3.03

Source: Xella Mexicana, according NOM-ONNCCE #2260, 2010

35.1.3 The Climate of Mexicali, Baja California, México

The city is located at latitude 32°N, 115°W at an elevation of 3 m above sea level. The local climate is an extreme, warm, dry type, with strong, daily and seasonal variations, predominantly sunny days and intense solar radiation. Its maximum annual average temperature is 31.4 °C, an annual average of 23.7 °C and a minimum annual average of 16.1 °C. In July, the average maximum temperature is 42.9 °C, with maximum monthly temperatures of 45.0 °C and maximum daily temperatures up to 52 °C. In August, the average maximum temperature is 42.0 °C, with maximum monthly temperatures of 44.2 °C and maximum daily temperatures up to 49.4 °C; while in December the average minimum temperature is 6.8 °C and 7.1 °C in January, with minimum monthly temperatures of 3.4 °C. The annual rainfall is 73.3 mm in the months of December and January [19]. Its warm period is from May to October and cold period is from November to April.

35.2 Method

The conditions of the case of study are presented; a longitudinal thermal monitoring was made in a bioclimatic economic dwelling model built with the AAC system, in an extreme hot-dry climate zone.

35.2.1 Case Study

The studied housing model is one of the obtained results from a research financed by the National Housing Commission [20]; its design is suitable for Mexicali's climatic conditions and uses guidelines and bioclimatic techniques, as well as an AAC constructive system.

The dwelling is located in El fraccionamiento Parajes de Puebla Neighborhood, in the 28th lot, block 32. In this division there are mostly predominate lots of 120 m² (6.86 m wide and 17.50 long). At the moment it is surrounded by commercial housing models. The orientation of the main façade is to the north; the dwelling is built in the north-south longitudinal axis. The housing has a construction area of 43.8 m², with a living room, dining room, kitchen, two bedrooms, and a bathroom. The interiors have sloping ceilings; there are heights superior to 2.40 m. It promotes the use of ventilation and daylighting, using high windows and the management of a bioclimatic patio and the protection of solar radiation, with horizontal and vertical shading elements (Fig. 35.2).

In Table 35.4 the technical specifications of AAC systems and commercial system are shown which includes overall heat transfer coefficient (U).

35.2.2   Longitudinal Thermal Monitoring

The field work consisted in measuring the environmental conditions inside the dwelling, during the warm period from September 23 to October 2, 2016. Dry-bulb temperature, relative humidity, and black globe temperature were registered. The measuring frequency of variables was 15 min. HOBO type sensors were put in common spaces and, in one of the bedrooms, placed in the geometrical center of the space, at a 1.30 m^2 height over the floor level according to the ISO 7726 [21].

The measuring equipment consisted of HOBO transducers U12-013 type; dry-bulb temperature and relative humidity were recorded. The measuring equipment consisted of HOBO transducers U12-013 with two external outputs and a USB communication port for data downloading; dry bulb temperature and relative humidity were recorded. Main specifications for this data logger are: Measurement range is −20 °C to 70 °C for temperature and 5% to 95% for relative humidity, accuracy of ±0.35 °C from 0 °C to 50 °C and ±2.5% from 10% to 90% to a maximum of ±3.5% for relative humidity., weight 0.046 kg and dimensions 0.058 m x 0.074 m x 0.022 m

For the conditions of the outside, the records of the meteorological station of the UABC were taken [22]. The consistencies of the records (outdoor and indoor) were checked in order to compare and determine hourly averages of all variables that were obtained.

A longitudinal monitoring of housing was carried out, with the houses' doors and windows closed; the house was not occupied.

35.3   Results

The Fig. 35.3 shows the behavior of outdoor dry-bulb temperature and outdoor relative humidity during the monitored period from September 23 to October 6, 2016.

Outdoors and indoor conditions are shown in Table 35.5. The effect of the envelope is observed in the conditions of temperature and relative humidity measured inside the spaces of the dwelling.

The envelope constructed with the AAC system dampens the daily temperature oscillations, while the daily external fluctuations oscillate between 10 °C and 16 °C; in the interior the oscillation was smaller, around 6 °C, as shown in Fig. 35.4.

Most of the time, DBTi and BGT were kept together; however, in the periods of high temperatures inside the house, the BGT was lower, which is indicating that the envelope built with the system AAC emits less radiation into the space (Fig. 35.5).

Fig. 35.2 Bioclimatic dwelling built with AAC, Mexicali

Table 35.4 Building systems' technical specifications and overall heat transfer coefficient (U), bioclimatic AAC housing and commercial housing, Mexicali

Element	Bioclimatic AAC housing	U (W/ m²°C)	Commercial housing	U(W/ m²°C)
Slabs	0.61 wide AAC-6 panels, with a thickness of 0.175 m	0.807	0.15 m beam with polystyrene 0.10 m vault, with a layer of 0.05 m compressed concrete	1.257
Walls	AAC-4 solid block 0.15 × 0.40 × 0.61 m, included "O" and "U" blocks, joints with adhesive mortar (0.015 m), castles with drowned rod	0.698	0.12 m common concrete block with cement-sand mortar,	3.460
Floors	0.10 m reinforced concrete slab, $f'c = 240$ kg/cm²	3.18	0.10 m reinforced concrete slab, $f'c = 240$ kg/cm²	3.18
Windows	Single pane 3 mm glass with aluminum frame	7.24	Single pane 3 mm glass with aluminum frame	7.24
Doors	Hollow core wooden door	2.78	Hollow core wooden door	2.78

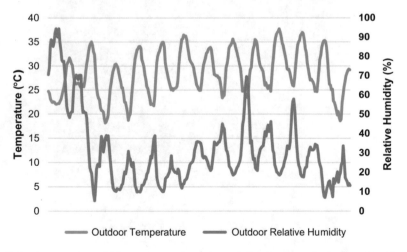

Fig. 35.3 Outdoor dry-bulb temperature and outdoor relative humidity, September 23 to October 6, Mexicali. (Source: Meteorology Department, UABC, 2017)

35.4 Conclusions

In hot-dry climate, the construction with the AAC system is thermally adequate, due to the low values of thermal conductivity, compared to the systems conventionally used for series of one single-family detached dwelling. This situation will cause less heat transfer to the interior of the house, however, in turn, can delay the cooling of the construction in summer.

Table 35.5 Thermal monitoring conditions, AAC housing, September 23 to October 6, Mexicali

	Outdoor		Indoor (AAC housing)			
	DBTo (°C)	RHo (%)	DBTi (°C)	RHi (%)	BGTi (°C)	DBTi_b (°C)
Maximum	37.717	94.333	36.898	56.926	35.616	33.888
Minimum	18.167	5.333	25.951	22.687	25.969	26.983
Mean	27.877	30.688	29.833	34.131	29.888	30.308
Oscillation	19.550		10.947		9.647	6.904

Note: *DBTo* Outdoor dry-bulb temperature, *RHo* outdoor relative humidity, *DBTi* indoor dry-bulb temperature, *RHi* indoor relative humidity, *BGTi* indoor black globe temperature, *DBTi_b* indoor dry-bulb temperature in bedroom

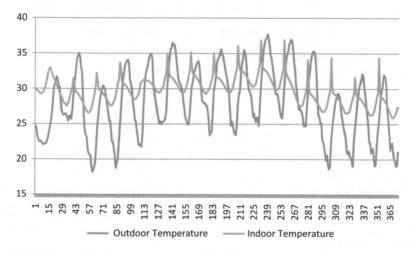

Fig. 35.4 Outdoor and indoor temperature, AAC housing, September 23 to October 6, Mexicali

Minor differences were found between the thermal conductivity values published in the literature, the manufacturer, and the laboratory studies performed in the area. However, it is important to know the behavior of the thermal conductivity of the AAC at temperatures around or above 40 °C, as presented in Mexicali; as well as porosity of the material, care must be taken with the effect of adsorption, absorption, and permeability of water vapor.

The management of the housing envelope alone does not ensure that it is in a condition of thermal comfort, even if the AAC system presents a better thermal performance than the current construction systems, so the use of air conditioning equipment is required but lower capacity. Finally, the design of a dwelling must respond to the needs of an adequate space and environmental habitability, so that the joint management of a suitable construction system associated to a project with other bioclimatic techniques, will help to improve the habitability conditions of the housing in extreme climates.

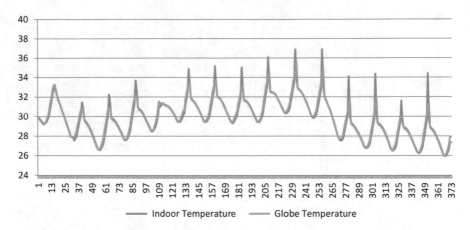

— Indoor Temperature   — Globe Temperature

Fig. 35.5 Indoor dry-bulb temperature and black globe temperature, AAC housing September 23 to October 6, Mexicali

Acknowledgments The authors would like to thank the National Council for Housing and the National Council for Science and Technology for the financial support provided for the "Thermal Comfort and Energy Savings in Low-income Dwellings in Hot Regions of Mexico, 2nd part CONAVI -01-20," besides "Environmental habitability in housing built to cities in Mexico" and "Energy Efficiency of Bioclimatic Models of Low-income housing in hot-dry climate" of Autonomous University of Baja California.

Also, thanks to the construction company RUBA for the support for the construction of bioclimatic dwelling and fieldwork development and Termoaislantes de México (Xella Mexicana) and Hebel for the donation of the material AAC (blocks and slabs) for the construction of housing with AAC. Finally, we thank the undergraduate and graduate students who have participated in the different research projects.

References

1. Secretaria de Energía (2017) Sector Eléctrico Nacional. https://www.gob.mx/sener/acciones-y- programas/estadisticas-del-sector-electrico-e-indicadores-de-cfe. Accessed 23 July 17
2. Instituto Nacional de Estadística y Geografía (2016) Anuario Estadístico y Geográfico de Baja California. México.
3. Instituto del Fondo Nacional de la Vivienda para los Trabajadores, Programa Nacional de Vivienda. Económica. http://www.infonavit.org.mx. México (2002)
4. Romero R, Vázquez E, Bojórquez G, Gómez G, Ochoa J, Marincic I, Resendiz O, Pérez M, García C (2009) Thermal comfort and energy savings in low-income dwelling in Mexico: Product 1, México: Final Technical Report CONAVI 2004-01-20
5. Secretaria de Desarrollo Social, Ley de Vivienda (2006) http://www.diputados.gob.mx/LeyesBiblio/ref/lviv.htm. Diario Oficial de la Federación, México
6. Comisión Nacional de Vivienda (2010) Código de Edificación de Vivienda, Segunda edición, México
7. Diario Oficial de la Federación (2014) Programa Nacional de Vivienda 2014–2018. http://www.dof.gob.mx/nota_detalle.php?codigo=5342865&fecha=30/04/2014

8. Instituto Nacional de Estadística y Geografía (2016) Anuario Estadístico y Geográfico de Baja California. México.
9. Comisión Nacional de Vivienda, Secretaria de Medio Ambiente y Recursos Naturales (2012) NAMA Apoyada para la Vivienda Sustentable en México – Acciones de Mitigación y Paquetes Financieros. México.
10. Xella Mexicana. http://www.hebel.mx/es/docs/Panel_Muro_-_v09.319(4).pdf
11. Jerman M, Keppert M, Jaroslav Vyborny J, Cerny R (2013) Hygric, thermal and durability properties of autoclaved aerated concrete. Constr Build Mater 41:352–359
12. Kocí V, Madera J, Cerný R (2013) Computer aided design of interior thermal insulation system suitable for autoclaved aerated concrete structures. Appl Therm Eng 58:165–172
13. Huang X, Ni W, Cui W, Wanga Z, Zhu L (2012) Cooper tailing preparation of autoclaved aerated concrete using copper tailings and blast furnace slag. Constr Build Mater 27:1–5
14. Drochytka R, Zach J, Azra Korjenic A, Hroudova J (2013) Improving the energy efficiency in buildings while reducing the waste using. Energ Buildings 58:319–232
15. Organismo Nacional de Normalización y Certificación de la Construcción y Edificación. Organismo (2010) Certificado SMM-017-002 y 003/2010. México.
16. Secretaría de Energía (2011) Mexican Standard NOM-020-ENER-2011. Energy efficiency in buildings.-envelope of buildings for housing. Diario Oficial de la Federación. August 9th
17. Secretaria de Energía (2016) Resolution modifying the values of global heat transfer coefficient (K) in Table 1, definitions are added and the verification of Official Mexican Standard NOM-020-ENER-2011. Diario Oficial de la Federación. October 4th
18. Gallegos R (2015) Determination of thermal properties of regional building materials for sustainable building, Final report research 11/566, Autonomous University of Baja California
19. National Weather Service (2014) Standard Climatological 1981–2010. http://smn.cna.gob.mx. Mexico
20. Autonomous University of Baja California (2013) Autonomous University of Baja California Sur, Autonomous University of Yucatan, University of Colima, University of Sonora Thermal comfort and energy savings in low-income dwelling in Mexico, Second Part Final Technical Report, Mexico
21. International Organization for Standardization (1998) Standard 7726: Ergonomics of the thermal environment—Instruments for measuring physical quantities, Geneva
22. National Weather Service (2017) Automatic Weather Stations. http://smn.cna.gob.mx/es/emas. Mexico.

Chapter 36
Renewables Are Commercially Justified to Save Fuel and Not for Storage

Donald T. Swift-Hook

Abstract Being capital intensive, renewables are not commercially appropriate for peak power plant. As they become cheaper than any other plants in many countries, subsidies are being eliminated, and then there is no commercial incentive to install renewables to reduce emissions or to save our planet. Their commercial justification is to *save fuel*.

Those countries with the biggest fuel production industries will have the biggest commercial incentive to save fuel. This is confirmed by noting that China, the USA and India, who between them produce two-thirds of the world's coal, generate more than half of the world's wind power.

Commercial storage of energy on a power system works by arbitrage, buying cheap electricity [typically in the middle of the night] and selling it when electricity is dear [during the daytime or evening]. If fuel saving renewables are stored, the round trip losses from putting their electricity into store and taking it out again waste some of the fuel already saved. There is no difference in fuel costs around the clock, so storing electricity from renewable fuel savers cannot be arbitraged or commercially justified.

36.1 Introduction

The renewable energy scene is changing rapidly. I have specialised in wind power, and I used to explain to my students half a dozen or so years ago that wind was ten times as cheap as photovoltaics which was why there was ten times as much of it. This meant that as much money was being spent on solar power as on wind, but

D. T. Swift-Hook (✉)
Kingston University, Kingston upon Thames, Surrey, UK

WREN, Brighton, UK

Bourne Place, Woking, Surrey, UK
e-mail: donald@swift-hook.com

© Springer International Publishing AG, part of Springer Nature 2019
A. Sayigh (ed.), *Sustainable Building for a Cleaner Environment*,
Innovative Renewable Energy, https://doi.org/10.1007/978-3-319-94595-8_36

Fig. 36.1 A typical domestic installation of photovoltaic solar panels

wind was almost entirely large-scale megawatt-sized industrial installations while solar was almost entirely domestic and on the rooftops of buildings; see Fig. 36.1.

I explained that wind was growing rapidly, doubling every 3 years, but that solar was growing even more rapidly doubling every 2 years.

As a snapshot, half a dozen or so years ago, all those things were true. Renewables represented only a few percent of electrical power generation, but the position was changing fast.

For instance, at the 2014 WREC at Kingston University in London, I was able to announce that wind power operating capacity worldwide had just overtaken nuclear.

In June 2014, the World Nuclear Authority redefined "operational" nuclear capacity to exclude long-term shutdown as well as permanent shutdown. They now exclude the Japanese reactors that have been shut down since the Fukushima disaster and may or may not be relicensed in the future. These represented around 10% of the installed capacity, so the new level of operational nuclear plant was much lower, 333 GW.

The wind capacity was growing rapidly from 318 GW at the end of 2013, and the 6-month figure available in July had just exceeded 340 GW, so wind capacity, which continues to grow rapidly year by year with little sign of any slowdown despite the economic crisis (see Fig. 36.2), had just overtaken nuclear, which is steadily declining.

Solar photovoltaic capacity was growing even more rapidly, and the majority of solar installations are now on the industrial scale. Figure 36.3 shows the 3 MW floating solar farm on Godley Reservoir, Tameside, Manchester, and Fig. 36.4 shows the 50 MW Shotwick Solar Park on Deeside in North Wales.

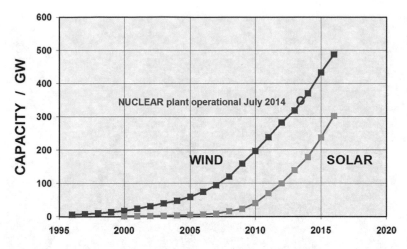

Fig. 36.2 Wind power generating capacity overtook nuclear by the end of June 2014

Fig. 36.3 Most solar pv capacity is now on the industrial scale. This is the 3 MW floating solar farm on Godley Reservoir, Tameside, Manchester, UK

36.2 Power Auctions

In many countries, power auctions [1] are now held regularly in which developers offer to provide a certain amount of capacity at a specific price [without necessarily saying what type of power plant they're planning to build]. Bids are listed from cheapest to most expensive, and distribution companies select the lowest-cost proposals available until reaching their target capacity.

Fig. 36.4 The 50 MW Shotwick Solar Park on Deeside in North Wales

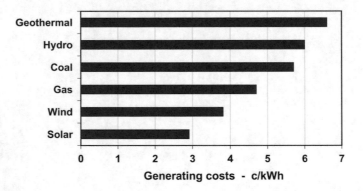

Fig. 36.5 The best prices bid in Chile in August 2016 for different types of generation [1]

Auction prices for both wind and solar have been falling fast. The best solar prices have fallen to where they are cheaper than any other type of generation; see Fig. 36.5.

These best prices are interesting and exciting, because they show what is feasible and achievable in the right circumstance, but average prices are more relevant and practical to show what is the present state of technology and economics around the world.

Figure 36.6 shows that wind is still significantly cheaper than solar on average, but prices of both of them have been falling significantly [1].

When wind capacity overtook nuclear in 2014, the prices of wind and solar were comparable with prices for coal and gas generation and about half those for nuclear.

In the last 2 years, the average prices for both wind and solar have virtually halved; see Fig. 36.6. This puts them both into a completely new situation. They are

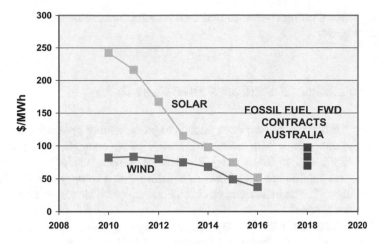

Fig. 36.6 The best prices bid for wind and solar at auctions around the world [1] compared with forward electricity contracts in Australia [almost entirely fossil]

both now cheaper than any other type of generation. Figure 36.6 shows the range of forward contract prices for generation in Australia, and these are all considerably above the current worldwide average of auction prices for both wind and solar. This is an appropriate comparison to make because Australia is one of the world's major coal producers and exporters, and more than 90% of the country's generation is from fossil fuels.

36.3   Subsidies

Governments around the world have been subsidising renewables to provide commercial incentives for generators to install them in order to avoid emissions. As prices have been falling, subsidies have been reduced. You don't need to subsidise activities that are cheap enough to be commercially profitable in their own right.

A good example is my own installation of pv solar panels shown in Fig. 36.1. I receive a feed-in tariff of £0.23 p for every unit of electricity that I generate. If I had installed my panels a year earlier, I would be receiving £0.46 p per unit (although the panels would have cost more than twice as much). If I had not completed my installation until the following month, I would only be getting £0.16 p/kWh, and if I installed the panels today, I would get little or no subsidy.

When they are cheaper than all other generation, then renewables are commercially viable in their own right without subsidies. It is inevitable that governments will then remove their feed-in tariffs and other subsidies and that is happening as rapidly as prices are falling.

Without subsidies, there are no commercial incentives for developers to cut their emissions or to save our planet. They will turn to renewables simply to make the greatest profits. Cutting emissions will then be a fortuitous side effect, just as it is for nuclear power.

36.4   No Longer Relentless Power Growth

One reason for installing new power plant is to provide more generating capacity.

Lenin famously said, "Communism is Soviet power plus electrification of the whole country", and in 2013 the UK regulator, Ofgen, said, "For 100 years, electricity consumption has been taken as equating with a country's well-being and standard of living". The increase in electricity consumption year by year has always been taken for granted.

How times change! In the last 10 years, UK electricity demand has fallen by 15% (it was 61.7 GW in 2005, but it had fallen to 52.7 GW by 2015). Other developed countries have had the similar experiences. So there is no longer a relentless increase in the power we use. We cannot assume that more generating capacity is needed.

Clearly, standards of living are being maintained and increased using less electricity. Energy efficiencies are increasing dramatically across the whole range of human activities but especially those which consume electricity.

However, many power stations are coming to the ends of their useful lives and need replacing. For example, very few nuclear power stations have been built in the last 30 years since the Chernobyl 1986 disaster, and, although many life extensions have been authorised, many reactors will have to be shut down in the next few years. At least some of them may need to be replaced.

36.5   To Stop the Lights Going Out

To maintain a secure supply, spare capacity is needed. On the UK power system, it used to be said that some 23% was sufficient spare generating capacity to ensure that the system failed to meet demand due to insufficient capacity only eight times in a century. In 2014, Ofgem estimated that there was "a one-in-31 chance of a black-out in the event of a harsh winter in 2015–2016 combined with little wind to propel turbines" and that is around three times in a century.

In the event, installed capacity in 2016 was 81 GW, with 54% of the capacity standing spare over and above that needed to meet peak demand. It seems clear that, if there is any danger of lights going out, that will not be avoided by installing any more spare capacity to add to the huge 54% surplus.

If peak capacity were required, it should be cheap (i.e. low capital cost) generating plant that needs not be particularly efficient because it would only need to operate for brief periods of peak demand. However, renewable and nuclear are both high capital cost (and little or no fuel cost) which is the last thing you need to provide peak capacity "to stop the lights going out".

In fact wind power does contribute to firm power a fraction of its capacity equal to the load factor, as I was the first to show analytically [2]. The capacity credit for any type of power plant on the system is in fact its load factor during the peak load periods. For wind this is well above 30% in the UK, and nuclear is around 60% (although wind is much less, and nuclear is much more on many other power systems).

In summary, there is no commercial justification for installing renewables (or nuclear) to provide peak generating capacity "to stop the lights going out".

36.6 Saving Fuel

What renewables (and nuclear) do is to save or replace fuel and that is something that almost all countries want.

The UK, for instance, was self-sufficient for a time with oil and gas from the North Sea but both these sources are running out and to import country now has the most of its fuel. It is keen to save the cost of imports by generating its own power from renewables.

However, Fig. 36.7 shows that the biggest coal producers are China, America and India [3], while Fig. 36.8 shows that China, America and India are among the world's biggest wind farm operators [4]. It is not surprising that those countries which have the greatest interest in saving or replacing fuel are the biggest fuel users.

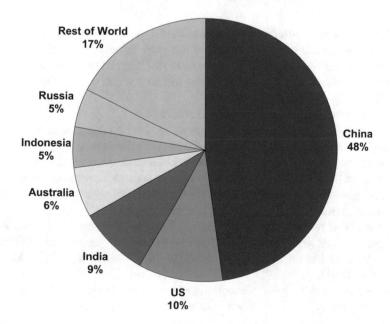

Fig. 36.7 The world's coal production [3]

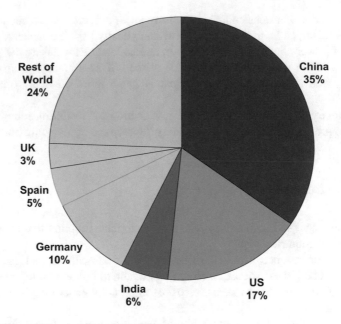

Fig. 36.8 The world's wind farmers 2016 [4]

36.7 Power System Storage

For electrical energy storage on a power system to be economic and commercially viable, the store must buy energy when it is cheap (at night) and sell it when it is dearer (during the day). The difference in price may then be sufficient to pay for the storage plant plus round trip energy losses incurred when putting the energy into store and taking it out again.

Making use of this difference is called arbitrage.

The first thing to note is that solar power cannot be stored economically on a typical power system [5]. By definition, there is no solar power at night.

A stand-alone solar generator that is not grid-connected has no alternative supply available, and then it usually has a battery to store enough power to cover any night-time requirements. Figure 36.9 shows a small garden light with a solar cell to collect sunlight and generate electricity which is stored internally in a battery.

On the power system scale, energy storage is always by pumped water associated with hydroelectric generation (apart from two small compressed air units in salt caverns built at Huntorf, Germany, 290 MW in 1979 and McIntosh, Alabama, 110 MW in 1991).

Pumped water storage is quite commonplace on power systems. In the EU, 5% of all generation is from hydroplant associated with pumped storage.

Fig. 36.9 Stand-alone pv usually has a battery to store one night's electrical energy

36.8 Fuel Savers Don't Want Storage

Renewables are fuel savers (see Sect. 36.6). Having saved fuel in producing electricity, it makes no sense at all to store that electrical energy. Losses in the storage and regeneration round trip waste some of the fuel that has already been saved, and there is nothing to be gained from retiming fuel, as there is for electricity.

Power in the middle of the day is worth more than power in the middle of the night but that is not true for fuel. Fuel is worth the same during the day (or when the wind blows) as it is in the middle of the night (or when the wind does not blow); there can be no arbitrage when fuel is saved.

So energy storage is not economic for a fuel saver [5].

36.9 Misleading Terminology

It is often suggested that the extra variability due to renewables needs energy storage to keep the system stable but that is a control problem, not energy storage.

Calling control and balancing plant "storage" are like calling a school bus a "method of fuel transportation"! Similarly, calling my mobile phone a method of "power system storage" or "a battery energy storage system" in this context would be simply misleading. Although it does actually do both those things on the power system, that is not what it is *for*.

References

1. IRENA (2017) Renewable energy auctions analysing 2016. IRENA, Abu Dhabi. 978-92-95111-089-0
2. Swift-Hook DT (1987) Firm power from the wind. In: Galt JM (ed) Wind energy conversion. MEP, London, p 33
3. British Petroleum (2016) Statistical review of world energy
4. Global Wind Energy Council (2016) Annual report
5. Swift-Hook DT (2013) Wind energy really is the last to be stored and solar energy cannot be stored economically. Renew Energy 50:971–976

Chapter 37
Climate Change Adaptation: Assessment and Simulation for Hot-Arid Urban Settlements – The Case Study of the Asmarat Housing Project in Cairo, Egypt

Mohsen Aboulnaga, Amr Alwan, and Mohamed R. Elsharouny

Abstract Urban areas in hot-arid climatic zones, especially in Egypt, are facing real challenges in responding to heat island effect, providing thermal comfort and adapt to climate change (CC) impacts. Such challenges are mounting due to CC risks that are manifested worldwide, e.g., severe storms that recently slashed the Gulf of Mexico, Texas, and Florida, USA. Metrological data indicate that the increase in hot summer days would result in rapid multiplication in heat stress, death cases, and economic impacts. A severe event was observed in Cairo, Egypt, in August 2015, where air temperature was recorded high 49 °C above the normal temperature for 10 days, hence resulting in 200 cases that were hospitalized from heat stress and 98 deaths. The CC direct risks are not only limited to urban areas and public health. Due to the fact that Egypt is highly dependent on fossil fuels to produce electricity, GHG emissions, mainly CO_2 will be significantly increasing. Therefore, sustainable and green measures and actions are vital to be considered and implemented in all sectors. Under such adverse CC impacts, it is necessary for all stakeholders to examine current urban projects in order to assess their ability to respond to CC adaptation measures. This paper presents the assessment of a low-income housing settlement that was recently built in Cairo. The Asmarat project is selected as the case study to simulate the long-term impact of CC scenarios by 2080 on one of the capital's urban settlements and to test the role of passive cooling configurations in mitigating CC effect in cities to identify possible countermeasures.

M. Aboulnaga (✉)
Sustainable Built Environment, Department of Architecture, Cairo University, Giza, Egypt

A. Alwan
Department of Architecture, Faculty of Engineering, Military Technical Collage, Cairo, Egypt

M. R. Elsharouny
M.Sc. in Environmental Design and Energy Efficiency, Department of Architecture, Faculty of Engineering, Cairo University, Giza, Egypt

© Springer International Publishing AG, part of Springer Nature 2019   437
A. Sayigh (ed.), *Sustainable Building for a Cleaner Environment*,
Innovative Renewable Energy, https://doi.org/10.1007/978-3-319-94595-8_37

Simulation programs ENVI-met and DesignBuilder were used to assess and measure the resilience and sustainability of the selected urban project. The study simulates the urban microclimate in terms of the urban form by 2016 and 2080 to evaluate CC impact. Six measures were tested including passive cooling design configurations, building elevation, buildings' envelops, vegetation, and water features, and orientation and high albedo were tested, and results were presented. These findings address adaptation policies, actions and measures, and simulations of the role of buildings' retrofitting and cities' upgrading in coping with CC mitigation/adaptation to narrow the information gap and yet understand the challenges facing the adaptation measures in hot-arid zones. The changes in climatic parameters resulted in an increased magnitude of thermal discomfort by 1 point on the PMV thermal sensation scale in the built environment within hot-arid climate zones. In addition, results indicate that adaptation measures through buildings' retrofitting and upgrading cities' strategies played a vital role in adapting with CC risks through the enhancement of outdoor and indoor thermal comfort and mitigating CO_2 emissions.

37.1   Introduction

The greenhouse gas (GHG) concentration has been rising steadily, and the mean global temperature is also elevating since the time of the industrial revolution. This is resulting from human activity, primarily the burning of fossil fuels and changes in land use [8]. Climate change is projected to affect the lives of billions of people around the world, and no region or country is immune to its impacts; however, the extent of vulnerability differs widely [18]. Climate change impacts threaten different sectors and activities in a complex series of reciprocal influence at multiple levels. According to the Intergovernmental Panel on Climate Change (IPCC), it was stated that even if emissions were completely halted, which is unexpected in the most optimistic scenarios, the severity of CC would continue to increase, so the climate change adaptation (CCA) is inevitable [19], which highlights the importance of adaptation in tackling CC consequences especially in the Delta, Egypt (Fig. 37.1a).

Urban areas are among the severely affected sectors that are facing and/or will face real challenges in responding to heat island effect, HIE (Fig. 37.1b), providing thermal comfort and mitigating climate change impacts [9]. During August 2015, Cairo, Egypt, witnessed a sudden increase in air temperature that was recorded high 49 °C for 10 days which is not normal according to weather records for similar months of August, and this heat wave has resulted in 200 cases that were hospitalized from heat stress and 98 deaths [15]. In addition, the increased pressure on the national electricity grid due to high energy demands for cooling is estimated at 88.7% in 2015. Increasing the dependence on fossil fuels to produce electricity followed by shooting in GHG emissions will increase CC risks if no adaptation measures and plans are developed and implemented [3].

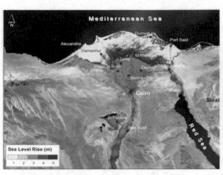 

Fig. 37.1 Climate change impacts in Egypt (**a**) Climate change impact in Delta (**b**) Heat island effect in urban areas, Cairo

Coupling methodology has been validated in many studies on coupling indoor and outdoor thermal comfort as part of climate change scenarios 2050 and 2080 in urban areas [1, 6]. These studies suggested that in order to integrate urban microclimate simulations, the coupling simulations, using software such as ENVI-met and other indoor energy simulation programs, namely DesignBuilder could be used [5]. This was conducted in investigating the effect of vegetation, water bodies, and urban geometry on outdoor thermal comfort. Also, Yi et al. conducted a study which stated that coupling methodology (ENVI-met and EnergyPlus) is important for evaluating the green roof impact on urban microclimate and building energy performance [2]. Therefore, there is a need for further studies about climate change assessment, simulation, and adaptation for hot-arid urban settlements in Cairo on long-term impact of CC scenarios.

37.2 Objectives

The objective of this study is to assess a selected low-income housing settlement that is newly built in Cairo in terms of climate change adaptation. The study simulates the urban microclimate in terms of the urban form by 2016 and 2080 to evaluate CC impacts. It also tests six measures, including passive cooling design configurations, building elevation, building envelop, vegetation, water features, and orientation and high albedo, in order to identify their impact on the adaptation process to heat stress and comfort indoor and outdoor to help fill the information gap and provide a deeper understanding of challenges and opportunities toward resilient cities.

The assessment includes current situation, adaptation policies, actions, and measures. Also, the study aims at simulating the role of buildings' retrofitting and cities' upgrading in coping with CC mitigation/adaptation in order to close the information gap and yet understand the challenges facing the adaptation measures in hot-arid zones. The Asmarat project is selected as the case study to simulate the long-term impact of CC scenarios by 2080 on one of the capital's urban settlements to test the role of cooling passive configurations in mitigating CC effect on cities to identify possible countermeasures to measure the resilience and sustainability of the selected urban project to envisage the scope of future impacts of CC to ensure thermal comfort is provided for better well-being and achieve livable sustainability, i.e., quality of life.

37.3 Methodology

The Asmarat project, one of the newest housing projects in Cairo, is selected as a case study. Upgrading informal risky housing areas is part of Egypt's Sustainable Development Strategy (SDS) 2030 to eliminate informal housing through transferring these slums into substitute residential areas that present novelty model for low-income and low-rise buildings in a new urban setting that is located in many new Egyptian cities [13].

This project is selected to enable authors to evaluate climate change impacts on hot-arid urban areas as well as to check how attainable such new projects with climate change adaptation measures. The significance of this study is that it highlights the importance of stakeholders (developers, architects, urban planners, and engineers) to examine the current projects in order to assess their ability to attain climate change mitigation and adaptation measures in hot-arid zones.

The research is based on coupling methodology, which refers to coupling ENVI-met with EnergyPlus and DesignBuilder to investigate the interaction between outdoor microclimate and indoor cooling energy consumption. This coupling simulation is mainly to provide more accuracy in the sensible results based on extracted morphed weather data of the microclimate of the selected project rather than that gathered from weather station data which is away from the project location [14]. Coupling methodology is validated in many studies [1, 6].

The program ENVI-met 3.1 (a three-dimensional microclimate model designed to simulate the surface-plant-air interactions in urban environment with a typical resolution down to 0.5 m in space and 1–5 s in time through using fundamental laws of fluid and thermodynamics) was used in simulations [12]. Leonardo 2014 program was also used to analyze and visualize ENVI-met model results and create 2D maps [11]. The CCWorldWeatherGen program was used to generate climate change weather scenarios based on the IPCC report [16, 17]. Climate calculator was used to estimate the water content as an essential input into the ENVI-met model data [10]. The PMV 2008 version 1.0 was also used to calculate the predicted mean vote – PMV [7]. The model used to simulate climate change scenarios is shown in Fig. 37.2. DesignBuilder program was utilized to assess energy efficiency (cooling loads) and predict carbon performance of different sets of the case study [4], is illustrated in Fig. 37.3.

Figure 37.3 shows the simulation methodology which is divided into two main stages: ENVI-met and DesignBuilder. In first stage, As-built of Asmarat project context and EPW weather files of Cairo are collected, and climate change 2080 scenario generated by CCWorldWeatherGen then used as input for ENVI-met to

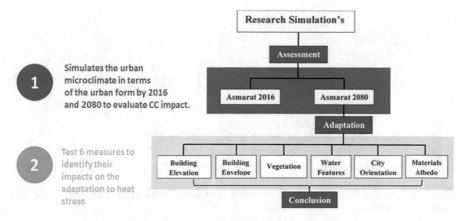

Fig. 37.2 The model used to simulate climate change scenarios

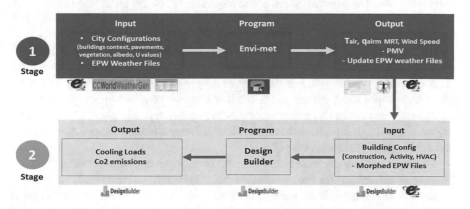

Fig. 37.3 Framework of simulations based on Google satellite image and a site survey

extract PMV and morphed weather files for project location; PMV evaluates outdoor thermal comfort performance. In the second stage, data inputs are building configurations of construction, activities, HVAC, and morphed weather files to calculate the cooling loads and CO_2 emissions of models.

37.4   Data Input

The Asmarat project layout was constructed by ENVI-met 3.1 software from a satellite image and a site survey, then the site location defined on Cairo's longitude and latitude, where the model area is on a (243x * 152y * 26z) grid as well as the size of a grid cell of (4.0 m × 4.0 m × 3.0 m). In the model, we started with the As-built 2016 that includes the urban form, soils, and pavement textures in addition

 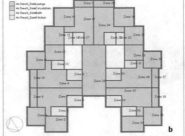

(a) Twenty-two receptors distributed in Envi-met model   (b) Floors typical plan of DesignBuilder model.

Fig. 37.4 Coupling data input (**a**) Twenty-two receptors distributed in ENVI-met model (**b**) Floors typical plan of DesignBuilder model

to specified plantation areas. Twenty-two receptors to monitor atmospheric parameters were added to cover all areas of the model and to obtain an average of climatic variables for more accurate and representative results as shown in Fig. 37.4a. The same process was repeated in all tested models.

For the simulation process, ENVI-met and DesignBuilder were used in the models and assessment process. ENVI-met simulations were constructed and run for each case on a typical summer day (1st of July) for 13 h (6:00–19:00). Thematic maps and graphs were extracted for the five climatic physical parameters that are influencing mostly the thermal comfort, including (a) dry-bulb temperature (DBT), (b) wind speed (*V*), (c) relative humidity (RH), (d) mean radiant temperature (MRT), and (e) the predicted mean vote (PMV).

These parameters are used to express the human perception of thermal comfort. Tests were carried out for the models: As-built 2016 and As-built 2080.

Results of As-built 2016 were compared with that of As-built 2080 to assess climate change, and six retrofitting proposals (Table 37.1) were finally compared with As-built 2080 to evaluate their role in climate change adaptation (CCA) through identifying the effect of each of the passive design techniques on the PMV for the outdoor environment and the indoor cooling loads of the dwelling shown in Fig. 37.4b (As-built 2080). In DesignBuilder, the building working hours were assumed to be from 06.00 to 19:00 similar to ENVI-met simulation hours of typical weather day (1st of July). Table 37.1 shows descriptions of simulated models.

37.5   Data Output

The Predicted Mean Vote (PMV) thematic maps' outputs of the Asmarat project As-built 2016 and As-built 2080 are presented in Fig. 37.5. Output data from ENVI-met are thematic maps and receptors' output that highlight the results of DBT, V, RH, MRT, and PMV for the 22 receptors for each the eight proposed models. The receptors' files show the state of the atmosphere and the surface at

Table 37.1 Descriptions of the simulation models

Models	Description
As-built 2016	As-built of the Asmarat under current weather of Cairo airport weather station
As-built 2080	As-built of the Asmarat under weather scenario 2080 based on the IPCC
Building elevation	Elevate all buildings' height 3.00 m above ground level by a clear storey
Building envelop	Retrofit buildings envelop for better insulation (walls = 0.57 W/m^2 K and roofs = 0.31 W/m^2 K)
Vegetation	Add grass on empty sandy areas, Ficus trees 4.00 m dense in streets, yellow Poinciana trees 20 m dense as buffer zone surround the city
Water feature	Add water features in empty sandy areas around the city, building's yards, and green areas
City orientation	Arrange urban settlement to orient streets toward main wind direction
High albedo	Select high albedo materials for the buildings' facades albedo = 50%, roofs' albedo = 50%, and street pavements albedo = 40%

selected points inside the model. The average values of the climatic factors were calculated hourly by an Excel spreadsheet for height of 1.5 m that is suitable to measure thermal comfort of pedestrians (Fig. 37.5).

The results extracted from the Asmarat As-built 2016 and the Asmarat As-built 2080 scenarios were analyzed and compared to each other to measure the impact of global warming on the urban microclimate. Also, the results extracted from the simulations of the six passive design configurations: a) building elevation, b) building envelop, c) vegetation, d) water features, e) city orientation, and f) high albedo proposals, were assessed and compared with the results of the Asmarat As-built 2080 in order to evaluate the role of cooling loads and passive design retrofitting configurations in adaptation scenarios. This is mainly to determine which of the six passive design configurations would lead to the best thermal comfort, plus developing design guidelines for a more sustainable retrofitting process in the context of CCA.

37.6   Results

The simulation results of the five parameters, including DBT, RH, *V*, MRT, and PMV, are presented in the below section.

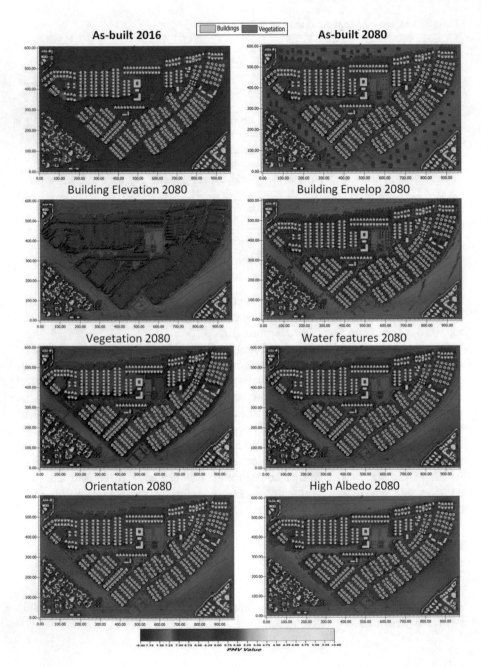

Fig. 37.5 Predicted mean vote thematic maps' outputs – Asmarat As-built 2016 and As-built 2080

37.6.1 Air Temperatures: DBT

Significant increase was observed in the air temperature (DBT) by 3.9 °C due to climate change impacts. The average air temperature increases from 33.8 °C to 37.7 °C. Results of the six retrofitting models: a) building elevation; b) building envelop; c) vegetation; d) water features; e) city orientation proposal; and f) high albedo were simulated. Significant increase was observed in the dry-bulb temperature (DBT) by 3.9 °C due to climate change impacts. The average air temperature increases from 33.8 °C to 37.7 °C. In the four tested retrofitting models: a) building envelop; b) water features; c) material albedo; and d) orientation) indicate that the differences between air temperatures in these four models in the simulations trials were insignificant. The air temperature was approximately at an average of 37.6 °C, whereas result of air temperature of the vegetation proposal (tested retrofitting model No. 5) is lowest at an average of 37.2 °C. This was followed by building elevation (tested retrofitting model No. 6) with an average of 37.3 °C.

37.6.2 Relative Humidity: RH

The relative humidity (RH) parameter was observed at the same value in both models of the Asmarat As-built 2016 and As-built 2080. The building elevation proposal has the highest RH ratio of 36% followed by vegetation proposal at 35 percent, whereas the proposals of building envelop, water features, and material albedo of As-built 2080 model were recorded at 34% compared to the Asmarat 2016. However, the orientation proposal has the lowest RH ratio equal to 33%.

37.6.3 Wind Speed: V

The average wind speed results of the Asmarat models, As-built 2016 and As-built 2080, were recorded at 1.2 m/s and 1.1 m/s, respectively. This indicates a very slight decrease, which could be insignificant. Results also indicate that the building elevation proposal has a slightly higher wind speed with an average of 1.2 m/s. The other proposals of building envelop, water features, and materials albedo indicate approximately an equal value for the As-built 2080 at an average of 1.1 m/s followed by the orientation proposal with 1.07 m/s. However, the vegetation proposal has a lower wind speed of 1.00 m/s.

37.6.4 Mean Radiant Temperature: MRT

The average mean radiant temperature (MRT) increased from 66.5 °C to 69 °C in the Asmarat As-built 2016 and As-built 2080, respectively, an increase of 2.5 °K. The vegetation proposal has the lowest MRT of 66 °C when compared to those of the elevation proposal at 68 °C, whereas the rest of the other proposal (building envelop, water features) recorded approximately same MRT with average of 69 °C, but the material albedo proposal has the highest MRT of 75 °C, an increase of 8.5 K. This indicates the significance of the MRT among other parameters.

37.6.5 Predicted Mean Vote: PMV

The simulations' results of the PMV indicate that climate change increase the feeling of thermal discomfort throughout the day. The PMV value of the Asmarat As-built 2016 and As-built 2080 models has increased from 5.42 to 6.40, respectively. The vegetation proposal shows the lowest PMV value at 6.12 followed by that of the elevation proposal at 6.27. The PMV of building envelop and water features proposal shows the same value of 6.40, whereas the orientation proposal is slightly higher than that of As-built 2080 base case; nevertheless, the material albedo proposal has the highest PMV value of 6.95.

37.6.6 Cooling Loads

In terms of cooling loads result, the simulation shows a significant increase in the cooling loads of the Asmarat project (As-built 2016 and As-built 2080) from 82.90 kW to 97.37 kW. These cooling loads results for the building envelop retrofitting model indicates the lowest value of cooling load of 70.80 kW. Nevertheless, the value of cooling load for the material albedo retrofitting model is followed by results of the orientation retrofitting models then by the vegetation model, and finally water features retrofitting mode – the lowest cooling load values when compared with those of the Asmarat As-built 2080 model. The building elevation retrofitting model shows the highest cooling load of 108.08 kW, at an increase of 11 per cent (25 kW).

37.6.7 GHG Footprint

Results show that there is a significant increase in greenhouse gas (GHG) emissions by 0.01 metric tons in the simulation of As-built 2080. The building envelop proposal has the lowest GHG footprint with a value 0.05 metric tons. The GHG value of the material albedo proposal was followed by that of orientation then that of the vegetation. The value of the water features has the lower GHG value compared with Asmarat As-built 2080 model. Finally, the building elevation proposal has the highest GHG footprint, which is 0.076 metric tons.

Coupling results of the models (As-built 2016 and As-built 2080) are presented in Table 37.2. The following table presents the values recorded for the eight

Table 37.2 Coupling results of the models (As-built 2016 and As-built 2080)

Models	Compared to	PMV	Delta (PMV)	Ratio of Delta %	Cooling Loads	Delta Cooling Loads)	Ratio of Delta %	GHG emissions	Delta (GHG)	Ratio of Delta %
As-built 2016	-	5.42	-	-	82.9	-	-	0.058	-	-
As-built 2080	As-built 2016	6.4	0.98	18	97.37	14.47	17	0.068	0.01	17
Building Elevation 2080	As-built 2080	6.27	-0.13	-2	108.08	10.71	11	0.076	0.008	12
Building Envelop 2080	As-built 2080	6.39	-0.01	-0.2	70.97	-26.4	-27	0.05	-0.018	-26
Vegetation 2080	As-built 2080	6.12	-0.28	-4	95.25	-2.12	-2	0.067	-0.001	-1
Water features 2080	As-built 2080	6.39	-0.01	-0.2	97.23	-0.14	-0.1	0.068	0	0
Orientation 2080	As-built 2080	6.47	0.07	1	92.59	-4.78	-5	0.065	-0.003	-4
High Albedo 2080	As-built 2080	6.95	0.55	9	90.52	-6.85	-7	0.064	-0.004	-6

Positive Impact	Negative Impact

proposals of the Asmarat models. This includes two As-built proposals (As-built 2016 and As-built 2080) and six environmental proposals: building elevation, buildings' envelops, vegetation and water features as well as orientation and high albedo (Table 37.2).

37.7 Discussion

It is clear from the above table and results that the vegetation model has the best performance on outdoor thermal comfort due to cooling effect through evaporation, transpiration, and shading that are the main processes through which vegetation affects climate. The vegetation model is considered the fourth in indoor thermal comfort performance due to the reduced air temperature. This is followed by reduction in the heat gain. The raised building's elevation model is the second in terms of performance due to continuity of air movement, which improves the cooling sensation and thus reduces the MRT and the worst performance in indoor due to increase heat gain from the ground floor.

The building envelop impact on the outdoor is negligible due to the fact that the main role of building envelop is to act as a thermal barrier regulating the interior temperature. This may have a slightly positive effect on the outdoor thermal comfort due to the decreased storage of sensible heat in the construction materials one of urban heat island causes, but it is the best proposal in terms of the indoor performance due to the insulation that reduces the heat gain and cooling loads.

The water features' impact on the outdoor thermal comfort is negligible due to that fact of the water bonds has a little effect on temperature. This is mainly due to fact that thermal stratification phenomenon that refers to vertical distribution of temperature with heights and the sun rays that warm the surface water is unlike the cooling effect of fountains spray that recirculates the water near the surface to prevent stratification. This water features model shows that the fifth for the indoor thermal comfort is due to reduced temperature and then it is followed by a decrease in the heat gain. The orientation retrofitting model is slightly negative despite that fact that air temperature decreases. Such reduction in the air temperature is encountered by a significant increase in the MRT due to the decrease in the buildings' shading. In term of ranking the retrofitting models, the orientation comes third in indoor thermal comfort performance. This is due to the reduction in air temperature and a reduction in the heat gain.

The high albedo material indicates that it is the worst proposal on outdoor thermal comfort despite it reduces air temperature, but the reduction has shown a significant increase in the MRT which is reflected by the short-wavelength radiation by the high albedo materials proposal. Thus, the conditions for thermal comfort index go warmer, but it shows the second level in the indoor results due to the reflected short-wavelength radiation that reduces the heat gain.

37.8 Conclusions and Recommendations

The simulation of the Asmarat models As-built 2016 and As-built 2080 was assessed and simulated. It is recommended to incorporate the four retrofitting proposals such as vegetation, raised building, building envelop, and water features in retrofitting or design to achieve better thermal comfort. Also it is recommended to incorporate high albedo material and orientation during design phase, but it is important to take into account the buildings' shading when defining the city orientation to maximize the shading and exclude the use of high albedo materials on pavements. This is mainly due to the negative impact of the reflected short-wavelength radiation on pedestrians which also depends on shading as well as the building walls, low albedo at the bottom part and a high albedo at the top of the building walls and roofs. This will have a lower impact on pedestrians' thermal comfort. It is also recommended that the ground floor is insulated when using raised floor building elevation strategy to reduce the heat gain into ground floor, thus lowering the cooling loads.

Despite the fact that passive cooling techniques play a key role in climate change adaptation (CCA) whether indoor and outdoor, the PMV outdoors are still out of thermal comfort zone as the current situation. Hence, it is highly recommended to limit warming to no more than current indicators and compile research recommendations with other tested sustainable techniques to reduce such gap for a better outdoor thermal comfort.

References

1. Elwan A et al (2014) An outdoor-indoor coupled simulation framework for climate change–conscious urban neighborhood design. Sage Publications ltd stm 90(8):874–891
2. Yi C et al (2014) Microclimate change outdoor and indoor coupled simulation for passive building adaptation design Elsevier B.V. Procedia Computer Science 32:691–698
3. Weisser D (2007) A guide to life-cycle greenhouse gas (GHG) emissions from electric supply technologies. Elsevier 32(9):1543–1559
4. DesignBuilder 2.1: user's manual (2009). Retrieved April 10, 2017, from DesignBuilder Ltd: http://www.designbuildersoftware.com/docs/designbuilder/DesignBuilder_2.1_Users-Manual_Ltr.pdf.
5. DesignBuilder Software Ltd (2017) DesignBuilder-simulation made easy. Retrieved 14 April 2017, from www.designbuilder.co.uk/
6. Fahmy M et al (2017) On the green adaptation of urban developments in Egypt; predicting community future energy efficiency using coupled outdoor-indoor simulations. Elsevier 153:241–261
7. Holmer I (2008) PMV 2008 ver 1.0. Retrieved 14 May 2017, from http://www.eat.lth.se/fileadmin/eat/Termisk_miljoe/PMV-PPD.html
8. Walsh J et al (2010) Ch. 2: our changing climate. Climate change impacts in the United States. In: The third national climate assessment. Global Change Research Program, pp 19–67. https://doi.org/10.7930/J0KW5CXT
9. Kaarin Taipale et al. (2012). Challenges and way forward in the urban sector: Sustainable Development in the 21st century (SD21). (Division for Sustainable Development of the United Nations) Retrieved April 20, 2017, from https://sustainabledevelopment.un.org/content/documents/challenges_and_way_forward_in_the_urban_sector_web.pdf
10. Bruse M (2009) ENVI-met 3.1 Help system. Retrieved 16 April 2017, from www.envi-met.com/documents/onlinehelpv3/helpindex.htm/
11. Bruse M (2014) LEONARDO 2014. Retrieved 12 April 2017, from http://www.model.envi-met.com/hg2e/doku.php?id=leonardo:start
12. Elnabawi M (2013) Use and evaluation of the ENVI-met model for two different urban forms in Cairo, Egypt: measurements and model simulations, Chambéry, France: 13th Conference of International Building Performance Simulation Association
13. MoIC (2016) National review sustainable development goals. Ministry of International Cooperation, Cairo
14. NREL (2017) Weather data by location. (National Renewable Energy Laboratory) Retrieved 4 March 2017, from https://energyplus.net/weather-location/africa_wmo_region_1/EGY//EGY_Cairo.Intl.Airport.623660_ETMY/
15. Phil Katzan et al. (2017). Protecting Health from Heat Stress in Informal Settlements of the Greater Cairo Region. (Deutsche Gesellschaft für, Internationale Zusammenarbeit (GIZ) GmbH) Retrieved March 15, 2017, from https://health.bmz.de/what_we_do/climate_health/Vulnerability_assessments/50_va_cairo/Qualitative_vulnerability_and_adaptation_assessment_Cairo_2016.pdf
16. SERG. (2013). Climate change world weather file generator. (Sustainable Energy Research Group, University of Southampton) Retrieved April 5, 2017, from http://www.energy.soton.ac.uk/files/2013/06/manual_weather_tool.pdf
17. SERG (2017) CCWeatherGen: climate change weather file generator for the UK. Retrieved 16 May 2017, from http://www.energy.soton.ac.uk/ccweathergen/
18. UNGA. (2008). Climate Change and The Most Vulnerable Countries: The Imperative to Act. (United Nations General Assembly) Retrieved May 2, 2017, from http://www.un.org/ga/president/62/ThematicDebates/ccact/vulnbackgrounder1July.pdf
19. WGII AR5. (2014). Climate Change 2014 Synthesis Report Summary for Policymakers. (Cambridge university press) Retrieved February 12, 2017, from https://www.ipcc.ch/pdf/assessment-report/ar5/syr/AR5_SYR_FINAL_SPM.pdf

Chapter 38
Ventilation Effectiveness of Residential Ventilation Systems and Its Energy-Saving Potential

Mohammad Reza Adili and Michael Schmidt

Abstract Ventilation systems can ensure the required minimum airflow to remove humidity and air pollutants from the building and maintain an acceptable Indoor Air Quality (IAQ). While determining these airflows, the type and the location of the supply and exhaust air outlets are less thoroughly investigated. As a result, the local ventilation effectiveness remains mainly unconsidered.

In this work, different types and positions of supply and exhaust air outlets for a living room were explored. The ventilation effectiveness of these systems was determined using computational fluid dynamics. The results indicate that an optimally distributed supply air makes it possible to improve the ventilation effectiveness. This leads to an improvement of the IAQ in the occupied zone or to a potential supply airflow reduction. This reduction has a positive impact on the energy demand of the air conveyance and air treatment subsystems.

38.1 Introduction

New residential buildings are designed to be airtight according to the requirements of new building regulations. The natural air exchange through infiltration is restricted, and sufficient ventilation cannot be guaranteed only via operable windows to remove humidity and air pollutants. This may lead to mould growth and negative effects on the IAQ. Ventilation systems can ensure the required minimum air change rate independent of user behaviour. Furthermore, by including a heat recovery unit, these systems can minimize the ventilation losses and reduce the heating demand. Assuming an ideal mixed airflow in the room, the design of systems is only made by balancing the airflows. However, the efficiency of air supply is not considered.

M. R. Adili (✉) · M. Schmidt
University of Stuttgart, Institute for Building Energetics, Stuttgart, Germany
e-mail: reza.adili@ige.uni-stuttgart.de

© Springer International Publishing AG, part of Springer Nature 2019
A. Sayigh (ed.), *Sustainable Building for a Cleaner Environment*,
Innovative Renewable Energy, https://doi.org/10.1007/978-3-319-94595-8_38

A precisely directed supply air maintains the desired ventilation effectiveness and leads to reduction of the required airflows. Thus, a potential energy saving with regard to the total energy demand can be achieved.

This work studies different types and positions of supply air outlets, exhaust air outlets and airflow patterns in the room. The ventilation effectiveness of these systems is determined by computational fluid dynamics (CFD). The energy-saving potential achieved by reducing the design airflow for systems with optimal ventilation effectiveness as well as by implementing heat recovery ventilation is described. A part of these systems is examined experimentally in the room airflow laboratory in order to validate the computational results.

38.2 Methodology

The study object is a typical living room with dimensions of 4.65 m × 4.20 m × 2.40 m ($L \times W \times H$) as presented in Fig. 38.1. It is assumed that the room is connected on the right side to a heated corridor with suspended ceiling. On the opposite side, there is an outside wall with a large window and a heating panel beneath the window. The occupied zone (grey volume) is specified according to EN 13779:2007 [1].

The supply outlet is positioned above the door which delivers the required flow rate of outside air into the room using a centralized ventilation system with heat recovery. A rectangular ventilation grille is considered at the bottom of the door to transfer air from the room to the corridor. At first, three typical supply outlets were investigated. These include a long throw jet nozzle (Fig. 38.2a), a radial supply valve (Fig. 38.2b) and a semi-radial supply valve (Fig. 38.2c).

In addition to these typical outlets, the following three novel solutions were also investigated: a combination displacement air diffuser for wall installation (Fig. 38.2d).

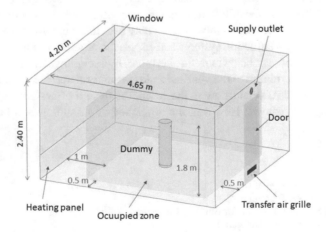

Fig. 38.1 3D representation of the model

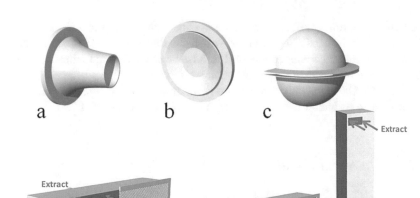

Fig. 38.2 The studied outlet types (**a**) long throw jet nozzle (**b**) radial supply valve (**c**) semi-radial supply valve (**d**) combination displacement air diffuser (**e**) combination slot air diffuser (**f**) decentralized vertical ventilation unit

This combination diffuser supplies the outside air through the lower section and extracts the room air through the top section. Due to the very low discharge velocity, these diffusers guarantee low turbulence air supply which leads to a displacement airflow. The next solution is a combination slot air diffuser (Fig. 38.2e), providing both supply and exhaust air in a single device. These combination slot diffuser supplies the outside air through the left section and extracts room air through the right section. The adjustable slots regulate the supply air angle so that a tangential, mixed or displacement ventilation can be achieved in the space. For the purpose of this study, four rows of slots for each supply and exhaust section were considered with a supply angle of 20° below the horizon. The third solution is a decentralized supply and extract air unit (Fig. 38.2f) with heat recovery for vertical installation on the outside wall adjacent to window. The supply air is discharged to the room as a displacement flow from the lower front part of the unit. Exhaust air is extracted from the upper part of the unit.

38.2.1 Boundary Conditions

A cylindrical dummy at the centre of the room represents a sitting occupant. The heat dissipation of the dummy and its CO_2 emission have been assumed as 85 W and 18 l/h, respectively. The supply airflow rate for all studied cases equals 72 m³/h according to the requirements of DIN EN 15251 [2]. The same amount of air will be extracted through the exhaust opening through a balanced ventilation system.

All enclosing room surfaces except the outside wall are assumed to be adiabatic. The outside wall includes a large window with a U-value of 1.3 W/m² K. Winter conditions with an outdoor air temperature of 0 °C have been considered for the simulations. A heating panel beneath the window maintains the average room air temperature around 20 °C. The supply air temperature to the room is 14 °C which represents the preheated outside air after the heat recovery unit with an efficiency of 70%.

38.2.2 Numerical Study

For the CFD simulations, the commercial software ANSYS Fluent has been implemented. The three-dimensional mesh for the fluid domain is generated using an unstructured grid with volumetric tetrahedral cells. Refinements were applied to areas with higher temperatures and air velocity gradients. An inflation layer has been considered for the enclosing surfaces especially the dummy, window and heating panel to capture the boundary layer effects. For turbulent flow calculations, the realizable k-ε model with enhanced wall treatment and full buoyancy effects was selected. For the prediction of radiant heat transfer, the discrete ordinates (DO) model was used. A fluid mixture composed of air and CO_2 was set in the species transport model. It was assumed that the CO_2 concentration in the supply air is zero. With this assumption the results indicate a room concentration distribution against the outdoor air CO_2 content. For the pressure-velocity coupling, the SIMPLE scheme was used. The simulations were performed for steady-state airflows. The infiltration effect was neglected for all simulations.

38.2.3 Ventilation Effectiveness

There are different ways and definitions to specify the effectiveness of the ventilation in a space. For the purpose of this study, the methods defined by Mundt et al. [3] were implemented. These include the contaminant removal effectiveness, local air quality index and local air change index. The corresponding equations are summarized in Table 38.1.

According to Mundt et al., the contaminant removal effectiveness (CRE), ε^c, is a measure of how quickly an airborne contaminant is removed from the room. It is defined as the ratio between the steady-state concentration of contaminant in the exhaust air, c_e, and the steady-state mean concentration of the room, $\langle c \rangle$. The air

Table 38.1 Summary of ventilation effectiveness equations

Contaminant removal effectiveness	Air quality index in the OZ	Local air change index
$\varepsilon^c = \dfrac{c_e}{\langle c \rangle}$	$\varepsilon_{oz}^c = \dfrac{c_e}{c_{oz}}$	$\varepsilon_P^a = \dfrac{\tau_n}{\overline{\tau}_P} \cdot 100 \ [\%]$

quality index for the occupied zone (OZ), ε_{oz}^c, is defined as the ratio between the steady-state concentration of contaminant at the exhaust air, c_e, and the average steady-state concentration in the occupied zone, c_{oz}. The local air change index, ε_P^a, is defined as the ratio between the nominal time constant, τ_n, and the local mean age of air, $\overline{\tau}_P$, at point P.

38.2.4   Building Energy Simulation

In order to determine the energy-saving potential of different cases, dynamic building simulations were carried out. For this purpose, the subject room of Fig. 38.1 with corresponding boundary conditions was studied in the building performance simulation program TRNSYS [4]. The location Stuttgart was selected for the variant study. The implemented ventilation system consists of a crossflow heat exchanger as a heat recovery (HR) unit, an electric preheater for frost protection of the heat exchanger, as well as a supply and an exhaust fan. In addition, a bypass of the heat exchanger has been considered, to enable a passive cooling operation during summer. In order to investigate the influence of different ventilation concepts on the heating or auxiliary energy demand, a parameter study was conducted with the input values airflow rate and efficiency of the heat recovery unit.

38.2.5   Experimental Set-Up

Measurements were carried out in a test room with a 1:1 scale of the simulation model as represented in Fig. 38.1. Tracer gas concentrations were acquired at a height of 1.1 m and 1.7 m, which represent the breathing level of a sitting and a standing person, respectively. Four measuring points were considered at each height with 1 m distance around the dummy, eight points for the concentration in total (Fig. 38.3a). The tracer gas concentration in the supply and exhaust ducts was also

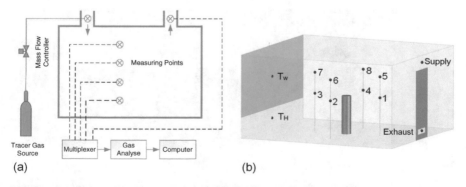

(a)   (b)

Fig. 38.3 Measuring points (**a**) and concentration measurement line-up (**b**)

measured. All boundary conditions such as airflow rate, supply temperature and dummy heat dissipation rate were set according to the simulation conditions.

For concentration measurements the Ansyco Gasmet™ DX4015 gas analyser which features a Fourier Transform Infrared (FTIR) spectrometer was implemented. As for tracer gas, nitrous oxide (N_2O) was used. N_2O has the same molar mass of the CO_2, and its background concentration in the ambient air is negligible. A multiplexer device was set to switch between different measuring points in the room. Figure 38.3b represents the line-up of the experiment. For the experimental studies, the supply outlets (a), (b) and (c) of Fig. 38.2 were used.

38.3 Results

The contours shown in Fig. 38.4 represent the CO_2 concentration distribution on a longitudinal surface through the middle of the room for the three first supply outlets of Fig. 38.2.

The supply airflow of the long throw jet nozzle (Fig. 38.4a) has enough momentum to compensate the thermal plume of the dummy. It reaches the outside wall, turns downward and flows over the floor towards the exhaust outlet resulting a perfect mixing in the room. The CO_2 distribution in room for this case is almost uniform. With the radial supply valve (Fig. 38.4b), the supply flow distributes radially over the back wall in all directions so that almost the whole wall and a small part of the ceiling is covered with a supply air layer. The flow is distributed radially in all directions and does not have enough momentum to reach the outside wall. The room airflow is less turbulent, and the thermal plume of the dummy rises unaffected towards the ceiling. From there it flows towards the back wall. The CO_2 distribution in the room is not uniform for this case. Some part of the supply air will be extracted through the exhaust opening which leads to a short circuit. The semi-radial supply valve also leads to a non-uniform CO_2 distribution as the radial supply valve (Fig. 38.4c).

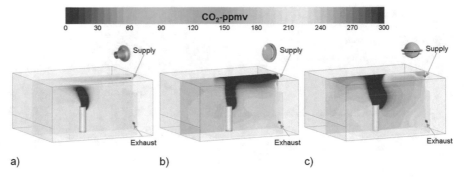

Fig. 38.4 CO_2 concentration distribution contours for the long throw jet nozzle (**a**) radial supply valve (**b**) and semi-radial supply valve (**c**)

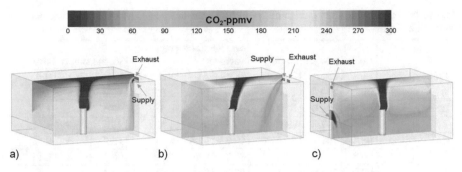

Fig. 38.5 CO_2 concentration distribution contours for the combination displacement air diffuser (**a**) combination slot air diffuser (**b**) and decentralized vertical ventilation unit (**c**)

Table 38.2 Ventilation effectiveness and CO_2 concentration in the occupied zone for different systems

Outlet type	c_{oz} (ppm)	ε^c	ε^c_{oz}
Long throw jet nozzle	228	1.00	0.99
Radial supply valve	248	0.90	0.91
Semi-radial supply valve	245	0.94	0.92
Combination displacement air diffuser	231	0.92	0.95
Combination slot air diffuser	179	1.23	1.28
Decentralized vertical ventilation unit	149	1.40	1.54

Figure 38.5 represents the CO_2 concentration distribution contours on a longitudinal surface through the middle of the room for the three last devices of Fig. 38.2. The discharge velocity of the combination displacement air diffuser and consequently the air supply turbulence are very low (Fig. 38.5a). The supply air falls down on the floor and moves towards the outside wall. There it will be heated by the heating panel, and then it rises up and meets the falling cold air on the window surface. Due to low turbulence, the thermal plume of the dummy rises unaffected towards the ceiling, and finally it will be extracted through the exhaust opening. The supply air of the combination slot air diffuser (Fig. 38.5b) has more turbulence than the displacement air diffuser but less than the mixing flow outlets of Fig. 38.2, so that a mixed/displacement airflow forms in the room. The supply air is directed with an angle to the occupied zone through four rows of slots but falls down on the floor, which leads to a stratification inside the room. The dummy thermal plume is slightly inclined towards the supply air and will be removed through the exhaust slots. With the decentralized vertical ventilation unit (Fig. 38.5c), the openings are integrated in the outside wall. The supply airflows with very low turbulence from the lower part to the room. The formation of a displacement flow is clearly observable.

Table 38.2 summarizes the average CO_2 concentration in the occupied zone, the CRE value as well as the air quality index in the occupied zone for different systems.

Based on the experimental studies, the local air change index, ε^a_p, was specified for the eight points represented in Fig. 38.3b. To obtain reproducible results, a cyclical sequence of "step-up" and "step-down" methods was used for the measurements

Table 38.3 Simulation and experimental results for local air change index with the corresponding relative deviations

Point number	Long throw jet nozzle			Semi-radial supply valve			Radial supply valve		
	Sim.	Exp.	Rel.dev.	Sim.	Exp.	Rel.dev.	Sim.	Exp.	Rel.dev.
1	99.0	100.0	1.00	98.6	96.6	2.07	92.8	95.3	2.62
2	98.9	98.0	0.92	97.5	97.2	0.31	92.6	94.2	1.70
3	98.6	97.8	0.82	97.5	98.8	1.32	93.4	96.7	3.41
4	98.4	101.8	3.34	97.2	97.6	0.41	93.8	94.2	0.42
5	99.3	98.8	0.51	97.7	97.6	0.10	95.0	97.4	2.46
6	99.0	101.6	2.56	97.2	97.0	0.21	93.4	94.6	1.27
7	98.5	98.2	0.31	98.5	96.3	2.28	94.8	96.3	1.56
8	98.1	97.9	0.20	97.4	96.7	0.72	95.2	95.1	0.11

All values are in percentage

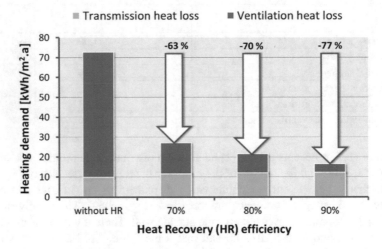

Fig. 38.6 Heating demand without and with heat recovery (72 m³/h)

[5]. The tracer gas was continuously released in the supply air duct at a constant rate, and its concentration at the eight measurement points was recorded (step-up). After the concentrations in the room and exhaust duct have reached their equilibrium values, the constant gas injection was stopped. Recording the concentration was continued after having stopped the tracer gas injection until no more changes were observed in the recorded concentrations (step-down). The mean age of air at each measurement point was determined from the recorded concentration versus time for each method [6]. Knowing the mean age of air at each point, the local air change index, ε_P^a, can be calculated. Table 38.3 compares the simulation and experimental results of local air change index at the eight points of Fig. 38.3b with the corresponding relative deviations. For the experimental values, the average of step-up and step-down measurements has been considered. Comparisons of measurements and simulations show good agreements.

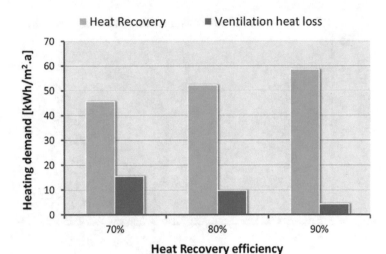

Fig. 38.7 Recovered heat and ventilation heat loss (72 m³/h)

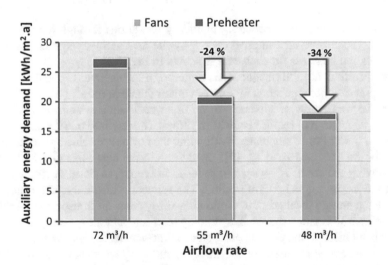

Fig. 38.8 Auxiliary energy demand of the fans and electric preheater

The deviations are due to uncertainties of simulations and measurements. Possible sources of uncertainties in the simulations are the numerical methods as well as the physical models used by the CFD code [7]. Uncertainties of measurements are caused by nonsymmetrical boundary conditions in comparison with simulations, leakage in the room and ducting, plus accuracy and calibration issues of the measurement devices [8].

Figure 38.6 represents the annual heating demand based on the building simulation results for a case without heat recovery as well as for cases with different heat recovery efficiencies of 70%, 80% and 90%. The outside airflow for all these cases is 72 m³/h. The results indicate that an energy saving of 63–77% is possible for the

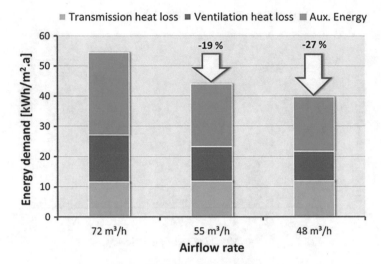

Fig. 38.9 Total energy demand (Heat recovery efficiency 70%)

simulated room with corresponding boundary conditions through the use of heat recovery. The regenerated thermal energy by the heat recovery system along with the ventilation heat loss for each case is shown in Fig. 38.7.

The specified 72 m³/h outside air is required in the occupied zone based on the assumption of a homogeneous mixing of contaminants in the air ($\varepsilon_{oz}^{c} = 1$). For this reason, the ventilation effectiveness of different ventilation systems can also influence the calculation of required outside airflows. The required outside airflow at the supply air outlet can be calculated by dividing the normative value by the air quality index for the occupied zone ε_{oz}^{c} [9]. In addition to the normative airflow for complete mixing (72 m³/h, $\varepsilon_{oz}^{c} = 1$), two reduced flow rates of 55 m³/h (for $\varepsilon_{oz}^{c} = 1.3$) and 48 m³/h (for $\varepsilon_{oz}^{c} = 1.5$) are studied. The reduction of the outside airflow also reduces the energy demand of the following subsystems air transport and air treatment. Figure 38.8 shows the auxiliary energy requirement for the supply and exhaust air fan in addition to the electric preheater for various outside airflows. Most of the auxiliary energy is required for the fans. By reducing airflows between 24% and 34% of auxiliary energy demand can be saved. The total energy demand of a system with a heat recovery efficiency of 70% for different airflow rates is represented in Fig. 38.9. Here, it can be seen that the reduction in the airflow reduces the auxiliary energy demand beside the heating demand, so that total energy requirement can be reduced between 19% and 27%.

38.4   Discussion

When reviewing the results for the three typical outlets of Fig. 38.2, it can be seen that they all achieve mixing ventilation in the room. The effectiveness of this ventilation depends on the geometry and momentum of the supply air jet and how well it is capable to induce the room air into itself. With the long throw jet nozzle, the CRE and air quality index in the occupied zone are near one (Table 38.2), which indicates a perfect mixing. For the radial and semi-radial supply valve, these values are slightly above 0.90 which represent a short circuit. This leads to a higher level of CO_2 concentration in the room.

With the combination displacement air diffuser, although a displacement flow is formed in the room, the values of CRE and the air quality index are below one, indicating a short circuit. This is resulted by close vicinity of supply and exhaust opening. Hence, part of the supply air will be instantly induced in the exhaust airflow and removed from the room without being used. Despite the short circuit, the level of CO_2 concentration in the occupied zone is lower than the levels with mixing ventilation outlets (Table 38.2). The air quality index in the occupied zone for the combination slot air diffuser and the decentralized vertical ventilation unit are 1.28 and 1.54, respectively. These values are greater than one and imply formation of displacement ventilation in the room. The CO_2 concentration in the occupied zone is 179 ppm and 149 ppm for these two systems, respectively, which is considerably lower than the values with mixing ventilation systems (Table 38.2).

The simulations are validated with experiments, and the results give an overview to examine both the ventilation effectiveness and the resulting IAQ of different systems.

38.5   Conclusions

There is a great interest in improving the energy efficiency of building ventilation systems while maintaining a desired IAQ. In this study, several CFD simulations supported by experimental studies were conducted in a typical room to investigate how the ventilation effectiveness of different ventilation systems is affected by the position and type of the air outlets in the room. Contaminant removal effectiveness and the air quality index have been considered as key indicators. Whereas typical outlets lead to mixing ventilation with CRE values near to one in best cases, the displacement ventilation outlets are capable to achieve stratification with CRE values higher than one. This indicates that, with displacement ventilation systems, there is a potential to improve the IAQ, or in other words, there is a potential to reduce the energy demand by reducing the supply airflow while maintaining the same IAQ. High care shall be taken into consideration to avoid any short circuit between supply and exhaust outlets. The outcome of this study contributes to the building designers at the basic phases to design a more efficient ventilation system by choosing and positioning of the system components in an optimum manner.

More studies shall be done to evaluate the effect of room geometry, occupancy and seasonal operating conditions.

Acknowledgement This project was funded by the Graduate and Research School Efficient use of Energy Stuttgart (GREES), Germany.

References

1. DIN EN (2007) DIN EN Standard 13779-2007. Ventilation for non-residential buildings-performance requirements for ventilation and room-conditioning systems. European Committee for Standardization, Brussels
2. DIN EN (2012) DIN EN Standard 15251-2012. Indoor environmental input parameters for design and assessment of energy performance of buildings- addressing indoor air quality, thermal environment, lighting and acoustics. European Committee for Standardization, Brussels
3. Mundt E, Mathisen HM, Nielsen PV, Moser A (2004) Ventilation effectiveness. REHVA guide-book, Finland
4. University of Wisconsin–Madison. Solar Energy Laboratory (1975) TRNSYS, a transient simulation program. The Laboratory, Madison
5. Jung A, Zeller M (1994) Analysis and testing of methods to determine indoor air quality and air change effectiveness. Rheinisch-Westfälische Technical University of Aachen, Aachen
6. Sandberg M, Sjöberg M (1983) The use of moments for assessing air quality in ventilated rooms. Build Environ 18(4):181–197
7. Nielsen PV, Allard F, Awbi HB, Davidson L, Schälin A (2007) Computational fluid dynamics in ventilation design. REHVA guidebook, Finland
8. ASHRAE (2002) ANSI/ASHRAE Standard 129-1997 (RA 2002): Measuring air-change effectiveness. American society of heating, refrigerating and air-conditioning engineers, Atlanta
9. ASHRAE (2010) ANSI/ASHRAE Standard 62.1. Ventilation for acceptable indoor air quality. American Society of Heating, Refrigerating and Air-Conditioning Engineers, Atlanta

Chapter 39
Assessment of Cardboard as an Environment-Friendly Wall Thermal Insulation for Low-Energy Prefabricated Buildings

Seyedehmamak Salavatian, M. D'Orazio, C. Di Perna, and E. Di Giuseppe

Abstract Exterior walls play a significant role in buildings thermal behaviour, and utilization of proper insulation materials with high thermal performance and low adverse environmental impacts is of great importance. The main aim of this study is to assess thermal and environmental benefits and drawbacks of honeycomb cardboard application in external wall configuration of prefabricated buildings. Its thermal conductivity was measured by guarded hot plate method, and its steady-state as well as periodic thermal transmittances were obtained. Furthermore, main construction junctions were simulated, and their linear thermal transmittance was studied. Secondly, with a life cycle assessment (LCA) approach, the adverse impacts of cardboard on environment in production phase and within a limited impact category set were studied. Additionally, the same procedures of thermal and environmental assessments were performed on a number of functionally equivalent insulation materials to be compared with cardboard. The results demonstrate an overall image of positive and negative consequences of cardboard application as insulation for wall envelopes.

S. Salavatian (✉)
Department of Architecture, Islamic Azad University, Rasht Branch, Rasht, Iran
e-mail: salavatian@iaurasht.ac.ir

M. D'Orazio · E. Di Giuseppe
Construction, Civil Engineering and Architecture Department,
Università Politecnica delle Marche, Ancona, Italy
e-mail: m.dorazio@univpm.it; e.digiuseppe@univpm.it

C. Di Perna
Industrial Engineering and Mathematics Sciences Department,
Università Politecnica delle Marche, Ancona, Italy
e-mail: c.diperna@univpm.it

© Springer International Publishing AG, part of Springer Nature 2019   463
A. Sayigh (ed.), *Sustainable Building for a Cleaner Environment*,
Innovative Renewable Energy, https://doi.org/10.1007/978-3-319-94595-8_39

39.1   Introduction

Buildings are the largest energy-consuming sector in the world and account for over one-third of total final energy consumption and an equally important source of CO_2 emissions [1]. These concerns have ended in tighter building regulations which increase use of insulation materials. External walls represent a significant share of the total envelope area of buildings and applying proper techniques in their insulation is influential in improving building energy performance. Although there are a large number of insulation products in building sector, it is important to use sustainable insulating materials in order to reduce the adverse environmental impact. A wealth of researches has conducted a life cycle assessment (LCA) approach in a comparative study of numerous building insulation materials [2–9]. Moreover, innovative insulation materials have recently emerged on the market and can result in considerable energy savings as well as reduction of CO_2 emission [7–10]. Further explorations must still be done in this area to identify insulation materials with the highest levels of thermal resistance and environmental sustainability.

In this paper, cardboard has been considered as the subject of study to realize if its use in wall envelopes would provide any thermal and environmental benefits. Cardboard is highly recycled and recyclable and a super light material, which combines lightness and strength [11] (average resistance to compression is about 6 Kg/cm^2 [12]). Because of these properties, it is predicted that cardboard could bring several advantages to wall envelopes of prefabricated buildings in terms of innovation, proper thermal properties and light-weight structures which are the major issues in prefabricated construction industry. Due to this adjustment, category of factory-made prefabricated homes was considered in this research. Although prefabricated housing has spread out dramatically recently, a noteworthy knowledge on low-energy construction methods in this field has not been established yet. Therefore there is a high level of importance to develop researches on advanced envelopes for these building types, and this work seeks to fill this gap.

In previous studies as [11, 13, 14], applicability of cardboard in architecture has been focused on its multifunctionality in buildings. To date, no study has examined its particular potentials as a building insulation material. In this paper, cardboard panels were investigated to provide more accurate data for a reliable perception on feasibility of its application as building insulation in prefabricated buildings.

39.2   Methodology

39.2.1   Thermal Properties of Cardboard

Cardboard panels under study are produced by an Italian firm specialized in honeycomb cardboard production made of 100% recycled paper. The samples utilized for experimental tests were composed of two panels of 500 by 500 mm made of two

Fig. 39.1 Honeycomb core of the cardboard panel

Fig. 39.2 Assembling the specimen in the hot plate apparatus

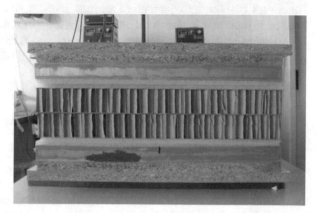

covering sheets of Kraft paper joined to a previously expanded and dried honeycomb core with the total thickness of 50 mm as shown in Fig. 39.1. As seen in Fig. 39.2, laboratory measurements were carried out by method of guarded hot plate and heat flow metre according to EN ISO 12667:2002 [15]. Based on this method, a temperature difference between the parallel faces of the specimen was generated by thermostatic baths, and then the heat flow was measured by means of heat flow probes placed on hot and cold sides of the specimen. The test lasted about 24 h in order to meet steady-state conditions, and surface temperatures and heat flux were registered every 60 s to be used as parameters in calculation of thermal conductivity of material.

39.2.2   Thermal Performance Assessment of Envelope

Configuration of proposed wall envelope is composed of four main elements: 1, structural framework, made of cold-formed steel profiles; 2, insulating material; 3, air/vapour barriers; and 4, facings, gypsum board as internal and PVC slats as external finishing. Two thermal parameters were investigated: firstly, steady-state thermal transmittance (U-value) of opaque wall envelope as a wintertime indicator in line with EN ISO 6946 [16] and, secondly, periodic thermal transmittance (Y_{mn}) which is mostly a summertime indicator according to EN ISO 13786 [17]. Local dispersions in thermal bridges were also calculated by linear thermal transmittance (Psi-value) to be complied with EN ISO 14683:2008 [18]. At the national level in Italy, Energy Performance of Buildings Directive (EPBD) was adopted with two legislative acts: Lgs. 192/2005 and Lgs. 311/2006 in which U-value limits for walls vary from 0.33 to 0.62 W/m^2 K depending on different climatic zones. In this study, target value for thermal transmittance of wall envelope was taken 0.33 W/m^2 K which is the lowest allowed value among six Italian climatic zones and could meet all over the territory. Moreover, considering a constant value provides the same functional basis for comparison of various choices. According to EN ISO 14683, four main junctions in which thermal bridges are likely to occur were modelled and investigated by the 2D thermal analysis software THERM version 7.3 [19]. The construction junctions include wall to wall, wall to roof, wall to floor and wall to openings (doors/windows).

39.2.2.1 Utilization of Other Replaceable Insulation Materials

In order to investigate cardboard in comparison with other common insulation materials in practice, the same thermal assessment procedure was performed for other competitive insulations in case of their replacement for cardboard. Selected insulation materials are based on European market and comprise three categories of natural, mineral and synthetic. As summarized in Table 39.1, the exact and practical thickness (based on its availability on the market) of insulations to obtain the transmittance limit was found.

Table 39.1 Properties of comparable insulation materials[a]

Insulation material	Density (kg/m^3)	Thermal conductivity (W/mK)	Required thickness for U-value = 0.33 W/ m^2 K (mm)	Practical thickness (mm)	U-value (W/ m^{2}K)	weight per f.u.[b] (kg)
Cardboard	43	0.125	250	250	0.308	10.75
Wood fibre	160	0.039	78	80	0.303	12.8
Cellulose	45	0.038	76	90	0.276	4.05
Glass wool	32	0.032	64	80	0.267	2.56
Rock wool	70	0.035	70	80	0.283	5.6
Polyurethane	35	0.028	56	60	0.295	2.1
EPS	35	0.036	72	80	0.288	2.8

[a]Physical properties are from technical datasheets of Italian manufacturers
[b]Functional unit

39.2.3   Environmental Impact Assessment of Cardboard

In this study, energy and mass flows and environmental impacts of cardboard have been assessed from the production of raw materials to the manufacture of end product. For this purpose, the software openLCA [20] which works according to ISO 14040 [21] and 14,044 [22] was utilized.

39.2.3.1   Scope Definition, Functional Unit and System Boundaries

Functional unit defines quantification of the identified functions of the product. The primary purpose of a functional unit is to provide a reference to which the inputs and outputs are related [21]. In this study, the functional unit (f.u.) would be the mass of insulation material that provides U-value of 0.33 W/m^2 K for the area of 1 m^2 of the wall. "Cradle to gate" study has been considered for this paper including product stages A1 to A3: A1, raw material extraction and processing, processing of secondary material input; A2, transport to the manufacturer; and A3, manufacturing [23]. Usage and disposal phases are excluded from the study due to time constraints and lack of required data.

39.2.3.2   Inventory Analysis and Setting Impact Categories

Life cycle inventory analysis includes quantification of input/output flows. Life cycle impact assessment, based on the inventory results, is qualified, quantified and compared. In this research, ProBas and Ecoinvent are the data sources used for inventory study of materials and by means of two impact assessment methods, CML (baseline) and cumulative energy demand, following impact categories, were assessed for cardboard and alternative insulations in openLCA: consumption of primary energy (MJ), global warming potential (GWP) (kg CO_2 eq), acidification potential (kg SO_2 eq), eutrophication (kg PO_4 eq) and photochemical oxidation (kg C_2H_4 eq).

39.3   Results and Discussion

39.3.1   Thermal Transmittance of Wall Envelope

The thermal conductivity for cardboard, experimentally obtained through the average method described in ISO 9869:1994, equals to 0.125 W/mK. Periodic thermal properties of cardboard and other insulations were analysed, and their compliance with Italian regulations is verified. All values meet required limits in regulations; however as seen in Fig. 39.3, higher values – except for time shift – belong to

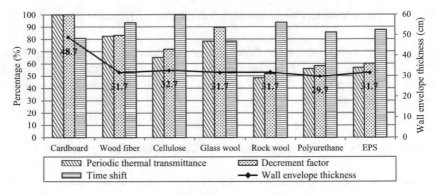

Fig. 39.3 Comparative periodic thermal properties of insulation materials

cardboard application. Y_{mn} and decrement factor in cardboard are quite two times greater than the minimum values among alternatives (rock wool). Time shift doesn't show any substantial variation for different alternatives. Total wall thickness of all other choices varies in a narrow range, while required thickness for cardboard is 55% greater than the average.

39.3.1.1 Linear Thermal Transmittance

Models of main junctions in THERM (as described in 39.2.2) show that typical wall structure is far from major anomalies because steel studs are placed frequently in the wall that their effect is negligible. Other building elements, e.g., roof, floor and openings, were assumed identical and designed in accordance with common current practices. Values of calculated y_e and y_i as described in EN 14683 [18] show that the use of cardboard doesn't cause any noticeable difference in linear thermal transmittance regarding application of other common insulations.

39.3.2   Environmental Impact Assessment

Environmental impacts of cardboard are compared with six other selected insulations in five impact categories. As shown in Fig. 39.4, although contribution of cardboard in total energy consumption compared to other insulations is high and ranked as the second one, the share of non-renewable energy goes to the fifth place among seven alternatives. Therefore in terms of non-renewable energy, it can compete with common insulations. Additionally, in the relative percentage results per functional unit (f.u.) for the selected insulation materials presented in Fig. 39.5, some results are observed regarding studied impact categories. In global warming

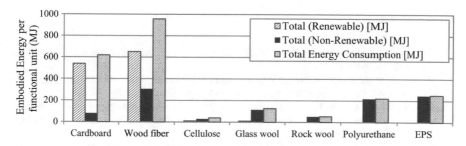

Fig. 39.4 Embodied energy of cardboard in comparison with other insulations per f.u

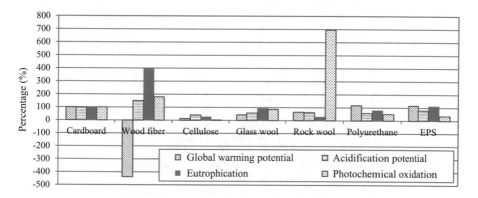

Fig. 39.5 Relative comparison of insulations in environmental impact categories

potential, cardboard is rather at the level of synthetic insulations such as polyurethane and EPS. In categories of eutrophication, acidification and photochemical oxidation, cardboard is placed as the third and fourth alternative.

39.4   Conclusion

Analysis and comparison of honeycomb cardboard with some of other common insulation materials in wall envelopes present a realistic perception of its feasibility in building sector. Its thermal conductivity was found out to be relatively three times higher than the average value of conventional insulations; higher thermal conductivity leads to utilization of thicker material for achievement of the equal function which increases weight of the wall despite low density of cardboard. LCA study results demonstrate that in terms of non-renewable energy consumption, cardboard is preferable to other insulations, while other investigated impact categories have not shown any significant superiority. Since its recyclability potential as a noticeable environmental benefit has not been considered in this study, it is predicted that by expanding system boundary of LCA to the end-of-life phase, other pluses of cardboard are brought into light and make it competitive in building insulation market.

References

1. International Energy Agency (IEA) (2018) Accessed 07 Apr 2018 at https://www.iea.org/etp/tracking2017/buildings/
2. Pargana N, Pinheiro MD, Silvestre JD, de Brito J (2014) Comparative environmental life cycle assessment of thermal insulation materials of buildings. Energ Buildings 82:466–481
3. Anastaselos D, Giama E, Papadopoulos AM (2009) An assessment tool for the energy, economic and environmental evaluation of thermal insulation solutions. Energ Buildings 41(11):1165–1171
4. Papadopoulos AM, Giama E (2007) Environmental performance evaluation of thermal insulation materials and its impact on the building. Build Environ 42(5):2178–2187
5. Azari R (2014) Integrated energy and environmental life cycle assessment of office building envelopes. Energ Buildings 82:156–162
6. Densley Tingley D, Hathway A, Davison B (2014) An environmental impact comparison of external wall insulation types. Build Environ 85:182–189
7. Ardente F, Beccali M, Cellura M, Mistretta M (2008) Building energy performance: a LCA case study of kenaf-fibres insulation board. Energ Buildings 40(1):1–10
8. Ricciardi P, Belloni E, Cotana F (2014) Innovative panels with recycled materials: thermal and acoustic performance and life cycle assessment. Appl Energy 134:150–162
9. Batouli SM, Zhu Y, Nar M, D'Souza NA (2014) Environmental performance of kenaf-fiber reinforced polyurethane: a life cycle assessment approach. J Clean Prod 66:164–173
10. La Rosa AD, Recca A, Gagliano A, Summerscales J, Latteri A, Cozzo G, Cicala G (2014) Environmental impacts and thermal insulation performance of innovative composite solutions for building applications. Constr Build Mater 55:406–414
11. Eekhout M, Verheijen F, Visser R (2008) Cardboard in architecture. IOS Press, Amsterdam
12. TVPLAST (2014) Accessed 01 May 2015 at www.tivuplast.it
13. Ayan O (2009) PhD thesis. Cardboard in architectural technology and structural engineering. ETH, Zurich
14. Pohl A (2009) PhD thesis. Strengthened corrugated paper honeycomb for application in structural elements. ETH, Zurich
15. European Standard (2002) EN 12667 thermal performance of building materials and products - determination of thermal resistance by means of guarded hot plate and he at flow meter methods. European Committee for Standardization, Brussels
16. European Standard (2008) EN ISO 6946 components and building elements - thermal resistance and thermal transmittance - calculation methods. European Committee for Standardization, Brussels
17. European Standard (2008) EN ISO 13786 thermal performance of building components - dynamic thermal characteristics - calculation method. European Committee for Standardization, Brussels
18. European Standard (2008) EN ISO 14683 thermal bridges in building construction - linear thermal transmittance - simplified methods and default values. European Committee for Standardization, Brussels
19. Lawrence Berkeley National Laboratory (LBNL) (2014) Accessed 05 Feb 2015 at https://windows.lbl.gov/software/therm/
20. OpenLCA (2015) Accessed 05 Feb 2015 at http://www.openlca.org/
21. European Standard (2006) EN ISO 14040 environment management - life cycle assessment - principles and framework. European Committee for Standardization, Brussels
22. European Standard (2006) EN ISO 14044 environmental management - life cycle assessment - requirements and guidelines. European Committee for Standardization, Brussels
23. European Standard (2012) EN 15804 sustainability of construction works - environmental product declaration - core rules for the product category of construction products. European Committee for Standardization, Brussels

Conclusions

In reviewing all the papers in this book, they were divided into the following categories: sustainable architecture, nine papers; building construction management and environment, six papers; ventilation and air movement in buildings, three papers; renewable energy in buildings and cities, nine papers; eco materials and technology, two papers; policy education and finance, eight papers; sustainable transport, two papers; and urban agriculture and soilless urban green space, three papers.

Most papers exhibit innovative ideas and concepts to reduce energy consumption yet maintain comfort and healthy environments. Some papers utilize indigenous local materials to create sustainable environment.

The majority of the papers came from countries within the Mediterranean region, and the remainder came from countries within a Mediterranean climate; 41 countries submitted papers.

The philosophy behind Mediterranean Green Buildings and Renewable Energy Forums (MED GREEN FORUM) is to encourage the region of Europe which has moderate climate with plentiful sunshine to start building their homes using totally renewable energy with features such as the use of daylight, natural ventilation, local material, and sustainable building design to achieve the required heating and cooling throughout the tear.

Since these forums commenced, most countries in the region have significantly increased their utilization of photovoltaic panels as building material and their construction offshore wind farms to generate more than 30% of required electricity.

© Springer International Publishing AG, part of Springer Nature 2019
A. Sayigh (ed.), *Sustainable Building for a Cleaner Environment*,
Innovative Renewable Energy, https://doi.org/10.1007/978-3-319-94595-8

Printed in the United States
By Bookmasters